Physical and Materials Constants

Boltzmann's constant	k	1.38×10^{-23}	J/K
Electron charge	q	1.6×10^{-19}	C
Thermal voltage	kT/q	0.026 (at $T = 300$ K)	V
Energy gap of silicon (Si)	E_g	1.12 (at $T = 300$ K)	eV
Intrinsic carrier concentration of silicon (Si)	n_i	1.45×10^{10} (at $T = 300$ K)	cm^{-3}
Dielectric constant of vacuum	ε_0	8.85×10^{-14}	F/cm
Dielectric constant of silicon (Si)	ε_{Si}	$11.7 \times \varepsilon_0$	F/cm
Dielectric constant of silicon dioxide (SiO_2)	ε_{ox}	$3.9 \times \varepsilon_0$	F/cm

Commonly Used Prefixes for Units

giga	G	10^9
mega	M	10^6
kilo	k	10^3
milli	m	10^{-3}
micro	μ	10^{-6}
nano	n	10^{-9}
pico	p	10^{-12}
femto	f	10^{-15}

CMOS Digital Integrated Circuits:
Analysis and Design

Sung-Mo (Steve) Kang
University of Illinois at Urbana-Champaign

Yusuf Leblebici
Istanbul Technical University

The McGraw-Hill Companies, Inc.
New York St. Louis San Francisco Auckland Bogotá
Caracas Lisbon London Madrid Mexico City Milan Montreal
New Delhi San Juan Singapore Sydney Tokyo Toronto

To Myoung-A, Jennifer and Jeffrey,
and Anil and Ebru

CMOS Digital Integrated Circuits: Analysis and Design

This book is printed on acid-free paper.

1 2 3 4 5 6 7 8 9 0   DOC DOC   9 0 9 8 7 6 5

ISBN 0-07-038046-5

The editor was Lynn Cox;
the production supervisor was Denise L. Puryear.
The cover was designed by Amy Becker/Designs & Visions.
R. R. Donnelley & Sons Company was printer and binder.

Library of Congress Cataloging-in-Publication Data

Kang, Sung-Mo, (date).
   CMOS digital integrated circuits: analysis and design / Sung-Mo
 (Steve) Kang, Yusuf Leblebici.
      p.   cm.
   ISBN 0-07-038046-5
   1. Metal oxide semiconductors, Complementary.
  2. Digital integrated circuits.   I. Leblebici, Yusuf.   II. Title.
 TK7871. 99. M44K36 1996
  621.39′5—dc20   95-21223

CONTENTS

PREFACE

Complementary metal oxide semiconductor (CMOS) digital integrated circuits are the enabling technology for the modern information age. Because of their intrinsic features in low-power consumption, large noise margins, and ease of design, CMOS integrated circuits have been widely used to develop random access memory (RAM) chips, microprocessor chips, digital signal processor (DSP) chips, and application-specific integrated circuit (ASIC) chips. The popular use of CMOS circuits will grow with the increasing demands for low-power, low-noise integrated electronic systems in the development of portable computers, personal digital assistants (PDAs), portable phones, and multimedia agents.

Since the field of CMOS integrated circuits alone is very broad, it is conventionally divided into digital CMOS circuits and analog CMOS circuits. This book is focused on the CMOS digital integrated circuits. At the University of Illinois at Urbana-Champaign, we have tried some of the available textbooks on digital MOS integrated circuits for our senior-level technical elective course, ECE382 - *Large Scale Integrated Circuit Design*. Students and instructors alike realized, however, that there was a need for a new book with more comprehensive treatment of CMOS digital circuits. Thus, our textbook project was initiated several years ago by assembling our own lecture notes. Since 1993, we have used evolving versions of this book both at the University of Illinois at Urbana-Champaign and at Istanbul Technical University. Both authors were very much encouraged by comments from their students, colleagues, and reviewers. In the meantime, the editorial staff of McGraw-Hill has arranged for rigorous reviews of the manuscript, and has been very supportive throughout the development phase.

This book, *CMOS Digital Integrated Circuits: Analysis and Design*, is primarily intended as a comprehensive textbook at the senior level and first-year graduate level, as well as a reference for practicing engineers in the areas of integrated circuit design, digital design, and VLSI. Recognizing that the area of digital integrated circuit design is evolving at an increasingly faster pace, we have made every possible effort to present up-to-date materials on all subjects covered. At the undergraduate level, coverage of the first ten chapters would probably provide sufficient material for a one-semester course on CMOS digital integrated circuits. Alternatively, this book can be used for a two-semester course, allowing a more detailed treatment of advanced issues, which are presented in the later chapters. At the graduate level, selected topics from the first ten chapters plus the last five chapters can be covered in one semester.

The first eight chapters of this book are devoted to a rigorous treatment of the MOS transistor with all its relevant aspects; to the static and dynamic operation principles, analysis and design of basic inverter circuits; and to the structure and operation of combinational and sequential logic gates. A separate chapter (Chapter 9) has been reserved for the treatment of *dynamic* logic circuits. Chapter 10 offers an in-depth presentation of semiconductor memory circuits. The relatively new field of BiCMOS digital circuit design is examined in Chapter 11, with a thorough coverage of bipolar transistor basics. Next, Chapter 12 provides a clear insight into the important subject of chip I/O design. Critical issues such as ESD protection, clock distribution, signal buffering, and latch-up are discussed in detail. General design methodologies and tools for very large scale integration (VLSI) are briefly presented in Chapter 13. Finally, the more advanced but very important topics of design for manufacturability and design for testability are covered in Chapters 14 and 15, respectively. The design and implementation of a 16-bit binary adder circuit, which was completed as a student team project in ECE382 - *Large Scale Integrated Circuit Design* course at the University of Illinois at Urbana-Champaign, are presented in detail in the Appendix.

The authors have long debated the coverage of nMOS circuits in this book. We have finally concluded that some coverage should be provided for pedagogical reasons. Studying nMOS circuits will better prepare readers for analysis of other field effect transistor (FET) circuits such as GaAs circuits, the topology of which is quite similar to that of depletion-load nMOS circuits. Thus, to emphasize the *load* concept, which is still widely used in many areas in digital circuit design, we present basic depletion-load nMOS circuits along with their CMOS counterparts in several places throughout the book.

The camera-ready manuscript, including the overwhelming majority of the figures, have been prepared by the authors on a Macintosh Quadra 700, using Aldus PageMaker 4.2, MacDraw Pro, and MathType 2.1. Although an immense amount of effort and attention to detail were expended to prepare the camera-ready manuscript, this book may still have some flaws and mistakes due to erring human nature. The authors would welcome and greatly appreciate suggestions and corrections from the readers, for the improvement of the technical content as well as the presentation style.

Our colleagues have provided many constructive comments and encouragement for the completion of this book. Professor Timothy N. Trick, former head of the department of electrical and computer engineering at the University of Illinois at Urbana-Champaign, has strongly supported our efforts from the very beginning. The appointment of Sung-Mo Kang as an associate in the Center for Advanced Study at the University of Illinois at Urbana-Champaign helped to start the process.

Yusuf Leblebici acknowledges the full support and encouragement from the department of electrical and electronics engineering at Istanbul Technical University, where he introduced a new digital integrated circuits course based on the early version of this book and received very valuable feedback from his students. Yusuf Leblebici also thanks the ETA Advanced Electronics Technologies Research and Development Foundation at Istanbul Technical University for their generous support.

Professor Elyse Rosenbaum and Professor Resve Saleh used the early versions of the manuscript as the textbook for ECE382 at Illinois and provided many helpful comments and corrections which have been fully incorporated with deep appreciation. Professor Elizabeth Brauer has also done the same at the University of Kentucky.

The authors would like to express sincere gratitude to Professor Janak Patel of the University of Illinois at Urbana-Champaign for generously mentoring the authors in writing Chapter 15, *Design for Testability*. Professor Patel has provided many constructive comments and many of his expert views on the subject are reflected in this chapter. Professor Prith Banerjee and Professor Farid Najm of the University of Illinois at Urbana-Champaign also provided many good comments. We would also like to thank Dr. Abhijit Dharchoudhury for his invaluable contribution to Chapter 14, *Design for Manufacturability*.

Professor Duran Leblebici of Istanbul Technical University, who, incidentally, is the father of Yusuf Leblebici, reviewed the entire manuscript in its early development phase, and provided very extensive and constructive comments, many of which are reflected in the final version. Both authors gratefully acknowledge his support during all stages of this venture. We also thank Professor Cem Göknar of Istanbul Technical University, who offered very detailed and valuable comments on *Design for Testability*, and Professor Uğur Çilingiroğlu of the same university, who offered many excellent suggestions for improving the manuscript, especially the chapter on semiconductor memories.

Many of the authors' former and current students at the University of Illinois at Urbana-Champaign also helped in the preparation of figures and verification of circuits using SPICE simulations. In particular, James Morikuni, Weishi Sun, Pablo Mena, Steve Ho, Sueng-Yong Park, and Jaewon Kim deserve special recognition. Ms. Lilian Beck and other staff members of the Publications Office in the department of electrical and computer engineering at the University of Illinois at Urbana-Champaign read the entire manuscript and provided excellent editorial comments.

The authors would also like to thank Dr. Masakazu Shoji of AT&T Bell Laboratories, Professor Gerold W. Neudeck of Purdue University, Professor Chin-Long Wey of Michigan State University, Professor Andrew T. Yang of the University of Washington, Professor Marwan M. Hassoun of Iowa State University, Professor Charles E. Stroud of the University of Kentucky, Professor Lawrence Pileggi of the University of Texas at Austin, and Professor Yu Hen Hu of the University of Wisconsin at Madison, who read all or parts of the manuscript and provided many valuable comments and encouragement. The editorial staff of McGraw-Hill has been an excellent source of strong support from the beginning of this textbook project. The venture was originally initiated with the enthusiastic encouragement from the previous electrical engineering editor, Ms. Anne (Brown) Akay. Mr. George Hoffman, in spite of his relatively short association, was extremely effective and helped settle the details of the publication planning. During the last stage, the new electrical engineering editor, Ms. Lynn Cox, and Mr. John Morriss, Mr. David Damstra, and Mr. Norman Pedersen of the Editing Department were superbly effective and we enjoyed dashing with them to finish the last mile.

Last, but not least, we would like to acknowledge the support from our families, in particular, from our spouses, Myoung-A (Mia) Kang and Anıl Leblebici, and our children, Jennifer and Jeffrey Kang and Deniz Ebru Leblebici, for tolerating many absences while we spent innumerable hours on this book over the last three years.

Sung-Mo (Steve) Kang   Yusuf Leblebici
Urbana, Illinois   *Istanbul, Turkey*
February 1995   *February 1995*

CHAPTER 1

INTRODUCTION

Digital CMOS (Complementary Metal Oxide Semiconductor) integrated circuits (ICs) have been the driving force behind Very Large Scale Integration (VLSI) for high-performance computing and other scientific and engineering applications. The demand for digital CMOS ICs will be continually strong due to salient features such as low power, reliable performance, circuit techniques for high speed such as using dynamic circuits, and ongoing improvement in processing technology.

Even as late as the early 1980s, designers and technologists were engaged in heated discussions on the merits and demerits of CMOS circuits in comparison with those of nMOS circuits and bipolar circuits. However, it has become very obvious that when the power consumption factor, which is closely related to device junction temperature, packaging cost, battery life, especially in portable systems, and reliability are considered, CMOS circuits are the clear choice.

Several years ago, a team of designers first developed a custom chip with nMOS technology to meet stringent timing requirements. However, the developed chip consumed too much power and the designers had to introduce innovative circuit techniques including automatic power-down to reduce the power consumption. Unfortunately, on-chip realization of additional circuitry usually requires additional silicon area and increases interconnection and fanout parasitics, thus slowing the circuit speed and also lowering the fabrication yield. When the chip was redesigned using CMOS technology, not only did the chip work at the required speed with much lower power consumption, but also the silicon area was smaller. The development of BiCMOS technology, which merges the merits of bipolar and CMOS circuits using essentially a CMOS process, has made CMOS technology even stronger.

It is now projected that the minimum feature size in CMOS ICs can decrease to 0.15 μm within a decade. With such a technology, the level of integration in a single chip can be on the order of several hundreds of millions of transistors for logic chips or even higher in the case of memory chips, which presents an immense challenge for chip developers in processing, design methodology, testing, and project management. Through the "divide-and-conquer" approach and more advanced design automation using computer-aided design (CAD) tools, ultra-large-scale problems should be handled.

Bipolar and gallium arsenide (GaAs) circuits have been used for very high speed circuits, and this practice may continue. For instance, in Monolithic Microwave Integrated Circuits (MMICs), GaAs MESFET (MEtal Semiconductor Field Effect Transistor) technology has been highly successful. However, they are still not efficient for VLSI or Ultra Large Scale Integration (ULSI) due to processing difficulties and high power consumption, although for special applications their use may continue. As long as the downward scaling of CMOS technology remains strong, other technologies are likely to remain the technology of tomorrow.

The objective of this book is to help readers develop in-depth analysis and design capability for digital CMOS circuits and chips. The development of VLSI chips requires an interdisciplinary team of architects, logic designers, circuit and layout designers, packaging engineers, test engineers, and process and device engineers. Also essential are the computer aids for design automation and optimization. It is not possible to discuss the full spectrum of development issues in any single book. Therefore, this book concentrates on digital circuits and also presents related materials in processing and device principles essential to in-depth understanding of CMOS digital circuits.

Often readers can become lost in details and fail to see the global picture. For VLSI circuit design, however, it is important that the design be done in the context of global optimization with proper boundary conditions. In fact, the beauty of integrated circuits is that the final design goal is the concerted performance of all transistors interconnected, and not of individual transistors. Therefore, the interconnect issues are almost as important as the issues of individual transistors. No matter how good the performance of an individual transistor is, if the technology fails to have equally good interconnects, the total performance can be very poor due to large parasitic capacitances and resistances, which translates into a large delay in the interconnection lines between transistors or logic gates.

This volume is intended as a comprehensive textbook for senior-level undergraduate students and first-year graduate students in an advanced course on digital circuit design. The material presented in this book should also be very useful to practicing VLSI design engineers. Most of the material presented has been taught over several years in undergraduate and graduate-level courses in the department of electrical and computer engineering at the University of Illinois at Urbana-Champaign. It is assumed that the readers of this book already have sufficient fundamental background on semiconductor devices, electronic circuit design and analysis, and logic theory. While the interactions among logic design, circuit design, and layout design are strongly emphasized throughout the text, the main focus is on transistor-level circuit design and analysis. This requires a fair amount of detailed current and voltage calculations, as well as a good understanding of how device characteristics affect overall circuit performances, such as propagation delay, noise margins, and power dissipation.

The relational ordering and extent of the topics covered in a typical digital integrated circuits course are depicted in Fig. 1.1. First, a fundamental knowledge of basic device physics is required to understand and use various MOSFET device models in circuit analysis. Following a review of device fundamentals, the emphasis will shift from single devices to simple two-transistor circuits, such as inverters, and then to more complex logic circuits. We will see that as we move to more advanced topics, the breadth of each subject also increases significantly. In fact, a large number of different variations may be considered for implementing complex circuits and systems. Consequently, we will examine representative examples for large-scale system implementations and compare their relative merits in terms of performance, reliability, and manufacturability.

The book begins with a review of fabrication-related issues. Representative integrated circuit fabrication techniques are summarized very briefly in the beginning, in order to establish a simple view of process flow and to provide the reader with the necessary terminologies related to processing. The level and the extent of MOS device physics covered in this book are specifically geared toward hands-on circuit design and analysis applications; hence, most of the device models used are relatively simple. The choice of simple device models imposes certain limitations on the accuracy; however, the emphasis is primarily on the clear understanding of basic design concepts and on the importance of generating meaningful estimates for circuit performance during the

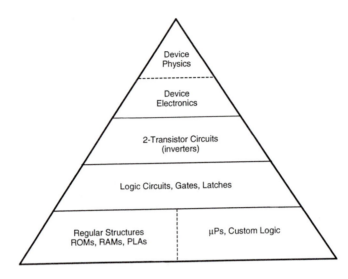

Figure 1.1.   The ordering of topics covered in a typical digital integrated circuits course.

early design stages. The very important role of computer-aided circuit simulation tools in VLSI design is also well recognized. The book contains a large number of computer simulation examples and exercise problems based on SPICE (Simulation Program with Integrated Circuit Emphasis), which has become a de facto standard in transistor-level circuit simulation in a wide range of computing platforms. An entire chapter is devoted to the examination and comparison of MOSFET models implemented in SPICE, including the identification of various device model parameters. Computer simulation is, and will continue to be, an essential part of the design process, both for performance verification and for fine-tuning of circuits. However, the emphasis on simulation must be well-balanced with the emphasis on hands-on design and analytical estimates, so that the significance of the latter is not overwhelmed by the extensive use of computer-aided techniques.

The main focus of this book is on CMOS digital integrated circuits, but a significant amount of material on nMOS digital circuits is also presented. Although CMOS has become the technology of choice in many applications in recent years, the fundamental concepts of nMOS logic provide a strong basis both for the conceptual understanding and for the development of CMOS designs. Chapters 5 through 9 are devoted exclusively to the analysis and design of basic CMOS and nMOS digital circuits. Fig. 1.2 shows a simple "family tree" for digital integrated circuits that clarifies the classification and relations among different types of circuits. Based on the utilization of periodic clock signals in circuit operation, the circuits are classified into two main categories, i.e., static circuits and dynamic circuits. The dynamic circuits are further divided into the categories of ratioless, ratioed, and domino logic.

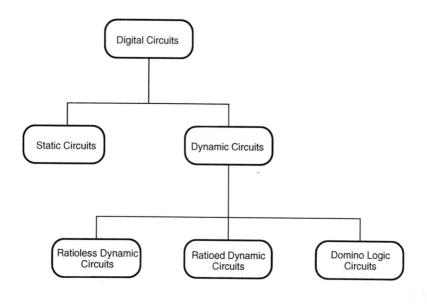

Figure 1.2.   Classification of digital circuit types.

Semiconductor memories are covered in detail in Chapter 10. Specific emphasis is given to the design and operation of different static and dynamic memory types and to comparisons of their performance characteristics. One chapter of the book is devoted to bipolar transistors and to bipolar/BiCMOS digital circuits, which continue to play an important role in the high-performance digital circuits arena. The inclusion of bipolar-based circuits in this book may be puzzling to some readers, but the significance of BiCMOS design techniques cannot be neglected in a comprehensive text on digital design. Various VLSI design styles and large-scale design considerations are discussed in Chapter 13. Finally, two chapters on design for manufacturability and design for testability cover many of the important topics, such as yield estimation, statistical design, and system testability, which deserve special attention in the context of large-scale integrated circuit design.

The chapters are organized in order to allow several different variations of course plans and self-study programs. A number of chapters can be grouped together to accommodate a specific course syllabus, and others can be skipped without a significant loss of continuity. Each chapter contains a large number of solved problems and examples, integrated into the text to enhance the understanding of the material at hand. Also, a collection of problems, some of which are geared specifically for computer-based SPICE simulation, is provided at the end of each chapter.

To help form a global picture of the digital circuit design cycle, in this chapter we begin with a "once over lightly" design exercise wherein we, as circuit designers, start from a logic diagram along with design specifications. The logic circuit is first translated into a CMOS circuit and the initial layout is done. From the layout, all of the important parasitics are calculated by using a circuit extraction program. Once a full circuit description is obtained from the initial layout, we analyze the circuit for DC and transient performance by using the circuit-level simulation program, SPICE, and then compare with the given design specifications. If the initial design fails to meet any one of the specifications, which is the case in this exercise, we devise an improved circuit design to meet the design objective. Then the improved design will be implemented into a new layout and the design-analysis cycle will be repeated until all of the design specifications are met. The simplified flow of this circuit design procedure is illustrated in Fig. 1.3. Note that the topics covered in this textbook concern primarily the two important steps enclosed in the dotted box, namely, *VLSI design* and *design verification.*

Example 1.1.

In the following example, we will design a one-bit binary full-adder circuit using 1-μm, n-well CMOS technology. The design specifications are

> Propagation delay times of sum and carry_out signals $<$ 5 ns
> Transition delay times of sum and carry_out signals $<$ 5 ns
> Circuit area $<$ 3000 μm^2
> Dynamic power dissipation (@ V_{DD} = 5 V and f_{max} = 20 MHz) $<$ 1 mW

We start our design by considering the Boolean description of the binary adder circuit. Let A and B represent the two input variables (addend bits), and let C represent the carry_in bit. The sum and carry_out signals are found as the following two combinational Boolean functions of the three input variables, A, B and C.

$$\text{sum} = A \oplus B \oplus C = ABC + A\overline{BC} + \overline{A}B\overline{C} + \overline{A}CB$$
$$\text{carry\_out} = AB + AC + BC$$

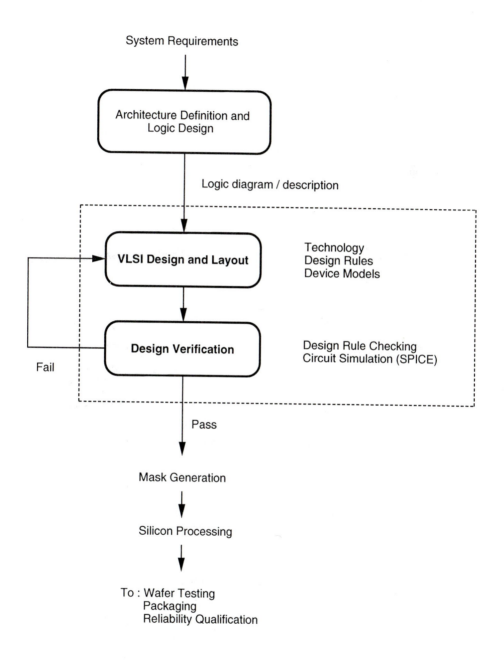

Figure 1.3.   The flow of circuit design procedures.

A gate-level realization of these two functions is shown in Fig. 1.4. Note that instead of realizing the two functions independently, we use the carry_out signal to generate the sum output. This implementation will ultimately reduce the circuit complexity and, hence, save chip area.

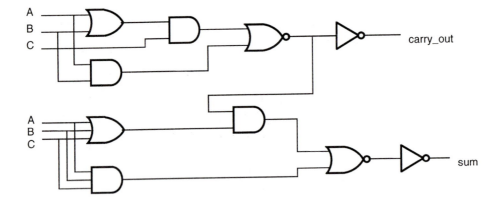

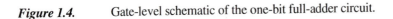

Figure 1.4.   Gate-level schematic of the one-bit full-adder circuit.

For translating the gate-level design into a transistor-level circuit description, we note that both the sum and the carry_out functions are represented by nested AND-OR-INVERT (AOI) structures in Fig. 1.4. Each AOI structure can be realized in CMOS as follows: the AND-terms are implemented by *series-connected* nMOS transistors, the OR-terms are implemented by *parallel-connected* nMOS transistors. The input variables are applied to the gates of the nMOS (and the complementary pMOS) transistors. Thus, the nMOS net may consist of nested series-parallel connections of nMOS transistors, between the output node and the ground. Once the nMOS part of a complex CMOS logic gate is realized, the corresponding pMOS net, which is connected between the output node and the power supply, is obtained as the exact dual of the nMOS net.

The transistor-level design of the CMOS full-adder circuit is shown in Fig. 1.5. Note that the circuit contains a total of 14 nMOS and 14 pMOS transistors. Initially, we will design all nMOS transistors with a (W/L) ratio of (2 μm/1 μm), and all pMOS transistors with a (W/L) ratio of (4 μm/1 μm). This initial sizing of transistors, which is obviously not an optimum solution, may be changed later depending on the performance characteristics of the adder circuit.

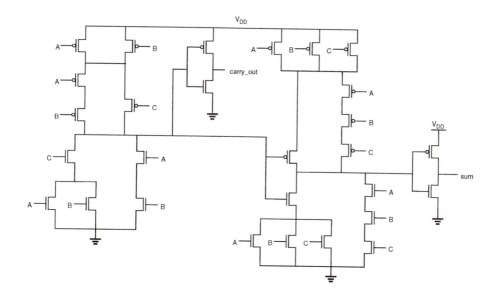

Figure 1.5.   Transistor-level schematic of the one-bit full-adder circuit.

Next, the initial layout of the full-adder circuit is generated. Here we use a regular, gate-matrix layout style in order to simplify the overall geometry and the signal routing. The initial layout (without the output inverters) is shown in Fig. 1.6.

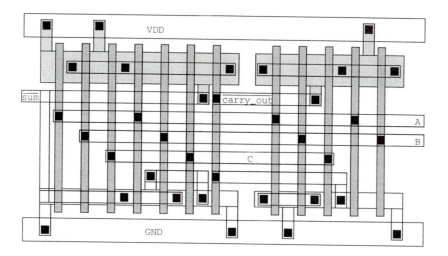

Figure 1.6.   Initial layout of the full-adder circuit.

Note that the horizontal power supply and the ground lines (metal) determine the upper and the lower boundaries of the adder cell layout. All polysilicon lines are laid out vertically, with equal spacing. The n-type and the p-type diffusion areas are placed at the bottom and the top part of the cell, respectively. The area between the two different diffusion regions is used for horizontally running metal interconnects (routing). Also note that the diffusion regions of neighboring transistors have been merged as much as possible, in order to save chip area. The gate-matrix layout style used in this example also has the inherent advantage of being easily adaptable to computer-aided design (CAD).

The designer must confirm that none of the physical layout design rules are violated in this adder layout, using an automatic *design rule checker* (DRC) tool. This is usually done concurrently during the graphical entry of the layout. The next step is to extract the parasitic capacitances and resistances from the initial layout, and then use detailed circuit simulation (SPICE) to estimate the dynamic performance of the adder circuit. Thus, we are now in the *design verification* stage of the design-flow diagram shown in Fig. 1.3. The parasitic extraction tool reads-in the physical layout file, analyzes the various mask layers to identify transistors, interconnects and contacts, calculates the parasitic capacitances and the parasitic resistances of these structures, and finally prepares a SPICE input file that accurately describes the circuit.

The circuit is now simulated using SPICE in order to determine its dynamic performance. Unfortunately, the simulation results of the initial full-adder design show that the circuit does not meet all of the design specifications. The propagation delay times of the sum and carry_out signals are found to violate the timing constraints.

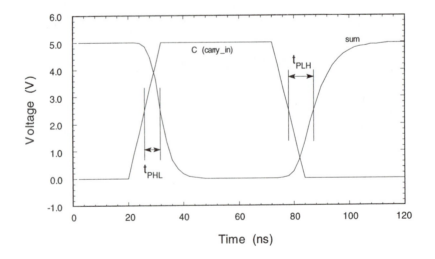

Figure 1.7.   Simulated (sum) output waveform of the full-adder circuit.

As an example, the simulated sum waveform is shown in Fig. 1.7 with $A = 0$, $B = 1$, and C (carry_in) switching. The propagation delay during rising output transition (t_{PLH}) is found to be 8.2 ns, whereas the design specification dictates a maximum propagation delay of $t_{PLH} = 5$ ns. Design modifications will be necessary to correct this problem. Thus, we go back to the design stage.

One approach to increase switching speed, and thus, to reduce delay times, would be to increase the (W/L) ratios of all transistors in the circuit. However, increasing the (W/L) ratios also increases the gate, source, and drain areas and, consequently, increases the parasitic capacitances loading the logic gates. Since the carry_out signal is used to generate the sum output, we will concentrate primarily on reducing the delay of the carry_out stage. Also consider that when several one-bit adder cells are used to build a multi-bit ripple-carry adder, the carry signal will have to propagate through the entire carry chain, causing significant delay. Therefore, reducing the delay in the carry_out stage may solve the performance problems of this full-adder circuit.

We resize the nMOS and pMOS transistors in the carry_out stage as follows. The (W/L) ratios of the nMOS transistors are increased from (2 μm/1 μm) to (4 μm/ 1 μm), and the (W/L) ratios of the pMOS transistors are increased from (4 μm/1 μm) to (7 μm/1 μm). The resulting cell layout is shown in Fig. 1.8. The modified transistors are highlighted with bold lines in the layout.

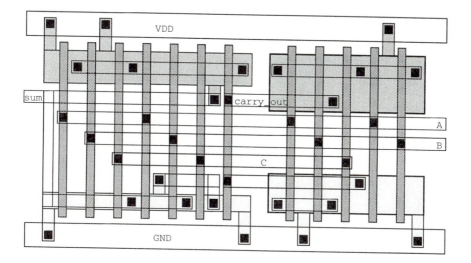

Figure 1.8.   Modified layout of the full-adder circuit.

The design verification steps must now be repeated for this new layout. Following parasitic extraction, we run a series of SPICE simulations on the extracted circuit file. We find that all propagation and transition (rise and fall) delay times are now within the specified limits, i.e., less than 5 ns. Also, the chip area is approximately 1540 μm^2, which is well below the specified upper limit of 3000 μm^2. The dynamic power dissipation of this circuit is estimated to be 670 μW. Thus, the circuit now satisfies the design specifications given in the beginning.

This example has shown us that the design of CMOS digital integrated circuits involves a wide range of issues, from Boolean logic to gate-level design, to transistor-level design, to physical layout design, and to parasitics extraction followed by detailed circuit simulation for design tuning and performance verification. In essence, the final output of integrated circuit design is the mask data from which the actual circuit is fabricated. Thus, it is important to design the layout and, hence, the mask set such that the fabricated integrated circuits meet test specifications with a high yield.

To achieve such a goal, designers perform extensive simulations using computer models extracted from the layout data and iterate the design until simulated results meet the specifications with sufficient margins.

In the following chapters, we will discuss the fabrication of MOS transistors using a set of masks, layout design rules, and electrical properties of MOS transistors and their computer models, before discussing the most basic CMOS inverter circuit.

CHAPTER 2

FABRICATION OF MOSFETs

2.1. Introduction

In this chapter, the fundamentals of MOS chip fabrication will be discussed and the major steps of the process flow will be examined. It is not the aim of this chapter to present a detailed discussion of silicon fabrication technology, which deserves separate treatment in a dedicated course. Rather, the emphasis will be on the general outline of the process flow and on the interaction of various processing steps, which ultimately determine the device and the circuit performance characteristics. The following chapters show that there are very strong links between the fabrication process, the circuit design process, and the performance of the resulting chip. Hence, circuit designers must have a working knowledge of chip fabrication to create effective designs and in order to optimize the circuits with respect to various manufacturing parameters. Also, the circuit designer must have a clear understanding of the roles of various masks used in the fabrication process, and how the masks are used to define various features of the devices on-chip.

The following discussion will concentrate on the well-established CMOS fabrication technology, which requires that both n-channel (nMOS) and p-channel (pMOS) transistors be built on the same chip substrate. To accommodate both nMOS and pMOS devices, special regions must be created in which the semiconductor type is opposite to the substrate type. These regions are called *wells* or *tubs*. A p-well is created in an n-type substrate or, alternatively, an n-well is created in a p-type substrate. In the simple n-well CMOS fabrication technology presented, the nMOS transistor is created in the p-type substrate, and the pMOS transistor is created in the n-well, which is built into the p-type substrate. In the twin-tub CMOS technology, additional tubs of the same type as the substrate can also be created for device optimization.

The simplified process sequence for the fabrication of CMOS integrated circuits on a p-type silicon substrate is shown in Fig. 2.1. The process starts with the creation of the n-well regions for pMOS transistors, by impurity implantation into the substrate. Then, a thick oxide is grown in the regions surrounding the nMOS and pMOS *active regions*. The thin gate oxide is subsequently grown on the surface through thermal oxidation. These steps are followed by the creation of n+ and p+ regions (source, drain, and channel-stop implants) and by final metallization (creation of metal interconnects).

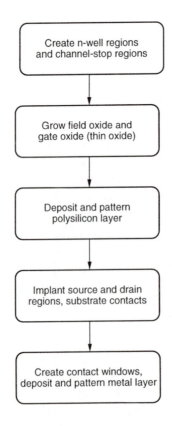

Figure 2.1.   Simplified process sequence for fabrication of the n-well CMOS integrated circuit with a single polysilicon layer, showing only major fabrication steps.

The process flow sequence pictured in Fig. 2.1 may at first seem to be too abstract, since detailed fabrication steps are not shown. To obtain a better understanding of the issues involved in the semiconductor fabrication process, we first have to consider some of the basic steps in more detail.

2.2. Fabrication Process Flow– Basic Steps

Note that each processing step requires that certain areas are *defined* on chip by appropriate *masks*. Consequently, the integrated circuit may be viewed as a set of patterned layers of doped silicon, polysilicon, metal, and insulating silicon dioxide. In general, a layer must be patterned before the next layer of material is applied on chip. The process used to transfer a pattern to a layer on the chip is called *lithography*. Since each layer has its own distinct patterning requirements, the lithographic sequence must be repeated for every layer, using a different mask.

To illustrate the fabrication steps involved in patterning silicon dioxide through optical lithography, let us first examine the process flow shown in Fig. 2.2. The sequence starts with the thermal oxidation of the silicon surface, by which an oxide layer of about 1-micrometer thickness, for example, is created on the substrate (Fig. 2.2(b)). The entire oxide surface is then covered with a layer of *photoresist*, which is essentially a light-sensitive, acid-resistant organic polymer, initially insoluble in the developing solution (Fig. 2.2(c)). If the photoresist material is exposed to ultraviolet (UV) light, the exposed areas become soluble so that they are no longer resistant to etching solvents. To selectively expose the photoresist, we have to cover some of the areas on the surface with a *mask* during exposure. Thus, when the structure with the mask on top is exposed to UV light, areas which are covered by the opaque features on the mask are shielded. In the areas where the UV light can pass through, on the other hand, the photoresist is exposed and becomes soluble (Fig. 2.2(d)).

The type of photoresist which is initially insoluble and becomes soluble after exposure to UV light is called *positive photoresist*. The process sequence shown in Fig. 2.2 uses positive photoresist. There is another type of photoresist which is initially soluble and becomes insoluble (hardened) after exposure to UV light, called *negative photoresist*. If negative photoresist is used in the photolithography process, the areas which are not shielded from the UV light by the opaque mask features become insoluble, whereas the shielded areas can subsequently be etched away by a developing solution. Negative photoresists are more sensitive to light, but their photolithographic resolution is not as high as that of the positive photoresists. Therefore, negative photoresists are used less commonly in the manufacturing of high-density integrated circuits.

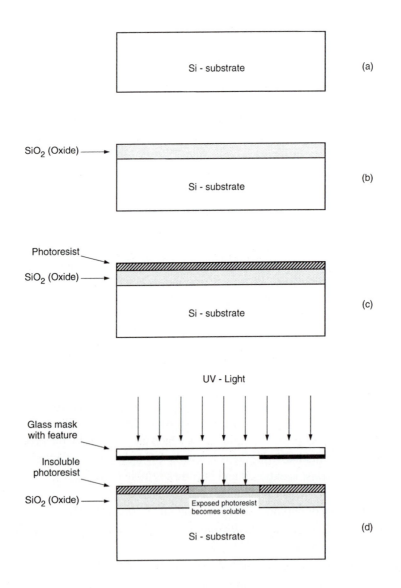

Figure 2.2.   Process steps required for patterning of silicon dioxide.

Following the UV exposure step, the unexposed portions of the photoresist can be removed by a solvent. Now, the silicon dioxide regions which are not covered by hardened photoresist can be etched away either by using a chemical solvent (HF

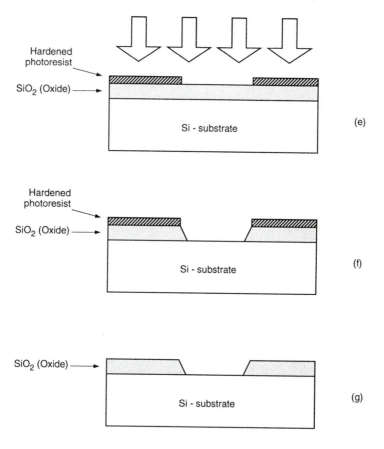

Figure 2.2.   Process steps required for patterning of silicon dioxide (*continued*).

acid) or by using a dry etch (plasma etch) process (Fig. 2.2(e)). Note that at the end of this step, we obtain an oxide window that reaches down to the silicon surface (Fig. 2.2(f)). The remaining photoresist can now be stripped from the silicon dioxide surface by using another solvent, leaving the patterned silicon dioxide feature on the surface as shown in Fig. 2.2(g).

The sequence of process steps illustrated in detail in Fig. 2.2 actually accomplishes a single pattern transfer onto the silicon dioxide surface, as shown in Fig.

2.3. The fabrication of semiconductor devices requires several such pattern transfers to be performed on silicon dioxide, polysilicon, and metal. The basic patterning process used in all fabrication steps, however, is quite similar to the one shown in Fig. 2.2. Also note that for accurate generation of high-density patterns required in sub-micron devices, *electron beam (E-beam) lithography* is used instead of optical lithography. In the following, the main processing steps involved in the fabrication of an n-channel MOS transistor on a p-type silicon substrate will be examined.

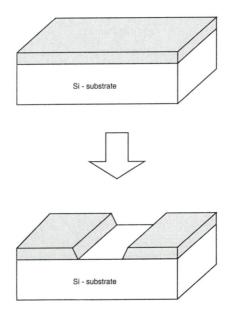

Figure 2.3.   The result of a single lithographic patterning sequence on silicon dioxide, without showing the intermediate steps. Compare the unpatterned structure (top) and the patterned structure (bottom) with Fig. 2.2(b) and Fig. 2.2(g), respectively.

The process starts with the oxidation of the silicon substrate (Fig. 2.4(a)), in which a relatively thick silicon dioxide layer, also called field oxide, is created on the surface (Fig. 2.4(b)). Then, the field oxide is selectively etched to expose the silicon surface on which the MOS transistor will be created (Fig. 2.4(c)). Following this step, the surface is covered with a thin, high-quality oxide layer, which will eventually form the gate oxide of the MOS transistor (Fig. 2.4(d)). On top of the thin oxide, a layer of polysilicon (polycrystalline silicon) is deposited (Fig. 2.4(e)). Polysilicon is used both as gate electrode material for MOS transistors and also as an interconnect medium in silicon integrated circuits. Undoped polysilicon has relatively high resistivity. The resistivity of polysilicon can be reduced, however, by doping it with impurity atoms.

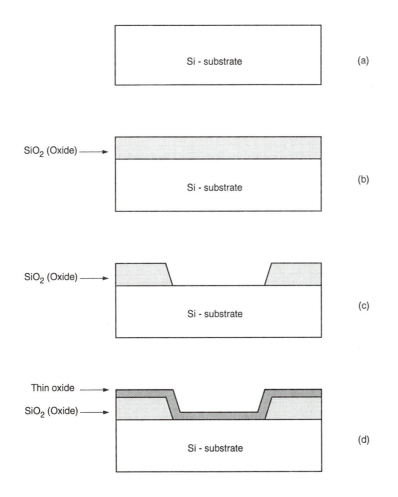

Figure 2.4.   Process flow for the fabrication of an n-type MOSFET on p-type silicon.

After deposition, the polysilicon layer is patterned and etched to form the interconnects and the MOS transistor gates (Fig. 2.4(f)). The thin gate oxide not covered by polysilicon is also etched away, which exposes the bare silicon surface on which the source and drain junctions are to be formed (Fig. 2.4(g)). The entire silicon surface is then doped with a high concentration of impurities, either through diffusion or ion implantation (in this case with donor atoms to produce n-type doping). Figure 2.4(h) shows that the doping penetrates the exposed areas on the silicon surface, ultimately creating two n-type regions (source and drain junctions) in the p-type substrate. The impurity doping also penetrates the polysilicon on the surface, reducing

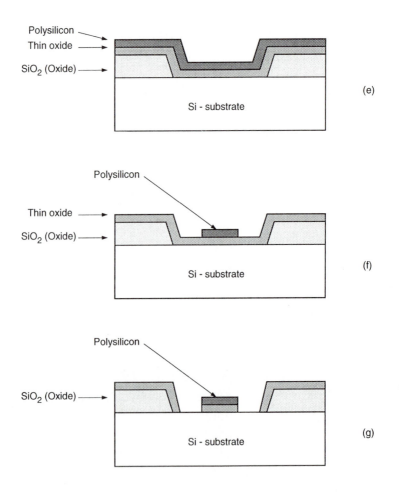

Figure 2.4.   Process flow for the fabrication of an n-type MOS transistor (*continued*).

its resistivity. Note that the polysilicon gate, which is patterned *before* doping, actually defines the precise location of the channel region and, hence, the location of the source and the drain regions. Since this procedure allows very precise positioning of the two regions relative to the gate, it is also called the *self-aligned process*.

Once the source and drain regions are completed, the entire surface is again covered with an insulating layer of silicon dioxide (Fig. 2.4(i)). The insulating oxide layer is then patterned in order to provide contact windows for the drain and source junctions (Fig. 2.4(j)). The surface is covered with evaporated aluminum which will form the interconnects (Fig. 2.4(k)). Finally, the metal layer is patterned and etched, completing the interconnection of the MOS transistors on the surface (Fig. 2.4(l)). Usually, a second (and third) layer of metallic interconnect can also be added on top of this structure by creating another insulating oxide layer, cutting contact (via) holes, depositing, and patterning the metal.

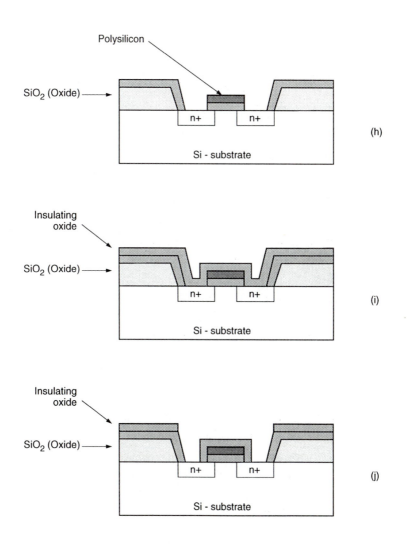

Figure 2.4. Process flow for the fabrication of an n-type MOS transistor (*continued*).

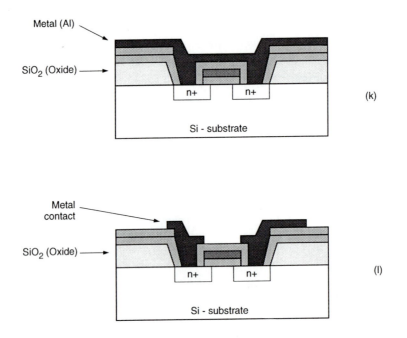

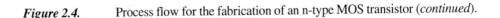

Figure 2.4.   Process flow for the fabrication of an n-type MOS transistor (*continued*).

2.3. The CMOS n-Well Process

Having examined the basic process steps for pattern transfer through lithography, and having gone through the fabrication procedure of a single n-type MOS transistor, we can now return to the generalized fabrication sequence of n-well CMOS integrated circuits, as shown in Fig. 2.1. In the following figures, some of the important process steps involved in the fabrication of a CMOS inverter will be shown by a top view of the lithographic masks and a cross-sectional view of the relevant areas.

The n-well CMOS process starts with a moderately doped (with impurity concentration typically less than 10^{15} cm^{-3}) p-type silicon substrate. Then, an initial oxide layer is grown on the entire surface. The first lithographic mask defines the n-well region. Donor atoms, usually phosphorus, are implanted through this window in the oxide. Once the n-well is created, the active areas of the nMOS and pMOS transistors can be defined. Figures 2.5 through 2.10 illustrate the significant milestones that occur during the fabrication process of a CMOS inverter.

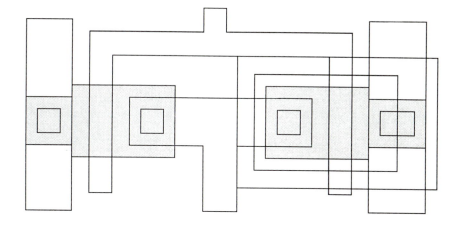

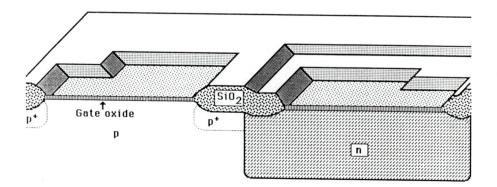

Figure 2.5. Following the creation of the n-well region, a thick field oxide is grown in the areas surrounding the transistor's active regions, and a thin gate oxide is grown on top of the active regions. The thickness and the quality of the gate oxide are two of the most critical fabrication parameters, since they strongly affect the operational characteristics of the MOS transistor, as well as its long-term reliability (Copyright © 1987 by Benjamin/Cummings Publishing Company, Inc.).

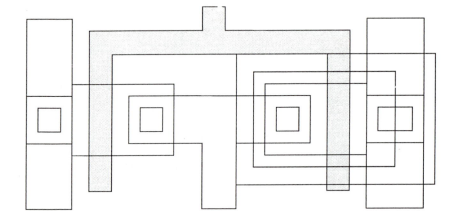

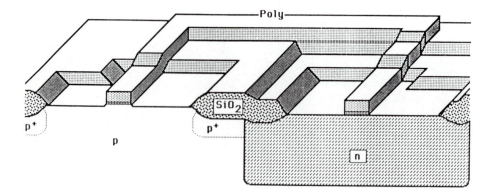

Figure 2.6. The polysilicon layer is deposited using chemical vapor deposition (CVD) and patterned by dry (plasma) etching. The created polysilicon lines will function as the gate electrodes of the nMOS and the pMOS transistors and their interconnects. Also, the polysilicon gates act as self-aligned masks for the source and drain implantations that follow this step (Copyright © 1987 by Benjamin/Cummings Publishing Company, Inc.).

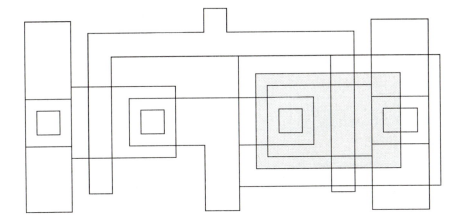

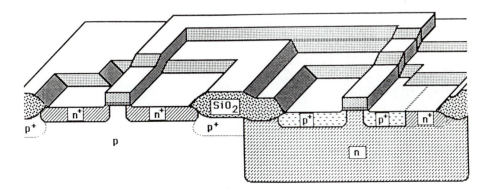

Figure 2.7. Using a set of two masks, the n+ and p+ regions are implanted into the substrate and into the n-well, respectively. Also, the ohmic contacts to the substrate and to the n-well are implanted in this process step (Copyright © 1987 by Benjamin/Cummings Publishing Company, Inc.).

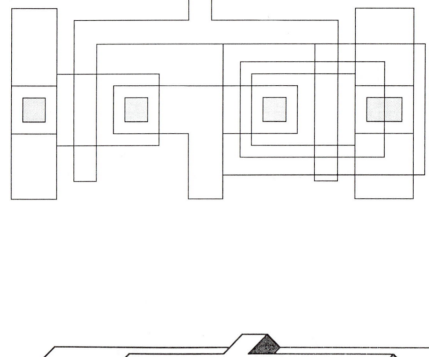

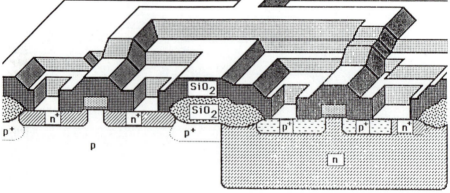

Figure 2.8. An insulating silicon dioxide layer is deposited over the entire wafer using CVD. Then, the contacts are defined and etched away to expose the silicon or polysilicon contact windows. These contact windows are necessary to complete the circuit interconnections using the metal layer, which is patterned in the next step (Copyright © 1987 by Benjamin/Cummings Publishing Company, Inc.).

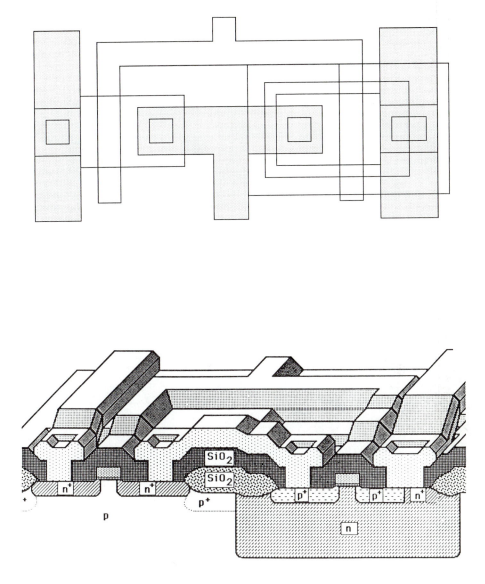

Figure 2.9. Metal (aluminum) is deposited over the entire chip surface using metal evaporation, and the metal lines are patterned through etching. Since the wafer surface is non-planar, the quality and the integrity of the metal lines created in this step are very critical and are ultimately essential for circuit reliability (Copyright © 1987 by Benjamin/Cummings Publishing Company, Inc.).

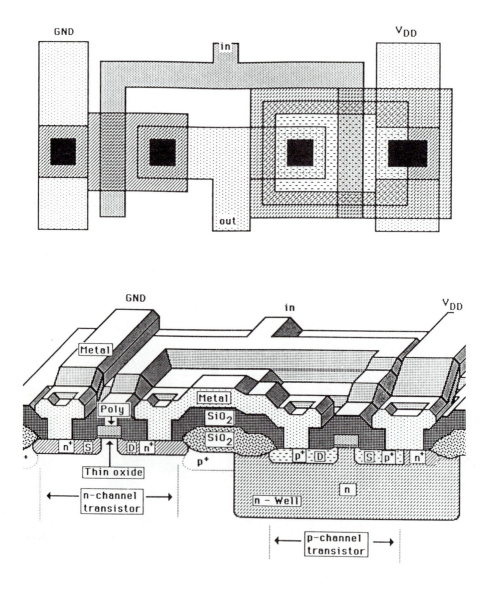

Figure 2.10. The composite layout and the resulting cross-sectional view of the chip, showing one nMOS and one pMOS transistor (built-in n-well), and the polysilicon and metal interconnections (Copyright © 1987 by Benjamin/Cummings Publishing Company, Inc.). The final step is to deposit the passivation layer (for protection) over the chip, except for wire-bonding pad areas.

2.4. Layout Design Rules

The physical mask layout of any circuit to be manufactured using a particular process must conform to a set of geometric constraints or rules, which are generally called layout design rules. These rules usually specify the minimum allowable line widths for physical objects on-chip such as metal and polysilicon interconnects or diffusion areas, minimum feature dimensions, and minimum allowable separations between two such features. If a metal line width is made too small, for example, it is possible for the line to break during the fabrication process or afterwards, resulting in an open circuit. If two lines are placed too close to each other in the layout, they may form an unwanted short circuit by merging during or after the fabrication process. The main objective of design rules is to achieve a high overall yield and reliability while using the smallest possible silicon area, for any circuit to be manufactured with a particular process.

Note that there is usually a trade-off between higher yield, which is obtained through conservative geometries, and better area efficiency, which is obtained through aggressive, high-density placement of various features on the chip. The layout design rules which are specified for a particular fabrication process normally represent a reasonable optimum point in terms of yield and density. It must be emphasized, however, that the design rules do not represent strict boundaries which separate "correct" designs from "incorrect" ones. A layout which violates some of the specified design rules may still result in an operational circuit with reasonable yield, whereas another layout observing all specified design rules may result in a circuit which is not functional and/or has very low yield. To summarize, we can say, in general, that observing the layout design rules significantly increases the probability of fabricating a successful product with high yield.

The design rules are usually described in two ways:

(a)   Micron rules, in which the layout constraints such as minimum feature sizes and minimum allowable feature separations, are stated in terms of absolute dimensions in micrometers, or,

(b)   Lambda rules, which specify the layout constraints in terms of a single parameter (λ) and thus allow linear, proportional scaling of all geometrical constraints.

Lambda-based layout design rules were originally devised to simplify the industry-standard micron-based design rules and to allow scaling capability for various processes. It must be emphasized, however, that most of the submicron CMOS process design rules do not lend themselves to straightforward linear scaling. The use of lambda-based design rules must therefore be handled with caution in submicron geometries. In the following, we present a sample set of the lambda-based layout design rules devised for the MOSIS (MOS Implementation System) CMOS process and illustrate the implications of these rules on a section of a simple layout which includes two transistors (Fig. 2.11). The MOSIS CMOS scalable design rules listed below are also illustrated in Fig. 2.12.

MOSIS Layout Design Rules (sample set)

Rule number	*Description*	*λ-Rule*
R1	Minimum active area width	3λ
R2	Minimum active area spacing	3λ
R3	Minimum poly width	2λ
R4	Minimum poly spacing	2λ
R5	Minimum gate extension of poly over active	2λ
R6	Minimum poly-active edge spacing (poly outside active area)	1λ
R7	Minimum poly-active edge spacing (poly inside active area)	3λ
R8	Minimum metal width	3λ
R9	Minimum metal spacing	3λ
R10	Poly contact size	2λ
R11	Minimum poly contact spacing	2λ
R12	Minimum poly contact to poly edge spacing	1λ
R13	Minimum poly contact to metal edge spacing	1λ
R14	Minimum poly contact to active edge spacing	3λ
R15	Active contact size	2λ
R16	Minimum active contact spacing (on the same active region)	2λ
R17	Minimum active contact to active edge spacing	1λ
R18	Minimum active contact to metal edge spacing	1λ
R19	Minimum active contact to poly edge spacing	3λ
R20	Minimum active contact spacing (on different active regions)	6λ

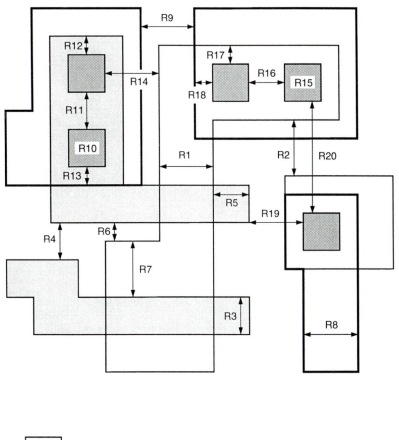

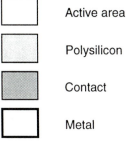

Active area

Polysilicon

Contact

Metal

Figure 2.11. Illustration of some of the typical MOSIS layout design rules listed above.

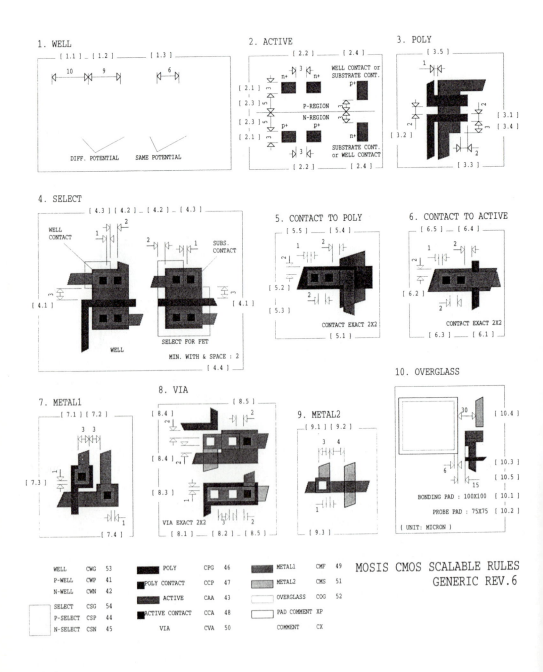

Figure 2.12.   MOSIS CMOS scalable design rules.

References

1.   W. Maly, *Atlas of IC Technologies*, Menlo Park, CA: Benjamin/Cummings, 1987.

2.   A. S. Grove, *Physics and Technology of Semiconductor Devices*, New York, NY: John Wiley & Sons, Inc., 1967.

3.   G. E. Anner, *Planar Processing Primer*, New York, NY: Van Nostrand Rheinhold, 1990.

4.   T. E. Dillinger, *VLSI Engineering*, Englewood Cliffs, NJ: Prentice-Hall, Inc., 1988.

5.   S.M. Sze, *VLSI Technology*, New York, NY: McGraw-Hill, 1983.

Exercise Problems

2.1   Design three masks for diffusion, window, and metal layers to realize a 1-kΩ resistor with n-type diffusion with sheet resistance of 100 Ω per square and lead metal lines. The minimum feature size allowed is 1 μm for line width and window opening. Also, 0.5-μm extension of metal feature and diffusion feature over the window opening is required.

2.2   Discuss whether the mask set in Problem 2.1 will realize the resistance exactly. What would be the effect of window design on the resistance? Discuss at least two other mechanisms in the processing that can make the processed resistance deviate from the target value even if the effect of window is neglected.

2.3   Let us now assume the variation in the resistance is affected only by the exact line feature sizes. Discuss the pros and cons of having the line width at minimum as compared to making the width larger than the minimum width.

2.4   A method for reducing interconnection resistance in the polysilicon lines is to use silicide material deposited on top to form polycide. This process can reduce the nominal sheet resistance from 20 Ω to 2 Ω or even less. For the same purpose, silicide material is also deposited on top of source and drain diffusions in MOS transistors. Thus, a single-step deposition of silicide can be achieved on polysilicon gates and source and drain regions of MOS transistors. Discuss how such deposition can be achieved without causing electrical shorts between the gate and source or drain of an individual transistor.

2.5   In VLSI technologies with multiple layers of metallic interconnects, one of the most yield-limiting processes is the patterning of metal lines, especially when the surface features before metal deposition are not flat. To achieve flat surfaces, chemical mechanical polishing (CMP) has been introduced. Discuss the side effects of CMP on the circuit performance and the process steps following CMP.

2.6   Discuss the difficulties in processing a window mask which contains a wide variation of window sizes, i.e., both very large rectangles and minimum-size square shapes. In particular, what problem would you expect when the process control is based on the monitoring of the smallest windows? How about the opposite case in which the process control is based on the largest window?

2.7   Photolithography has been the driving force behind massive processing of MOS chips at low cost. Despite significant improvements in both photolithography and photoresist materials, it has become increasingly more difficult to process very small, deep submicron feature sizes. Alternatives can be X-ray lithography or direct electron-beam writing. Discuss the difficulties inherent in such alternatives.

2.8   Consider a chip design using 10 mask levels. Suppose that each mask can be made with 98% yield. Determine the composite mask yield for the set of 10 masks. Would the processed chip yield be lower or higher than this composite yield? If your results are inconclusive, explain the reason.

CHAPTER 3

MOS TRANSISTOR

The MOS *Field Effect Transistor* (MOSFET) is the fundamental building block of MOS and CMOS digital integrated circuits. Compared to the bipolar junction transistor (BJT), the MOS transistor occupies a relatively smaller silicon area, and its fabrication involves fewer processing steps. These technological advantages, together with the relative simplicity of MOSFET operation, have helped make the MOS transistor the most widely used switching device in LSI and VLSI circuits. In this chapter, we will examine the basic structure and the electrical behavior of nMOS (n-channel MOS), as well as pMOS (p-channel MOS) devices. The nMOS transistor is used as the primary switching device in virtually all digital circuit applications, whereas the pMOS transistor is used mostly in conjunction with the nMOS device in CMOS circuits. However, the basic operation principles of both nMOS and pMOS transistors are very similar to each other.

This chapter starts with a detailed investigation of the basic electrical and physical properties of *Metal Oxide Semiconductor* (MOS) systems, upon which the MOSFET structure is based. We will consider the effects of external bias conditions on charge distribution in the MOS system and on the conductance of free carriers. It will be shown that, in field effect devices, the current flow is controlled by externally applied electric fields, and that the operation depends only on the majority carrier flow between two device terminals. Next, the current-voltage characteristics of MOS transistors will be examined in detail, including physical limitations imposed by small device geometries and various second-order effects observed in MOSFETs. Note that these considerations will be particularly important for the overall performance of large-scale digital circuits built by using small-geometry MOSFET devices.

3.1. The Metal Oxide Semiconductor (MOS) Structure

We will start our investigation by considering the electrical behavior of the simple two-terminal MOS structure shown in Fig. 3.1. Note that the structure consists of three layers: The metal *gate electrode*, the insulating oxide (SiO_2) layer, and the p-type bulk semiconductor (Si), called the *substrate*. As such, the MOS structure forms a capacitor, with the gate and the substrate acting as the two terminals (plates), and the oxide layer as the dielectric. The thickness of the silicon dioxide layer is usually between 10 nm and 50 nm. The carrier concentration and its local distribution within the semiconductor substrate can now be manipulated by the external voltages applied to the gate and substrate terminals. A basic understanding of the bias conditions for establishing different carrier concentrations in the substrate will also provide valuable insight into the operating conditions of more complicated MOSFET structures.

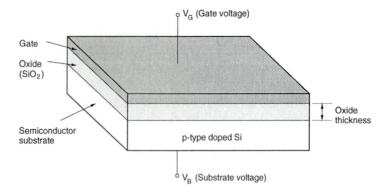

Figure 3.1.   Two-terminal MOS structure.

Consider first the basic electrical properties of the semiconductor (Si) substrate, which acts as one of the electrodes of the MOS capacitor. The equilibrium concentrations of mobile carriers in a semiconductor always obey the *Mass Action Law* given by

$$n \cdot p = n_i^2 \qquad\qquad (3.1)$$

Here, n and p denote the mobile carrier concentrations of electrons and holes, respectively, and n_i denotes the intrinsic carrier concentration of silicon, which is a function of the temperature T. At room temperature, i.e., $T = 300$ K, n_i is approximately equal to 1.45×10^{10} cm^{-3}. Assuming that the substrate is uniformly doped with an acceptor (e.g., Boron) concentration N_A, the equilibrium electron and hole concentrations in the p-type substrate are approximated by

$$n_{po} \cong \frac{n_i^2}{N_A}$$

$$p_{po} \cong N_A$$

(3.2)

The doping concentration N_A is typically on the order of 10^{15} to 10^{16} cm^{-3}; thus, it is much greater than the intrinsic carrier concentration n_i. Note that the bulk electron and hole concentrations given in (3.2) are valid in the regions farther away from the *surface*, where the semiconductor substrate and the oxide layer meet. The conditions on the surface, however, are far more significant for the electrical behavior and the operation of the MOS system, and we will discuss these conditions in more detail.

The energy band diagram of the p-type substrate is shown in Fig. 3.2. The band-gap between the *conduction band* and the *valence band* is approximately 1.1 eV in silicon. The location of the *equilibrium Fermi level* E_F within the band-gap is determined by the doping type and the doping concentration in the silicon substrate. The *Fermi potential* ϕ_F, which is a function of temperature and doping, denotes the difference between the intrinsic Fermi level E_i, and the Fermi level E_F.

$$\phi_F = \frac{E_F - E_i}{q}$$

(3.3)

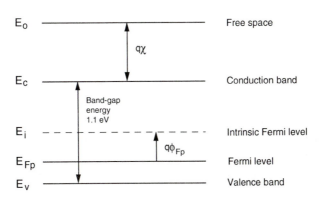

Figure 3.2.   Energy band diagram of a p-type silicon substrate.

For a p-type semiconductor, the Fermi potential can be approximated by

$$\phi_{Fp} = \frac{kT}{q} \ln \frac{n_i}{N_A} \tag{3.4}$$

whereas for an n-type semiconductor (doped with a donor concentration N_D), the Fermi potential is given by

$$\phi_{Fn} = \frac{kT}{q} \ln \frac{N_D}{n_i} \tag{3.5}$$

Here, k denotes the Boltzmann constant, and q denotes the unit (electron) charge. Note that the definitions given in (3.4) and (3.5) result in a positive Fermi potential for n-type material, and a negative Fermi potential for p-type material. We will use this convention throughout the text. The *electron affinity* of silicon, which is the potential difference between the conduction band level and the vacuum (free-space) level, is denoted by $q\chi$ in Fig. 3.2. The energy required for an electron to move from the Fermi level into free space is called the *work function* $q\Phi_S$, and is given by

$$q\Phi_S = q\chi + (E_c - E_F) \tag{3.6}$$

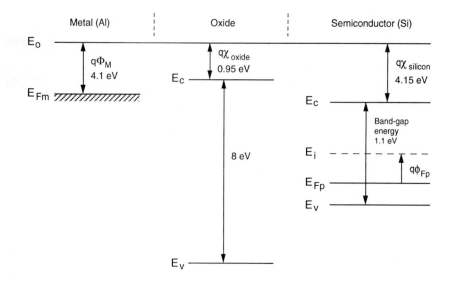

Figure 3.3.   Energy band diagrams of the components that make up the MOS system.

The insulating silicon dioxide layer between the silicon substrate and the gate has a large band-gap of about 8 eV and an electron affinity of about 0.95 eV. On the other hand, the work function $q\Phi_M$ of an aluminum gate is about 4.1 eV. Figure 3.3 shows the energy band diagrams of metal, oxide, and semiconductor layers in a MOS system as three separate components.

Now consider that the three components of the ideal MOS system are brought into physical contact. The Fermi levels of all three materials must line up, as they form the MOS capacitor shown in Fig. 3.1. Because of the work-function difference between the metal and the semiconductor, a voltage drop occurs across the MOS system. Part of this built-in voltage drop occurs across the insulating oxide layer. The rest of the voltage drop (potential difference) occurs at the silicon surface next to the silicon-oxide interface, forcing the energy bands of silicon to bend in this region. The resulting combined energy band diagram of the MOS system is shown in Fig. 3.4. Notice that the equilibrium Fermi levels of the semiconductor (Si) substrate and the metal gate are at the same potential. The bulk Fermi level is not significantly affected by the band bending, whereas the surface Fermi level moves closer to the intrinsic Fermi (mid-gap) level. The Fermi potential at the surface, also called *surface potential* ϕ_s, is smaller in magnitude than the bulk Fermi potential ϕ_F.

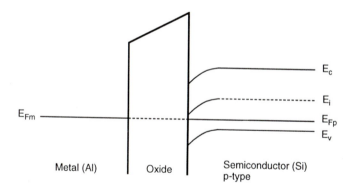

Figure 3.4.   Energy band diagram of the combined MOS system.

Example 3.1

Consider the MOS structure that consists of a p-type doped silicon substrate, a silicon dioxide layer, and a metal (aluminum) gate. The equilibrium Fermi potential of the doped silicon substrate is given as $q\phi_{Fp} = 0.2$ eV. Using the electron affinity for silicon and the work function for aluminum given in Fig. 3.3, calculate the built-in potential difference across the MOS system. Assume that the MOS system contains no other charges in the oxide or on the silicon-oxide interface.

First, we have to calculate the work function for the doped silicon, which is given by (3.6). Since the electron affinity of silicon is 4.15 eV, the work function $q\Phi_S$ is found as

$$q\Phi_S = 4.15 \text{ eV} + 0.75 \text{ eV} = 4.9 \text{ eV}$$

Now calculate the work function difference between the silicon substrate and the aluminum gate. Note that the work function of aluminum is given as 4.1 eV in Fig. 3.3. Thus, the built-in potential difference across this MOS system is

$$q\Phi_M - q\Phi_S = 4.1 \text{ eV} - 4.9 \text{ eV} = -0.8 \text{ eV}$$

If a voltage corresponding to this potential difference is applied externally between the gate and the substrate, the bending of the energy bands near the surface can be compensated, i.e., the energy bands become "flat." Thus, the voltage defined by

$$V_{FB} = \Phi_M - \Phi_S$$

is called the *flat-band* voltage.

3.2. The MOS System under External Bias

We now turn our attention to the electrical behavior of the MOS structure under externally applied bias voltages. Assume that the substrate voltage is set at $V_B = 0$, and let the gate voltage be the controlling parameter. Depending on the polarity and the magnitude of V_G, three different operating regions can be observed for the MOS system: *accumulation, depletion,* and *inversion.*

If a negative voltage V_G is applied to the gate electrode, the holes in the p-type substrate are attracted to the semiconductor-oxide interface. The majority carrier concentration near the surface becomes larger than the equilibrium hole concentration in the substrate; hence, this condition is called carrier *accumulation* on the surface (Fig. 3.5). Note that in this case, the oxide electric field is directed towards the gate electrode. The negative surface potential also causes the energy bands to bend upward near the surface. While the hole density near the surface increases as a result of the applied negative gate bias, the electron (minority carrier) concentration decreases as the negatively charged electrons are pushed deeper into the substrate.

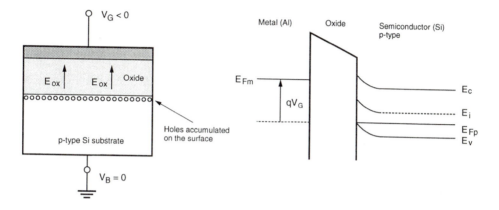

Figure 3.5.   The cross-sectional view and the energy band diagram of the MOS structure operating in accumulation region.

Now consider the next case in which a small positive gate bias V_G is applied to the gate electrode. Since the substrate bias is zero, the oxide electric field will be directed towards the substrate in this case. The positive surface potential causes the energy bands to bend downward near the surface, as shown in Fig. 3.6. The majority carriers, i.e., the holes in the substrate, will be repelled back into the substrate as a result of the positive gate bias, and these holes will leave negatively charged fixed acceptor

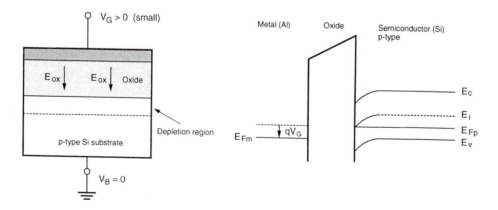

Figure 3.6.   The cross-sectional view and the energy band diagram of the MOS structure operating in depletion mode, under small gate bias.

ions behind. Thus, a *depletion* region is created near the surface. Note that under this bias condition, the region near the semiconductor-oxide interface is nearly devoid of all mobile carriers.

The thickness x_d of this depletion region on the surface can easily be found as a function of the surface potential ϕ_s. Assume that the mobile hole charge in a thin horizontal layer parallel to the surface is

$$dQ = -q \cdot N_A \cdot dx \tag{3.7}$$

The *change* in surface potential required to displace this charge sheet dQ by a distance x_d away from the surface can be found by using the Poisson equation.

$$d\phi_s = -x \cdot \frac{dQ}{\varepsilon_{Si}} = \frac{q \cdot N_A \cdot x}{\varepsilon_{Si}} dx \tag{3.8}$$

Integrating (3.7) along the vertical dimension (perpendicular to the surface) yields

$$\int_{\phi_F}^{\phi_s} d\phi_s = \int_{0}^{x_d} \frac{q \cdot N_A \cdot x}{\varepsilon_{Si}} dx \tag{3.9}$$

$$\phi_s - \phi_F = \frac{q \cdot N_A \cdot x_d^2}{2 \varepsilon_{Si}} \tag{3.10}$$

Thus, the depth of the depletion region is

$$x_d = \sqrt{\frac{2 \varepsilon_{Si} \cdot |\phi_s - \phi_F|}{q \cdot N_A}} \tag{3.11}$$

and the depletion region charge density, which consists solely of fixed acceptor ions in this region, is given by the following expression

$$Q = -q \cdot N_A \cdot x_d = -\sqrt{2q \cdot N_A \cdot \varepsilon_{Si} \cdot |\phi_s - \phi_F|} \tag{3.12}$$

The amount of this depletion region charge plays a very important role in the analysis of threshold voltage, as we will examine shortly.

To complete our qualitative overview of different bias conditions and their effects upon the MOS system, consider next that the positive gate bias is further increased. As a result of the increasing surface potential, the downward bending of the energy bands will increase as well. Eventually, the mid-gap energy level E_i becomes smaller than the Fermi level E_{Fp} on the surface, which means that the substrate semiconductor in this region becomes n-type. Within this thin layer, the electron density is larger than the majority hole density, since the positive gate potential attracts additional minority carriers (electrons) from the bulk substrate, to the surface (Fig. 3.7). The n-type region created near the surface by the positive gate bias is called the *inversion layer*, and this condition is called *surface inversion*. It will be seen that the thin inversion layer on the surface with a large mobile electron concentration can be utilized for conducting current between two terminals of the MOS transistor.

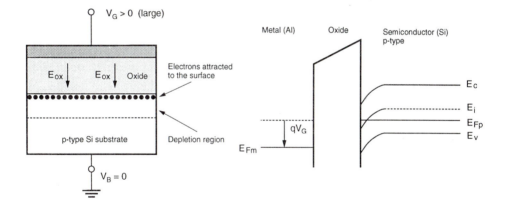

Figure 3.7.   The cross-sectional view and the energy band diagram of the MOS structure in surface inversion, under larger gate bias voltage.

As a practical definition, the surface is said to be *inverted* when the density of mobile electrons on the surface becomes equal to the density of holes in the bulk (p-type) substrate. This condition requires that the surface potential has the same magnitude, but the reverse polarity, as the bulk Fermi potential ϕ_F. Once the surface is inverted, any further increase in the gate voltage leads to an increase of mobile electron concentration on the surface, but not to an increase of the depletion depth. Thus, the depletion region depth achieved at the onset of surface inversion is also equal to the maximum depletion depth, x_{dm}, which remains constant for higher gate voltages. Using the inversion condition $\phi_s = -\phi_F$, the maximum depletion region depth at the onset of surface inversion can be found from (3.1!) as follows.

$$x_{dm} = \sqrt{\frac{2 \cdot \varepsilon_{Si} \cdot |2\phi_F|}{q \cdot N_A}} \qquad (3.13)$$

The creation of a conducting surface inversion layer through externally applied gate bias is an essential phenomenon for current conduction in MOS transistors. In the following section, we will examine the structure and the operation of the MOS Field Effect Transistor (MOSFET).

3.3. Structure and Operation of MOS Transistor (MOSFET)

The basic structure of an n-channel MOS Field Effect Transistor (MOSFET) is shown in Fig. 3.8. This four-terminal device consists of a p-type substrate, in which two n$^+$ diffusion regions, the drain and the source, are formed. The surface of the substrate region between the drain and the source is covered with a thin oxide layer, and the metal (or polysilicon) gate is deposited on top of this gate dielectric. The midsection of the device can easily be recognized as the basic MOS structure which was examined in the previous sections. The two n$^+$ regions will be the current-conducting terminals of this device. Note that the device structure is completely symmetrical with respect to the drain and source regions; the different roles of these two regions will be defined only in conjunction with the applied terminal voltages and the direction of the current flow.

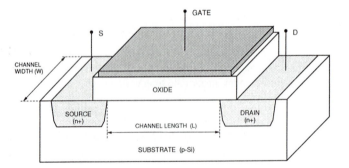

Figure 3.8.   The physical structure of an n-channel enhancement-type MOSFET.

A conducting *channel* will eventually be formed through applied gate voltage in the section of the device between the drain and the source diffusion regions. The distance between the drain and source diffusion regions is the *channel length L,* and the lateral extent of the channel (perpendicular to the length dimension) is the *channel width W.* Both the channel length and the channel width are important parameters which can be used to control some of the electrical properties of the MOSFET. The thickness of the oxide layer covering the channel region, t_{ox}, is also an important parameter.

A MOS transistor which has no conducting channel region at zero gate bias is called an *enhancement-type* (or *enhancement-mode*) MOSFET. If a conducting channel already exists at zero gate bias, on the other hand, the device is called a *depletion-type* (or *depletion-mode*) MOSFET. In a MOSFET with p-type substrate, and with n$^+$ source and drain regions, the channel region to be formed on the surface is n-type. Thus, such a device with p-type substrate is called an *n-channel MOSFET*. In a MOSFET with n-type substrate and with p$^+$ source and drain regions, on the other hand, the channel is p-type, and the device is called a *p-channel MOSFET*.

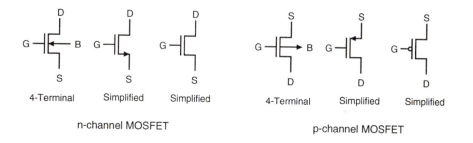

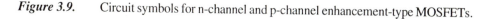

Figure 3.9.   Circuit symbols for n-channel and p-channel enhancement-type MOSFETs.

The abbreviations used for the device terminals are: G for the gate, D for the drain, S for the source, and B for the substrate (or body). In an n-channel MOSFET, the source is defined as the n$^+$ region which has a *lower* potential than the other n$^+$ region, the drain. By convention, all terminal voltages of the device are defined with respect to the source potential. Thus, the gate-to-source voltage is denoted by V_{GS}, the drain-to-source voltage is denoted by V_{DS}, and the substrate-to-source voltage is denoted by V_{BS}. Circuit symbols for both n-channel and p-channel enhancement-type MOSFETs are shown in Fig. 3.9. While the four-terminal symbolic representation shows all external terminals of the device, the simple three-terminal representation will also be used extensively. Note that in the simple MOSFET circuit symbol, the small arrow always marks the source terminal.

Consider first the n-channel enhancement-type MOSFET shown in Fig. 3.8. The simple operation principle of this device is: *controlling the current conduction between the source and the drain, using the electric field generated by the gate voltage as a control variable.* Since the current flow in the channel is also controlled by the drain-to-source voltage and by the substrate voltage, the current can be considered a function of these external terminal voltages. We will examine in detail the functional relationships between the channel current (also called the *drain current*) and the terminal voltages. In order to start current flow between the source and the drain regions, however, we have to form a conducting channel first.

The simplest bias condition that can be applied to the n-channel enhancement-type MOSFET is shown in Fig. 3.10. The source, the drain, and the substrate terminals are all connected to ground. A positive gate-to-source voltage V_{GS} is then applied to the gate in order to create the conducting channel underneath the gate. With this bias arrangement, the channel region between the source and the drain diffusions behaves exactly the same as for the simple MOS structure we examined in Section 3.2. For small gate voltage levels, the majority carriers (holes) are repelled back into the substrate, and the surface of the p-type substrate is depleted. Since the surface is devoid of any mobile carriers, current conduction between the source and the drain is not possible.

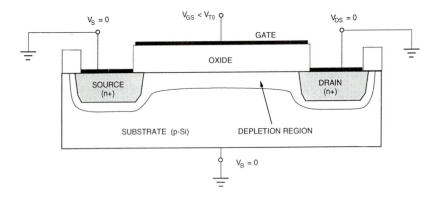

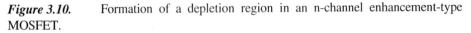

Figure 3.10.   Formation of a depletion region in an n-channel enhancement-type MOSFET.

Now assume that the gate-to-source voltage is further increased. As soon as the surface potential in the channel region reaches $-\phi_{Fp}$, surface inversion will be established, and a conducting n-type layer will form between the source and the drain diffusion regions (Fig. 3.11). This channel now provides an electrical connection between the two n^+ regions, and it allows current flow, as long as there is a potential difference between the source and the drain terminal voltages (Fig. 3.12). The bias conditions for the onset of surface inversion and for the creation of the conducting channel are therefore very significant for MOSFET operation.

The value of the gate-to-source voltage V_{GS} needed to cause surface inversion (to create the conducting channel) is called the *threshold voltage V_{T0}*. Any gate-

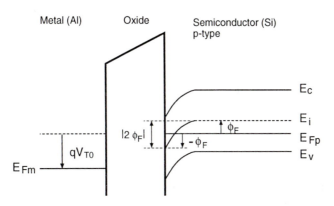

Figure 3.11.   Band diagram of the MOS structure underneath the gate, at surface inversion. Notice the band bending by $|2\phi_F|$ at the surface.

to-source voltage smaller than V_{T0} is not sufficient to establish an inversion layer; thus, the MOSFET can conduct no current between its source and drain terminals unless V_{GS} $>V_{T0}$. For gate-to-source voltages larger than the threshold voltage, on the other hand, a larger number of minority carriers (electrons) are attracted to the surface, which ultimately contribute to channel current conduction. Also note that increasing the gate-to-source voltage above and beyond the threshold voltage will not affect the surface potential and the depletion region depth. Both quantities will remain approximately constant and equal to their values attained at the onset of surface inversion.

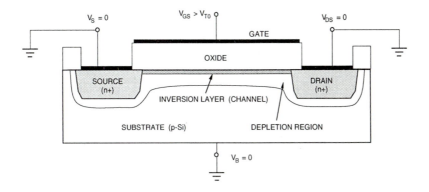

Figure 3.12.   Formation of an inversion layer (channel) in an n-channel enhancement-type MOSFET.

The Threshold Voltage

In the following, physical parameters affecting the threshold voltage of a MOS structure will be examined by considering the various components of V_{T0}. For all practical purposes, we can identify four physical components of the threshold voltage: (i) the work function difference between the gate and the channel, (ii) the gate voltage component to change the surface potential, (iii) the gate voltage component to offset the depletion region charge, and (iv) the voltage component to offset the fixed charges in the gate oxide and in the silicon-oxide interface. The analysis will be carried out for an n-channel device, but the results are applicable to p-channel devices as well, with minor modifications.

The work function difference Φ_{GC} between the gate and the channel reflects the built-in potential of the MOS system, which consists of the p-type substrate, the thin silicon dioxide layer, and the gate electrode. Depending on the gate material, the work function difference is

$$\Phi_{GC} = \phi_F(substrate) - \phi_M \qquad \text{for metal gate} \qquad (3.14)$$
$$\Phi_{GC} = \phi_F(substrate) - \phi_F(gate) \qquad \text{for polysilicon gate} \qquad (3.15)$$

This first component of the threshold voltage accounts for part of the voltage drop across the MOS system that is built-in. Now, the externally applied gate voltage must be changed to achieve surface inversion, i.e., to change the surface potential by $-2\phi_F$. This will be the second component of the threshold voltage.

Another component of the applied gate voltage is necessary to offset the depletion region charge, which is due to the fixed acceptor ions located in the depletion region near the surface. We can calculate the depletion region charge density at surface inversion ($\phi_s = -\phi_F$) using (3.12).

$$Q_{B0} = -\sqrt{2q \cdot N_A \cdot \varepsilon_{Si} \cdot |-2\phi_F|} \qquad (3.16)$$

Note that if the substrate (body) is biased at a different voltage level than the source, which is at ground potential (reference), then the depletion region charge density can be expressed as a function of the source-to-substrate voltage V_{SB}.

$$Q_B = -\sqrt{2q \cdot N_A \cdot \varepsilon_{Si} \cdot |-2\phi_F + V_{SB}|} \qquad (3.17)$$

The component that offsets the depletion region charge is then equal to $-Q_B/C_{ox}$, where C_{ox} is the gate oxide capacitance per unit area.

$$C_{ox} = \frac{\varepsilon_{ox}}{t_{ox}} \tag{3.18}$$

Finally, we must consider the influence of a nonideal physical phenomenon which we have neglected until now. There always exists a fixed positive charge density Q_{ox} at the interface between the gate oxide and the silicon substrate, due to impurities and/or lattice imperfections at the interface. The gate voltage component that is necessary to offset this positive charge at the interface is $-Q_{ox}/C_{ox}$. Now, we can combine all of these voltage components to find the threshold voltage.

For zero substrate bias, the threshold voltage V_{T0} is expressed as follows:

$$\tag{3.19}$$
$$V_{T0} = \Phi_{GC} - 2\phi_F - \frac{Q_{B0}}{C_{ox}} - \frac{Q_{ox}}{C_{ox}}$$

For nonzero substrate bias, on the other hand, the depletion charge density term must be modified to reflect the influence of V_{SB} upon that charge, resulting in the following generalized threshold voltage expression.

$$V_T = \Phi_{GC} - 2\phi_F - \frac{Q_B}{C_{ox}} - \frac{Q_{ox}}{C_{ox}} \tag{3.20}$$

The generalized form of the threshold voltage can also be written as

$$V_T = \Phi_{GC} - 2\phi_F - \frac{Q_{B0}}{C_{ox}} - \frac{Q_{ox}}{C_{ox}} - \frac{Q_B - Q_{B0}}{C_{ox}} = V_{T0} - \frac{Q_B - Q_{B0}}{C_{ox}} \tag{3.21}$$

Note that in this case, the threshold voltage differs from V_{T0} only by an additive term. This substrate-bias term is a simple function of the material constants and of the source-to-substrate voltage V_{SB}.

$$\frac{Q_B - Q_{B0}}{C_{ox}} = -\frac{\sqrt{2q \cdot N_A \cdot \varepsilon_{Si}}}{C_{ox}} \cdot \left(\sqrt{|-2\phi_F + V_{SB}|} - \sqrt{|2\phi_F|} \right) \tag{3.22}$$

Thus, the most general expression of the threshold voltage V_T can be found as follows.

$$V_T = V_{T0} + \gamma \cdot \left(\sqrt{|-2\phi_F + V_{SB}|} - \sqrt{|2\phi_F|} \right) \tag{3.23}$$

where the parameter γ

$$\gamma = \frac{\sqrt{2q \cdot N_A \cdot \varepsilon_{Si}}}{C_{ox}} \tag{3.24}$$

is the *substrate-bias* (or *body-effect*) coefficient.

The threshold voltage expression given in (3.23) can be used both for n-channel and p-channel MOS transistors. One must be careful, however, since some of the terms and coefficients in this equation have different polarities for the n-channel (nMOS) case and for the p-channel (pMOS) case. The reason for this polarity difference is that the substrate semiconductor is p-type in an n-channel MOSFET and n-type in a p-channel MOSFET. Specifically,

*   the substrate Fermi potential ϕ_F is *negative in nMOS, positive in pMOS,*
*   the depletion region charge densities Q_{B0} and Q_B are *negative in nMOS, positive in pMOS,*
*   the substrate bias coefficient γ is *positive in nMOS, negative in pMOS,*
*   the substrate bias voltage V_{SB} is *positive in nMOS, negative in pMOS.*

Typically, the threshold voltage of an enhancement-type n-channel MOSFET is a positive quantity, whereas the threshold voltage of a p-channel MOSFET is negative.

Example 3.2

Calculate the threshold voltage V_{T0} at $V_{SB} = 0$, for a polysilicon gate n-channel MOS transistor, with the following parameters: substrate doping density $N_A = 10^{16}$ cm^{-3}, polysilicon gate doping density $N_D = 2 \times 10^{20}$ cm^{-3}, gate oxide thickness $t_{ox} = 500$ Å, and oxide-interface fixed charge density $N_{ox} = 4 \times 10^{10}$ cm^{-2}.

First, calculate the Fermi potentials for the p-type substrate and for the n-type polysilicon gate:

$$\phi_F(substrate) = \frac{kT}{q} \ln\left(\frac{n_i}{N_A}\right) = 0.026 \text{ V} \cdot \ln\left(\frac{1.45 \cdot 10^{10}}{10^{16}}\right) = -0.35 \text{ V}$$

Since the doping density of the polysilicon gate is very high, the heavily doped n-type gate material is expected to be degenerate. Thus, we may assume that the Fermi potential of the polysilicon gate is approximately equal to the conduction band potential, i.e., $\phi_F(gate) = 0.55$ V. Now, calculate the work function difference between the gate and the channel:

$$\Phi_{GC} = \phi_F(substrate) - \phi_F(gate) = -0.35 \text{ V} - 0.55 \text{ V} = -0.90 \text{ V}$$

The depletion region charge density at $V_{SB} = 0$ is found as follows:

$$Q_{B0} = -\sqrt{2 \cdot q \cdot N_A \cdot \varepsilon_{Si} \cdot |-2\phi_F(substrate)|}$$
$$= -\sqrt{2 \cdot 1.6 \cdot 10^{-19} \cdot 10^{16} \cdot 11.7 \cdot 8.85 \cdot 10^{-14} |-2 \cdot 0.35|} = -4.82 \cdot 10^{-8} \text{C/cm}^2$$

The oxide-interface charge is :

$$Q_{ox} = q \cdot N_{ox} = 1.6 \cdot 10^{-19} \text{C} \times 4 \cdot 10^{10} \text{cm}^{-2} = 6.4 \cdot 10^{-9} \text{C/cm}^2$$

The gate oxide capacitance per unit area is calculated using the dielectric constant of silicon dioxide and the oxide thickness t_{ox}.

$$C_{ox} = \frac{\varepsilon_{ox}}{t_{ox}} = \frac{3.97 \cdot 8.85 \cdot 10^{-14} \text{ F/cm}}{500 \cdot 10^{-8} \text{ cm}} = 7.03 \cdot 10^{-8} \text{ F/cm}^2$$

Now, we can combine all components and calculate the threshold voltage.

$$V_{T0} = \Phi_{GC} - 2\phi_F(substrate) - \frac{Q_{B0}}{C_{ox}} - \frac{Q_{ox}}{C_{ox}}$$
$$= -0.90 - (-0.70) - (-0.69) - 0.09 = 0.40 \text{ V}$$

In this simplified analysis, the doping concentrations of the source and the drain diffusion regions and the geometry (physical dimensions) of the channel region have no influence upon the threshold voltage V_{T0}.

Note that the exact value of the threshold voltage of an actual MOS transistor cannot be determined using (3.23) in most practical cases, due primarily to uncertainties and variations of the doping concentrations, the oxide thickness, and the fixed oxide-interface charge. The nominal value and the statistical range of the threshold voltage for any MOS process are ultimately determined by direct measurements, which will be described later in Section 3.4. In most MOS fabrication processes, the threshold voltage can be adjusted by selective dopant ion implantation into the channel region of the MOSFET. For n-channel MOSFETs, the threshold voltage is *increased* (made more positive) by adding extra p-type impurities (acceptor ions). Alternatively, the threshold voltage of the n-channel MOSFET can be *decreased* (made more negative) by implanting n-type impurities (dopant ions) into the channel region.

The amount of change in the threshold voltage as a result of extra implants can be approximated as follows. Let the density of implanted impurities be represented by N_I [cm^{-2}]. Assume that all implanted ions are electrically active, i.e., each ion contributes to the depletion region charge. Then, the threshold voltage V_{T0} at zero substrate bias ($V_{SB} = 0$) will be shifted by an amount of qN_I/C_{ox}. This approximation obviously neglects the variation of the substrate Fermi level ϕ_F as the result of extra implants, but it nevertheless provides a fair estimate for the threshold voltage shift.

Exercise 3.1

Consider the following p-channel MOSFET process:

Substrate doping $N_D = 10^{15}$ cm^{-3}, polysilicon gate doping density $N_D = 10^{20}$ cm^{-3}, gate oxide thickness $t_{ox} = 650$ Å, and oxide-interface charge density $N_{ox} = 2 \times 10^{10}$ cm^{-2}. Use $\varepsilon_{Si} = 11.7\varepsilon_0$ and $\varepsilon_{ox} = 3.97\varepsilon_0$ for the dielectric coefficients of silicon and silicon-dioxide, respectively.

(a)   Calculate the threshold voltage V_{T0}, for $V_{SB} = 0$.

(b)   Determine the type and the amount of channel ion implantation which are necessary to achieve a threshold voltage of $V_{T0} = -2$ V.

Note that, using selective ion implantation into the channel, the threshold voltage of an n-channel MOSFET can also be made negative. This means that the resulting nMOS transistor will have a conducting channel at $V_{GS} = 0$, enabling current flow between its source and drain terminals as long as V_{GS} is larger than the negative threshold voltage. Such a device is called a *depletion-type* (or *normally-on*) n-channel MOSFET. We will see several practical applications for depletion-type nMOS transistors in the design of MOS digital circuits. Except for its negative threshold voltage, the depletion-type n-channel MOSFET exhibits the same electrical behavior as the enhancement-type n-channel MOSFET. Figure 3.13 shows the conventional circuit symbols used for depletion-type n-channel MOSFETs.

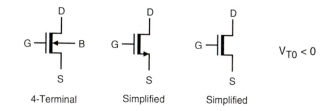

| 4-Terminal | Simplified | Simplified |

Figure 3.13. Circuit symbols for n-channel depletion-type MOSFETs.

Example 3.3

Consider the n-channel MOSFET process given in Example 3.2. In several digital circuit applications, the condition $V_{SB} = 0$ cannot be guaranteed for all transistors. We will examine in this example how a nonzero source-to-substrate voltage V_{SB} affects the threshold voltage of the MOS transistor.

First, we must calculate the substrate-bias coefficient γ using the process parameters given in Example 3.2.

$$\gamma = \frac{\sqrt{2 \cdot q \cdot N_A \cdot \varepsilon_{Si}}}{C_{ox}} = \frac{\sqrt{2 \cdot 1.6 \cdot 10^{-19} \cdot 10^{16} \cdot 11.7 \cdot 8.85 \cdot 10^{-14}}}{7.03 \cdot 10^{-8}} = 0.82 \text{ V}^{\frac{1}{2}}$$

Now we compute and plot the threshold voltage V_T as a function of the source-to-substrate voltage V_{SB}. The voltage V_{SB} will be assumed to vary between zero and 5 V.

$$V_T = V_{T0} + \gamma \cdot \left(\sqrt{\left| -2\phi_F + V_{SB} \right|} - \sqrt{\left| 2\phi_F \right|} \right) = 0.40 + 0.82 \cdot \left(\sqrt{0.7 + V_{SB}} - \sqrt{0.7} \right)$$

It is seen that the threshold voltage variation is about 1.5 V over this range, which could present serious design problems if neglected. We will see in the following chapters that the substrate-bias effect is unavoidable in most digital circuits and that the circuit designer usually must take appropriate measures to account for and/or to compensate for the threshold voltage variations.

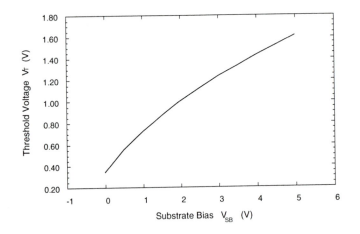

Variation of the threshold voltage as a function of the source-to-substrate voltage.

MOSFET Operation: A Qualitative View

The basic structure of the n-channel MOS (nMOS) transistor built on a p-type substrate was shown in Fig. 3.8. The MOSFET consists of a MOS capacitor with two p-n junctions placed immediately adjacent to the *channel* region that is controlled by the MOS gate. The carriers, i.e., electrons in an nMOS transistor, enter the structure through the source contact (S), leave through the drain (D), and are subject to the control of the gate (G) voltage. To ensure that both p-n junctions are reverse-biased initially, the substrate potential is kept lower than the other three terminal potentials.

We have seen that when $0 < V_{GS} < V_{T0}$, the gated region between the source and the drain is depleted; no carrier flow can be observed in the channel. As the gate voltage is increased beyond the threshold voltage ($V_{GS} > V_{T0}$), however, the mid-gap energy level at the surface is pulled below the Fermi level, causing the surface potential ϕ_s to turn positive and to *invert* the surface (Fig. 3.12). Once the inversion layer is established on the surface, an n-type conducting channel forms between the source and the drain, which is capable of carrying the drain current.

Next, the influence of drain-to-source bias V_{DS} and different modes of drain current flow will be examined for an nMOS transistor with $V_{GS} > V_{T0}$. At $V_{DS} = 0$, thermal equilibrium exists in the inverted channel region, and the drain current I_D is equal to zero (Fig. 3.14(a)). If a small drain voltage $V_{DS} > 0$ is applied, a drain current

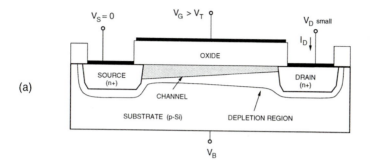

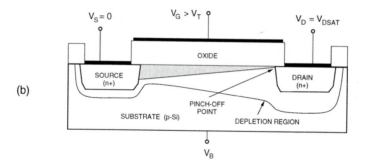

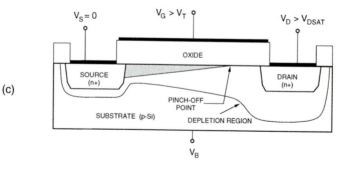

Figure 3.14. Cross-sectional view of an n-channel (nMOS) transistor, (a) operating in the linear region, (b) operating at the edge of saturation, and (c) operating beyond saturation.

proportional to V_{DS} will flow from the source to the drain through the conducting channel. The inversion layer, i.e., the channel, forms a continuous current path from the source to the drain. This operation mode is called the *linear mode*, or the *linear region*. Thus, in linear region operation, the channel region acts as a voltage-controlled resistor. The electron velocity in the channel for this case is usually much lower than the drift velocity limit. Note that as the drain voltage is increased, the inversion layer charge and the channel depth at the drain end start to decrease. Eventually, for $V_{DS} = V_{DSAT}$, the inversion charge at the drain is reduced to zero, which is called the *pinch-off* point (Fig. 3.14(b)).

Beyond the pinch-off point, i.e., for $V_{DS} > V_{DSAT}$, a depleted surface region forms adjacent to the drain, and this depletion region grows toward the source with increasing drain voltages. This operation mode of the MOSFET is called the *saturation mode* or the *saturation region*. For a MOSFET operating in the saturation region, the effective channel length is reduced as the inversion layer near the drain vanishes, while the channel-end voltage remains essentially constant and equal to V_{DSAT} (Fig. 3.14(c)). Note that the pinched-off (depleted) section of the channel absorbs most of the excess voltage drop ($V_{DS} - V_{DSAT}$) and a high-field region forms between the channel-end and the drain boundary. Electrons arriving from the source to the channel-end are injected into the drain-depletion region and are accelerated toward the drain in this high electric field, usually reaching the drift velocity limit. The pinch-off event, or the disruption of the continuous channel under high drain bias, characterizes the saturation mode operation of the MOSFET.

The influence of these operating conditions upon the external (terminal) current-voltage characteristics of the MOS transistor will be examined in the following section. A good understanding of these relationships, and of the factors involved therein, will be essential for the design and analysis of MOS digital circuits.

3.4. MOSFET Current-Voltage Characteristics

The analytical derivation of the MOSFET current-voltage relationships for various bias conditions requires that several approximations be made to simplify the problem. Without these simplifying assumptions, analysis of the actual three-dimensional MOS system would become a very complex task and would prevent the derivation of closed-form current-voltage equations. In the following, we will use the *gradual channel approximation* (GCA) for establishing the MOSFET current-voltage relationships, which will effectively reduce the analysis to a one-dimensional current-flow problem. This will allow us to devise relatively simple current equations that agree well with experimental results. As in every approximate approach, however, the GCA also has its limitations, especially for small-geometry MOSFETs. We will investigate the most significant limitations and examine some of the possible remedies.

Gradual Channel Approximation

To begin with the current-flow analysis, consider the cross-sectional view of the n-channel MOSFET operating in the linear mode, as shown in Fig. 3.15.

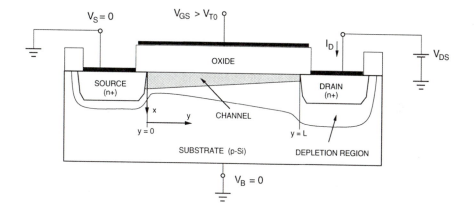

Figure 3.15.   Cross-sectional view of an n-channel transistor, operating in linear region.

Here, the source and the substrate terminals are connected to ground, i.e., $V_S = V_B = 0$. The gate-to-source voltage (V_{GS}) and the drain-to-source voltage (V_{DS}) are the external parameters controlling the drain (channel) current I_D. The gate-to-source voltage is set to be larger than the threshold voltage V_{T0} to create a conducting inversion layer between the source and the drain. We define the coordinate system for this structure such that the x-direction is perpendicular to the surface, pointing down into the substrate, and the y-direction is parallel to the surface. The y-coordinate origin ($y = 0$) is at the source end of the channel. The *channel voltage* with respect to the source will be denoted by $V_c(y)$. Now assume that the threshold voltage V_{T0} is constant along the entire channel region, between $y = 0$ and $y = L$. In reality, the threshold voltage changes along the channel since the channel voltage is not constant. Next, assume that the electric field component E_y along the y-coordinate is *dominant* compared to the electric field component E_x, along the x-coordinate. This assumption will allow us to reduce the current-flow problem in the channel to the y-dimension only. Note that the boundary conditions for the channel voltage V_c are:

$$V_c(y = 0) = V_S = 0$$
$$V_c(y = L) = V_{DS} \tag{3.25}$$

Also, it is assumed that the entire channel region between the source and the drain is inverted, i.e.,

$$V_{GS} \geq V_{T0}$$
$$V_{GD} = V_{GS} - V_{DS} \geq V_{T0} \tag{3.26}$$

The channel current (drain current) I_D is due to the electrons in the channel region traveling from the source to the drain under the influence of the lateral electric field component E_y. Since the current flow in the channel is primarily governed by the lateral drift of the mobile electron charge in the surface inversion layer, we will consider the amount and the bias-voltage dependence of this inversion layer in more detail.

Let $Q_I(y)$ be the total mobile electron charge in the surface inversion layer. This charge can be expressed as a function of the gate-to-source voltage V_{GS} and of the channel voltage $V_c(y)$, as follows:

$$Q_I(y) = -C_{ox} \cdot \left[V_{GS} - V_c(y) - V_{T0} \right] \tag{3.27}$$

Figure 3.16 shows the spatial geometry of the surface inversion layer and indicates its significant dimensions. Note that the thickness of the inversion layer tapers off as we move from the source to the drain, since the gate-to-channel voltage causing surface inversion is smaller at the drain end.

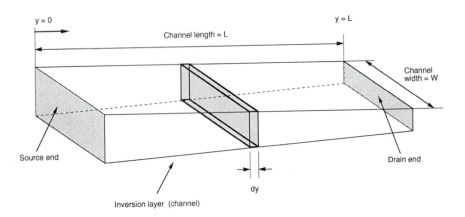

Figure 3.16.   Simplified geometry of the surface inversion layer (channel region).

Now consider the incremental resistance dR of the differential channel segment shown in Fig. 3.16. Assuming that all mobile electrons in the inversion layer have a constant *surface mobility* μ_n, the incremental resistance can be expressed as follows. Note that the minus sign is due to the negative polarity of the inversion layer charge Q_I.

$$dR = -\frac{dy}{W \cdot \mu_n \cdot Q_I(y)} \tag{3.28}$$

The *electron surface mobility* μ_n used in (3.28) depends on the doping concentration of the channel region, and its magnitude is typically about one-half of that of the bulk electron mobility. We will assume that the channel current density is uniform across this segment. According to our one-dimensional model, the channel (drain) current I_D flows between the source and the drain regions in the y-coordinate direction. Applying Ohm's law for this segment yields the voltage drop along the incremental segment dy, along the y-direction.

$$dV_c = I_D \cdot dR = -\frac{I_D}{W \cdot \mu_n \cdot Q_I(y)} \cdot dy \tag{3.29}$$

This equation can now be integrated along the channel, i.e., from $y = 0$ to $y = L$, using the boundary conditions given in (3.25).

$$\int_0^L I_D \cdot dy = -W \cdot \mu_n \int_0^{V_{DS}} Q_I(y) \cdot dV_c \tag{3.30}$$

The left-hand side of this equation is simply equal to $L\,I_D$. The integral on the right-hand side is evaluated by replacing $Q_I(y)$ with (3.27). Thus,

$$I_D \cdot L = W \cdot \mu_n \cdot C_{ox} \int_0^{V_{DS}} \left(V_{GS} - V_c - V_{T0}\right) \cdot dV_c \tag{3.31}$$

Assuming that the channel voltage V_c is the only variable in (3.31) that depends on the position y, the drain current is found as follows.

$$I_D = \frac{\mu_n \cdot C_{ox}}{2} \cdot \frac{W}{L} \cdot \left[2 \cdot \left(V_{GS} - V_{T0}\right)V_{DS} - V_{DS}^2\right] \tag{3.32}$$

Equation (3.32) represents the drain current I_D as a simple second-order function of the two external voltages, V_{GS} and V_{DS}. This current equation can also be rewritten as

$$I_D = \frac{k'}{2} \cdot \frac{W}{L} \cdot \left[2 \cdot (V_{GS} - V_{T0}) V_{DS} - V_{DS}^2\right] \tag{3.33}$$

or

$$I_D = \frac{k}{2} \cdot \left[2 \cdot (V_{GS} - V_{T0}) V_{DS} - V_{DS}^2\right] \tag{3.34}$$

where the parameters k and k' are defined as

$$k' = \mu_n \cdot C_{ox} \tag{3.35}$$

and

$$k = k' \cdot \frac{W}{L} \tag{3.36}$$

  The drain current equation given in (3.33) is the simplest analytical approximation for the MOSFET current-voltage relationship. Note that, in addition to the process-dependent constants k' and V_{T0}, the current-voltage relationship is also affected by the device dimensions, W and L. In fact, we will see that the ratio of W/L is one of the most important design parameters in MOS digital circuit design. Now, we must determine the *region of validity* for this equation and what this means for the practical use of the equation.

======

Example 3.4

For an n-channel MOS transistor with $\mu_n = 600 \text{ cm}^2/\text{V·s}$, $C_{ox} = 7 \cdot 10^{-8} \text{ F/cm}^2$, $W = 20$ μm, $L = 2$ μm and $V_{T0} = 1.0$ V, examine the relationship between the drain current and the terminal voltages.

First, calculate the parameter k:

$$k = \mu_n \cdot C_{ox} \cdot \frac{W}{L} = 600 \text{ cm}^2/\text{V·s} \times 7 \cdot 10^{-8} \text{ F/cm}^2 \times \frac{20 \text{ μm}}{20 \text{ μm}} = 0.42 \text{ mA/V}^2$$

Now, the current-voltage equation (3.34) can be written as follows.

$$I_D = 0.21 \text{ mA/V}^2 \left[2 \cdot (V_{GS} - 1.0) \cdot V_{DS} - V_{DS}^2\right]$$

To examine the effect of the gate-to-source voltage and the drain-to-source voltage upon the drain current, we will plot I_D as a function of V_{DS}, for different (constant) values of V_{GS}. It can easily be seen that the second-order current-voltage equation given above produces a set of inverted parabolas for each constant V_{GS} value.

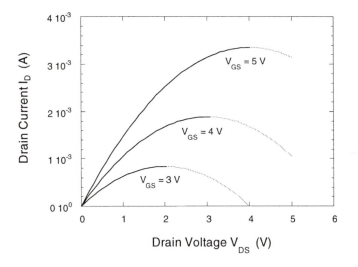

The drain current-drain voltage curves shown above reach their peak value for $V_{DS} = V_{GS} - V_{T0}$. Beyond this maximum, each curve exhibits a *negative* differential conductance, which is not observed in actual MOSFET current-voltage measurements (section shown by the dashed lines). We must remember now that the drain current equation (3.32) has been derived under the following voltage assumptions,

$$V_{GS} \geq V_{T0}$$
$$V_{GD} = V_{GS} - V_{DS} \geq V_{T0}$$

which guarantee that the entire channel region between the source and the drain is inverted. This condition corresponds to the *linear* operating mode for the MOSFET, which was examined qualitatively in Section 3.4. Hence, the current equation (3.32) is valid only for the linear mode operation. Beyond the linear region boundary, i.e., for V_{DS} values *larger* than $V_{GS} - V_{T0}$, the MOS transistor will be assumed to be in *saturation*. A different current-voltage expression will be necessary for the MOSFET operating in this region.

Example 3.4 shows that the current equation (3.32) is not valid beyond the linear region/saturation region boundary, i.e., for

$$V_{DS} \geq V_{DSAT} = V_{GS} - V_{T0} \qquad (3.37)$$

Also, drain current measurements with constant V_{GS} show that the current I_D does not show much variation as a function of the drain voltage V_{DS} beyond the saturation boundary, but rather remains approximately constant around the peak value reached for $V_{DS} = V_{DSAT}$. This saturation drain current level can be found simply by substituting (3.37) for V_{DS} in (3.32).

$$I_D(sat) = \frac{\mu_n \cdot C_{ox}}{2} \cdot \frac{W}{L} \cdot \left[2 \cdot (V_{GS} - V_{T0}) \cdot (V_{GS} - V_{T0}) - (V_{GS} - V_{T0})^2\right]$$

$$= \frac{\mu_n \cdot C_{ox}}{2} \cdot \frac{W}{L} \cdot (V_{GS} - V_{T0})^2 \qquad (3.38)$$

Thus, the drain current I_D becomes a function only of the gate-to-source voltage V_{GS}, beyond the saturation boundary. Note that this constant saturation current approximation is not very accurate in reality, and that the saturation-region drain current continues to have a certain dependence on the drain voltage. For simple hand calculations, however, (3.38) provides a sufficiently accurate approximation of the MOSFET drain (channel) current in saturation.

Figure 3.17 shows the typical drain current vs. drain voltage characteristics of an n-channel MOSFET, as described by the current equations (3.32) and (3.38). The

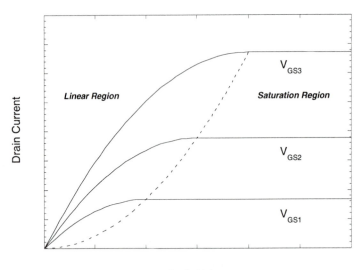

Figure 3.17.   Basic current-voltage characteristics of an n-channel MOS transistor.

parabolic boundary between the linear and the saturation regions is indicated here by the dashed line. The current-voltage characteristics of the MOS transistor can also be visualized by plotting the drain current as a function of the gate voltage, as shown in Fig. 3.18. This $I_D - V_{GS}$ transfer characteristic in saturation mode ($V_{DS} > V_{DSAT}$) provides a simple view of the drain current increasing as a second-order function of the gate-to-source voltage (cf. Equation (3.38)). The current is obviously equal to zero for any gate voltage smaller than the threshold voltage V_{T0}.

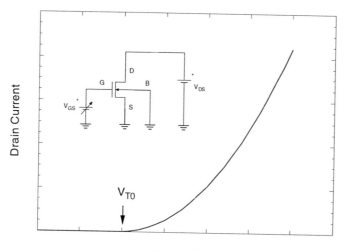

Figure 3.18.   Drain current of the n-channel MOS transistor as a function of the gate-to-source voltage V_{GS}, with $V_{DS} > V_{DSAT}$ (transistor in saturation).

Channel Length Modulation

Next, we will examine the mechanisms of channel pinch-off and current flow in saturation mode in more detail. Consider the inversion layer charge Q_I that represents the total mobile electron charge on the surface, given by (3.27). The inversion layer charge at the source end of the channel is

$$Q_I(y = 0) = -C_{ox} \cdot (V_{GS} - V_{T0})$$
(3.39)

and the inversion layer charge at the drain end of the channel is

$$Q_I(y = L) = -C_{ox} \cdot (V_{GS} - V_{T0} - V_{DS})$$
(3.40)

Note that at the edge of saturation, i.e., when the drain-to-source voltage reaches V_{DSAT},

$$V_{DS} = V_{DSAT} = V_{GS} - V_{T0} \tag{3.41}$$

the inversion layer charge at the drain end becomes zero, according to (3.40). In reality, the channel charge does not become exactly equal to zero (remember that the GCA is just a simple approximation of the actual conditions in the channel), but it indeed becomes very small.

$$Q_I(y = L) \approx 0 \tag{3.42}$$

Thus, we can state that under the bias condition given in (3.41), the channel is *pinched-off* at the drain end, i.e., at $y = L$. The onset of the saturation mode operation in the MOSFET is signified by this pinch-off event. If the drain-to-source voltage V_{DS} is increased even further beyond the saturation edge so that $V_{DS} > V_{DSAT}$, an even larger portion of the channel becomes pinched-off.

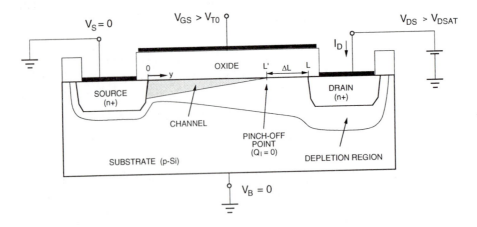

Figure 3.19.   Channel length modulation in an n-channel MOSFET operation in saturation mode.

Consequently, the *effective channel length* (the length of the inversion layer where GCA is still valid) is reduced to

$$L' = L - \Delta L \tag{3.43}$$

where ΔL is the length of the channel segment with $Q_I = 0$ (Fig. 3.19). Hence, the pinch-off point moves from the drain end of the channel toward the source with increasing drain-to-source voltages. The remaining portion of the channel between the pinch-off point and the drain will be in depletion mode. Since $Q_I(y)=0$ for $L'<y<L$, the channel voltage at the pinch-off point remains equal to V_{DSAT}:

$$V_c(y=L')=V_{DSAT} \tag{3.44}$$

The electrons traveling from the source toward the drain traverse the inverted channel segment of length L', and then they are injected into the depletion region of length ΔL that separates the pinch-off point from the drain edge. As seen in Fig. 3.19, we can represent the inverted portion of the surface by a shortened channel, with a channel-end voltage of V_{DSAT}. The gradual channel approximation is valid in this region; thus, the channel current can be found using (3.38).

$$I_D(sat) = \frac{\mu_n \cdot C_{ox}}{2} \cdot \frac{W}{L'} \cdot (V_{GS} - V_{T0})^2 \tag{3.45}$$

Note that this current equation corresponds to a MOSFET with effective channel length L', operating in saturation. Thus, (3.45) accounts for the actual shortening of the channel, also called *channel length modulation*. Since $L' < L$, the saturation current calculated by using (3.45) will be larger than that found by using (3.38), under the same bias conditions. As L' decreases with increasing V_{DS}, the saturation mode current $I_D(sat)$ will also increase with V_{DS}. By approximating the effective channel length L' $=L-\Delta L$ as a function of the drain bias voltage, we can modify (3.45) to reflect this drain voltage dependence. First, rewrite the saturation current as follows.

$$I_D(sat) = \left(\frac{1}{1 - \dfrac{\Delta L}{L}} \right) \cdot \frac{\mu_n \cdot C_{ox}}{2} \cdot \frac{W}{L} \cdot (V_{GS} - V_{T0})^2 \tag{3.46}$$

The first term of this saturation current expression accounts for the channel modulation effect, while the rest of this expression is identical to (3.38). It can be shown that the channel length shortening ΔL is actually proportional to the square root of (V_{DS} – V_{DSAT}).

$$\Delta L \propto \sqrt{V_{DS} - V_{DSAT}} \tag{3.47}$$

To simplify the analysis even further, we will use the following empirical relation between ΔL and the drain-to-source voltage instead.

$$1 - \frac{\Delta L}{L} \approx 1 - \lambda \cdot V_{DS} \tag{3.48}$$

Here, λ is an empirical model parameter, and is called the *channel length modulation coefficient*. Assuming that $\lambda V_{DS} \ll 1$, the saturation current given in (3.45) can now be written as:

$$I_D(sat) = \frac{\mu_n \cdot C_{ox}}{2} \cdot \frac{W}{L} \cdot (V_{GS} - V_{T0})^2 \cdot (1 + \lambda \cdot V_{DS}) \tag{3.49}$$

This simple current equation prescribes a linear drain-bias dependence for the saturation current in MOS transistors, determined by the empirical parameter λ. Although this rough approximation does not accurately reflect the physical relationship between the channel length shortening ΔL and the drain bias, (3.49) can be used with sufficient confidence for most first-order hand calculations. The drain current vs. drain-to-source voltage characteristics of an n-channel MOSFET, obtained by using (3.32) for the linear region and (3.49) for the saturation region, are shown in Fig. 3.20. The saturation mode current increases linearly with V_{DS}, instead of remaining constant. The slope of the current-voltage curve in the saturation region is determined by the channel length modulation coefficient λ.

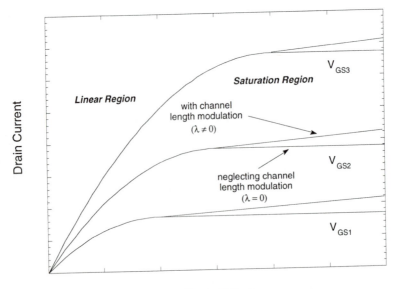

Figure 3.20.   Current-voltage characteristics of an n-channel MOS transistor, including the channel length modulation effect.

Substrate Bias Effect

Note that the derivation of linear-mode and saturation-mode current-voltage characteristics in the previous pages has been done under the assumption that the substrate potential is equal to the source potential, i.e., $V_{SB} = 0$. Consequently, the zero-substrate bias threshold voltage V_{TO} has been used in the current equations. In many digital circuit applications, on the other hand, the source potential of an nMOS transistor can be larger than the substrate potential, which results in a positive source-to-substrate voltage $V_{SB} > 0$. In this case, the influence of the nonzero V_{SB} upon the current characteristics must be accounted for. Recall that the general expression (3.23) for the threshold voltage V_T already includes the substrate bias term and, hence, it reflects the influence of the nonzero source-to-substrate voltage upon the device characteristics.

$$V_T(V_{SB}) = V_{TO} + \gamma \cdot \left(\sqrt{|2\phi_F| + V_{SB}} - \sqrt{|2\phi_F|} \right) \tag{3.50}$$

We can simply replace the threshold voltage terms in linear-mode and saturation-mode current equations with the more general $V_T(V_{SB})$ term.

$$I_D(lin) = \frac{\mu_n \cdot C_{ox}}{2} \cdot \frac{W}{L} \cdot \left[2 \cdot (V_{GS} - V_T(V_{SB})) V_{DS} - V_{DS}^2 \right] \tag{3.51}$$

$$I_D(sat) = \frac{\mu_n \cdot C_{ox}}{2} \cdot \frac{W}{L} \cdot (V_{GS} - V_T(V_{SB}))^2 \cdot (1 + \lambda \cdot V_{DS}) \tag{3.52}$$

In general, we will use only the term V_T instead of $V_T(V_{SB})$ to express the general (substrate-bias dependent) threshold voltage. As already demonstrated in Example 3.3, the substrate-bias effect can significantly change the value of the threshold voltage and, hence, the current capability of the MOSFET. With this modification, we finally arrive at a complete first-order characterization of the drain (channel) current as a nonlinear function of the terminal voltages.

$$I_D = f(V_{GS}, V_{DS}, V_{BS}) \tag{3.53}$$

In the following, we will repeat the current-voltage equations derived under the first-order gradual channel approximation (GCA), both for n-channel and for p-channel MOS transistors. Figure 3.21 shows the polarities of applied terminal voltages and the drain current directions. Note that the threshold voltage V_T and the terminal voltages V_{GS}, V_{DS}, and V_{SB} are all *negative* for the pMOS transistor. The parameter μ_p denotes the surface hole mobility in the pMOSFET.

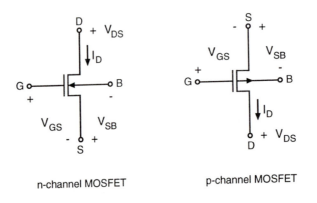

n-channel MOSFET   p-channel MOSFET

Figure 3.21.   Terminal voltages and currents of the nMOS and the pMOS transistor.

Current-voltage equations of the n-channel MOSFET :

$$I_D = 0, \quad for \quad V_{GS} < V_T \tag{3.54}$$

$$I_D(lin) = \frac{\mu_n \cdot C_{ox}}{2} \cdot \frac{W}{L} \cdot \left[2 \cdot (V_{GS} - V_T) V_{DS} - V_{DS}^2 \right] \quad for \quad V_{GS} \geq V_T$$
$$and \quad V_{DS} < V_{GS} - V_T \tag{3.55}$$

$$I_D(sat) = \frac{\mu_n \cdot C_{ox}}{2} \cdot \frac{W}{L} \cdot (V_{GS} - V_T)^2 \cdot (1 + \lambda \cdot V_{DS}) \quad for \quad V_{GS} \geq V_T$$
$$and \quad V_{DS} \geq V_{GS} - V_T \tag{3.56}$$

Current-voltage equations of the p-channel MOSFET :

$$I_D = 0, \quad for \quad V_{GS} > V_T \tag{3.57}$$

$$I_D(lin) = \frac{\mu_p \cdot C_{ox}}{2} \cdot \frac{W}{L} \cdot \left[2 \cdot (V_{GS} - V_T) V_{DS} - V_{DS}^2 \right] \quad for \quad V_{GS} \leq V_T$$
$$and \quad V_{DS} > V_{GS} - V_T \tag{3.58}$$

$$I_D(sat) = \frac{\mu_p \cdot C_{ox}}{2} \cdot \frac{W}{L} \cdot (V_{GS} - V_T)^2 \cdot (1 + \lambda \cdot V_{DS}) \quad for \quad V_{GS} \leq V_T$$
$$and \quad V_{DS} \leq V_{GS} - V_T \tag{3.59}$$

Measurement of Parameters

The MOSFET current-voltage equations (3.54)-(3.59), together with the general threshold voltage expression (3.50), are very useful for simple, first-order calculations of the currents and voltages in the nMOS and pMOS transistors. Because of several simplifications and approximations involved in their derivation, however, the accuracy of these current-voltage equations is fairly limited. To exploit the simplicity of the equations and to achieve the maximum possible accuracy in calculations, the parameters appearing in the current equations must be determined carefully, through experimental measurements. The model parameters that are used in (3.50) and in (3.54)-(3.59) are the zero-bias threshold voltage V_{T0}, the substrate-bias coefficient γ, the channel length modulation coefficient λ, and the transconductance parameters

$$k_n = \mu_n \cdot C_{ox} \cdot \frac{W}{L} \tag{3.60}$$

$$k_p = \mu_p \cdot C_{ox} \cdot \frac{W}{L} \tag{3.61}$$

In the following section, some simple measurements for an enhancement-type n-channel MOSFET will be described for the determination of these parameters. First, consider the test circuit setup shown in Fig. 3.22(a). The source-to-substrate voltage V_{SB} is set at a constant value, and the drain current is measured for different values of the gate-to-source voltage V_{GS}. Since the drain and the gate of the transistor are at the same potential, $V_{DS} = V_{GS}$. Hence, the saturation condition $V_{DS} > V_{GS} - V_T$ is always satisfied, i.e., the nMOS transistor shown in Fig. 3.22(a) operates in saturation mode. Neglecting the channel length modulation effect for simplicity, the drain current is described by

$$I_D(sat) = \frac{k_n}{2} \cdot (V_{GS} - V_{T0})^2 \tag{3.62}$$

Now, the square root of the drain current can be written as a *linear* function of the gate-to-source voltage.

$$\sqrt{I_D} = \sqrt{\frac{k_n}{2}} \cdot (V_{GS} - V_{T0}) \tag{3.63}$$

If the square root of the measured drain current values is plotted against the gate-to-source voltage, the slope and the voltage-axis intercept of the resulting curve(s) can determine the parameters k_n, V_{T0}, and γ. Figure 3.22(b) shows the measured drain current vs. gate voltage curves, obtained for different values of substrate bias. By extrapolating the curves to zero-drain-current (voltage-axis intercept point), we can find the threshold voltage V_T that corresponds to each V_{SB} value. The voltage-axis intercept of the curve with $V_{SB} = 0$ gives the zero-bias threshold voltage, V_{T0}. Note that these *extrapolated threshold voltage* values do not exactly match the threshold voltage values usually measured in the production environment, at a certain nonzero drain current. They can rather be viewed as fitting parameters for the current-voltage equations. The slope of each curve is equal to the square root of k_n, the transconductance parameter. Thus, k_n can simply be calculated from this slope.

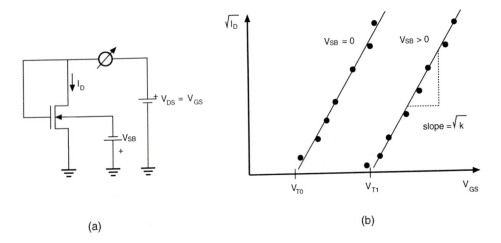

(a)   (b)

Figure 3.22.   (a) Test circuit arrangement and (b) measured data for experimental determination of the parameters k_n, V_{T0}, and γ.

Next, consider the extrapolated threshold voltage values, obtained from voltage axis intercepts at nonzero substrate bias voltage. Using one of the available V_{SB} values, the substrate bias coefficient γ can be found from

$$\gamma = \frac{V_T(V_{SB}) - V_{T0}}{\sqrt{|2\phi_F| + V_{SB}} - \sqrt{|2\phi_F|}} \tag{3.64}$$

The experimental measurement of the channel length modulation coefficient λ requires a different test circuit setup, as shown in Fig. 3.23(a).

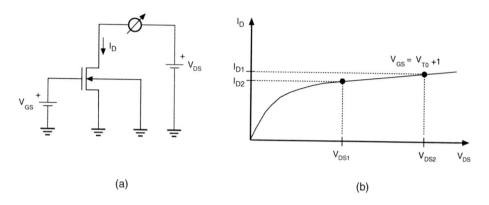

Figure 3.23. (a) Test circuit arrangement and (b) measured data for experimental determination of the channel length modulation coefficient λ.

The gate-to-source voltage V_{GS} is set to $V_{T0} + 1$. The drain-to-source voltage is chosen sufficiently large $(V_{DS} > V_{GS} - V_{T0})$ that the transistor operates in the saturation mode. The saturation drain current is then measured for two different drain voltage values, V_{DS1} and V_{DS2}. Note that the drain current in the saturation mode is given by

$$I_D(sat) = k_n \cdot (V_{GS} - V_{T0})^2 \cdot (1 + \lambda \cdot V_{DS}) \tag{3.65}$$

Since $V_{GS} = V_{T0} + 1$, the ratio of the measured drain current values I_{D1} and I_{D2} is

$$\frac{I_{D2}}{I_{D1}} = \frac{1 + \lambda \cdot V_{DS2}}{1 + \lambda \cdot V_{DS1}} \tag{3.66}$$

which can be used to calculate the channel length modulation coefficient λ. This in fact corresponds to calculating the *slope* of the drain current vs. drain voltage curve in the saturation region, as shown in Fig. 3.23(b).

Example 3.5

Measured voltage and current data for a MOSFET are given below. Determine the type of the device, and calculate the parameters k_n, V_{T0}, and γ. Assume $\phi_F = -0.3$ V.

$V_{GS}(V)$	$V_{DS}(V)$	$V_{SB}(V)$	$I_D(\mu A)$
3	3	0	97
4	4	0	235
5	5	0	433
3	3	3	59
4	4	3	173
5	5	3	347

First, the MOS transistor is on ($I_D > 0$) for $V_{GS} > 0$ and $V_{DS} > 0$. Thus, the transistor must be an n-channel MOSFET. Assume that the transistor is enhancement-type, and therefore, operating in saturation mode for $V_{GS} = V_{DS}$. Neglecting the channel length modulation effect, the saturation mode current is written as

$$I_D = \frac{k_n}{2} \cdot (V_{GS} - V_T)^2 \quad \Leftrightarrow \quad \sqrt{I_D} = \sqrt{\frac{k_n}{2}} \cdot (V_{GS} - V_T)$$

Let (V_{GS1}, I_{D1}) and (V_{GS2}, I_{D2}) be any two current-voltage pairs obtained from the table. Then, the square-root of the transconductance parameter k_n can be calculated.

$$\sqrt{\frac{k_n}{2}} = \frac{\sqrt{I_{D1}} - \sqrt{I_{D2}}}{V_{GS1} - V_{GS2}} = \frac{\sqrt{433\ \mu A} - \sqrt{97\ \mu A}}{5\ V - 3\ V} = 5.48 \times 10^{-3}\ A^{1/2}/V$$

Thus, the transconductance parameter of this n-channel MOSFET is:

$$k_n = 2 \cdot (5.48 \times 10^{-3})^2 = 60 \times 10^{-6}\ A/V^2 = 60\ \mu A/V^2$$

The *extrapolated threshold voltage* V_{T0} at zero substrate bias can be found by calculating the x-axis intercept of the *sqrt(I_D)* vs. V_{GS} curve.

$$V_{T0} = V_{GS} - \sqrt{\frac{2 \cdot I_D}{k_n}} = 1.2\ V$$

To find the substrate bias coefficient γ, we must first determine the threshold voltage

V_T at the source-to-substrate voltage of 3 V. Using one of the current-voltage data pairs corresponding to $V_{SB} = 3$ V, V_T can be calculated as follows.

$$V_T\left(V_{SB} = 3 \text{ V}\right) = V_{GS} - \sqrt{\frac{2 \cdot I_D}{k_n}} = 4 \text{ V} - \sqrt{\frac{2 \cdot 173 \text{ }\mu\text{A}}{60 \text{ }\mu\text{A}/\text{V}^2}} = 1.6 \text{ V}$$

Finally, the substrate bias coefficient is found as:

$$\gamma = \frac{V_T\left(V_{SB} = 3 \text{ V}\right) - V_{T0}}{\sqrt{|2\phi_F| + V_{SB}} - \sqrt{|2\phi_F|}} = \frac{1.6 \text{ V} - 1.2 \text{ V}}{\sqrt{0.6 \text{ V} + 3 \text{ V}} - \sqrt{0.6 \text{ V}}} = 0.36 \text{ V}^{1/2}$$

3.5. MOSFET Scaling and Small-Geometry Effects

The design of high-density chips in MOS VLSI (Very Large Scale Integration) technology requires that the packing density of MOSFETs used in the circuits is as high as possible and, consequently, that the sizes of the transistors are as small as possible. The reduction of the size, i.e., the dimensions of MOSFETs, is commonly referred to as *scaling*. It is expected that the operational characteristics of the MOS transistor will change with the reduction of its dimensions. Also, some physical limitations eventually restrict the extent of scaling that is practically achievable. There are two basic types of size-reduction strategies: *full scaling* (also called constant-field scaling) and *constant-voltage scaling*. Both types of scaling approaches will be shown to have unique effects upon the operating characteristics of the MOS transistor. In the following, we will examine in detail the scaling strategies and their effects, and we will also consider some of the physical limitations and small-geometry effects that must be taken into account for scaled MOSFETs.

Scaling of MOS transistors is concerned with systematic reduction of overall dimensions of the devices as allowed by the available technology, while preserving the geometric ratios found in the larger devices. The proportional scaling of all devices in a circuit would certainly result in a reduction of the total silicon area occupied by the circuit, thereby increasing the overall functional density of the chip. To describe device scaling, we introduce a constant *scaling factor* $S > 1$. All horizontal and vertical dimensions of the *large-size* transistor are then divided by this scaling factor to obtain the scaled device. The extent of scaling that is achievable is obviously determined by the fabrication technology, and more specifically, by the minimum feature size. Table 3.1 below shows the recent history of reducing feature sizes for the typical CMOS gate-array process. It is seen that a new generation of manufacturing technology replaces the previous one about every two or three years, and the down-scaling factor S of the minimum feature size from one generation to the next is about 1.4 - 1.5.

Year	1980	1983	1985	1987	1989	1991	1993	1995
Feature size (μm)	5.0	3.5	2.5	1.75	1.25	1.0	0.8	0.6

Table 3.1.   Reduction of the minimum feature size (minimum dimensions that can be defined and manufactured on chip) over the years, for a typical CMOS gate-array process.

We consider the proportional scaling of all three dimensions by the same scaling factor S. Figure 3.24 shows the reduction of key dimensions on a typical MOSFET, together with the corresponding increase of the doping densities.

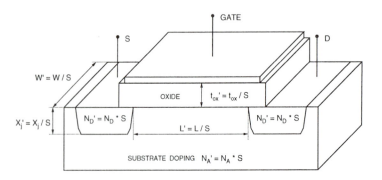

Figure 3.24.   Scaling of a typical MOSFET by a scaling factor of S.

The primed quantities in Fig. 3.24 indicate the scaled dimensions and doping densities. It is easy to recognize that the scaling of all dimensions by a factor of $S > 1$ leads to the reduction of the area occupied by the transistor by a factor of S^2. To better understand the effects of scaling upon the current-voltage characteristics of the MOSFET, we will examine two different scaling options in the following section.

Full Scaling (Constant-Field Scaling)

This scaling option attempts to preserve the magnitude of internal electric fields in the MOSFET, while the dimensions are scaled down by a factor of S. To achieve this goal, all potentials must be scaled down proportionally, by the same scaling factor. Note that this potential scaling also affects the threshold voltage V_{T0}. Finally, the Poisson equation describing the relationship between charge densities and electric fields dictates that the charge densities must be *increased* by a factor of S in order to maintain the field conditions. Table 3.2 lists the scaling factors for all significant dimensions, potentials, and doping densities of the MOS transistor.

Quantity	Before Scaling	After Scaling
Channel length	L	$L' = L/S$
Channel width	W	$W' = W/S$
Gate oxide thickness	t_{ox}	$t_{ox}' = t_{ox}/S$
Junction depth	x_j	$x_j' = x_j/S$
Power supply voltage	V_{DD}	$V_{DD}' = V_{DD}/S$
Threshold voltage	V_{T0}	$V_{T0}' = V_{T0}/S$
Doping densities	N_A N_D	$N_A' = S \cdot N_A$ $N_D' = S \cdot N_D$

Table 3.2.   Full scaling of MOSFET dimensions, potentials, and doping densities.

Now consider the influence of full scaling described here upon the current-voltage characteristics of the MOS transistor. It will be assumed that the surface mobility μ_n is not significantly affected by the scaled doping density. The gate oxide capacitance per unit area, on the other hand, is changed as follows.

$$C_{ox}' = \frac{\varepsilon_{ox}}{t_{ox}'} = S \cdot \frac{\varepsilon_{ox}}{t_{ox}} = S \cdot C_{ox} \tag{3.67}$$

The aspect ratio W/L of the MOSFET will remain unchanged under scaling. Consequently, the transconductance parameter k_n will also be scaled by a factor of S. Since all terminal voltages are scaled down by the factor S as well, the linear-mode drain current of the scaled MOSFET can now be found:

$$I_D'(lin) = \frac{k_n'}{2} \cdot \left[2 \cdot (V_{GS}' - V_T') \cdot V_{DS}' - V_{DS}'^2 \right]$$

$$= \frac{S \cdot k_n}{2} \cdot \frac{1}{S^2} \cdot \left[2 \cdot (V_{GS} - V_T) \cdot V_{DS} - V_{DS}^2 \right] = \frac{I_D(lin)}{S} \tag{3.68}$$

Similarly, the saturation-mode drain current is also reduced by the same scaling factor.

$$I_D'(sat) = \frac{k_n'}{2} \cdot \left(V_{GS}' - V_T' \right)^2 = \frac{S \cdot k_n}{2} \cdot \frac{1}{S^2} \cdot \left(V_{GS} - V_T \right)^2 = \frac{I_D(sat)}{S} \qquad (3.69)$$

Now consider the power dissipation of the MOSFET. Since the drain current flows between the source and the drain terminals, the instantaneous power dissipated by the device (before scaling) can be found as:

$$P = I_D \cdot V_{DS} \qquad (3.70)$$

Notice that full scaling reduces both the drain current and the drain-to-source voltage by a factor of S; hence, the power dissipation of the transistor will be reduced by the factor S^2.

$$P' = I_D' \cdot V_{DS}' = \frac{1}{S^2} \cdot I_D \cdot V_{DS} = \frac{P}{S^2} \qquad (3.71)$$

This significant reduction of the power dissipation is one of the most attractive features of full scaling. Note that with the device area reduction by S^2 discussed earlier, we find the *power density* per unit area remaining virtually unchanged for the scaled device.

Finally, consider the gate oxide capacitance defined as $C_g = W L C_{ox}$. It will be shown later in Section 3.6 that charging and discharging of this capacitance plays an important role in the transient operation of the MOSFET. Since the gate oxide capacitance C_g is scaled down by a factor of S, we can predict that the transient characteristics, i.e., the charge-up and charge-down times, of the scaled device will improve accordingly. In addition, the proportional reduction of all dimensions on-chip will lead to a reduction of various parasitic capacitances and resistances as well, contributing to the overall performance improvement. Table 3.3 summarizes the changes in key device characteristics as a result of full (constant-field) scaling.

Quantity	Before Scaling	After Scaling
Oxide capacitance	C_{ox}	$C_{ox}' = S \cdot C_{ox}$
Drain current	I_D	$I_D' = I_D / S$
Power dissipation	P	$P' = P / S^2$
Power density	$P / Area$	$P'/Area' = P / Area$

Table 3.3.   Effects of full scaling upon key device characteristics.

Constant-Voltage Scaling

While the full scaling strategy dictates that the power supply voltage and all terminal voltages be scaled down proportionally with the device dimensions, the scaling of voltages may not be very practical in many cases. In particular, the peripheral and interface circuitry may require certain voltage levels for all input and output voltages, which in turn would necessitate multiple power supply voltages and complicated level-shifter arrangements. For these reasons, constant-voltage scaling is usually preferred over full scaling.

In constant-voltage scaling, all dimensions of the MOSFET are reduced by a factor of S, as in full scaling. The power supply voltage and the terminal voltages, on the other hand, remain unchanged. The doping densities must be increased by a factor of S^2 in order to preserve the charge-field relations. Table 3.4 shows the constant-voltage scaling of key dimensions, voltages, and densities. Under constant-voltage scaling, the changes in device characteristics are significantly different compared to those in full scaling, as we will demonstrate. The gate oxide capacitance per unit area C_{ox} is increased by a factor of S, which means that the transconductance parameter is also increased by S. Since the terminal voltages remain unchanged, the linear mode drain current of the scaled MOSFET can be written as:

$$I_D'(lin) = \frac{k_n'}{2} \cdot \left[2 \cdot \left(V_{GS}' - V_T' \right) \cdot V_{DS}' - V_{DS}'^2 \right]$$

$$= \frac{S \cdot k_n}{2} \cdot \left[2 \cdot \left(V_{GS} - V_T \right) \cdot V_{DS} - V_{DS}^2 \right] = S \cdot I_D(lin) \tag{3.72}$$

Quantity	Before Scaling	After Scaling
Dimensions	W, L, t_{ox}, x_j	reduced by S ($W' = W/S$, ...)
Voltages	V_{DD}, V_T	remain unchanged
Doping densities	N_A, N_D	increased by S^2 ($N_A' = S^2 \cdot N_A$, ...)

Table 3.4. Constant-voltage scaling of MOSFET dimensions, potentials, and doping densities.

Also, the saturation-mode drain current will be increased by a factor of S after constant-voltage scaling. This means that the drain current *density* (current per unit area) is increased by a factor of S^3, which may cause serious reliability problems for the MOS transistor.

$$I_D'(sat) = \frac{k_n'}{2} \cdot \left(V_{GS}' - V_T'\right)^2 = \frac{S \cdot k_n}{2} \cdot \left(V_{GS} - V_T\right)^2 = S \cdot I_D(sat) \qquad (3.73)$$

Next, consider the power dissipation. Since the drain current is increased by a factor of S while the drain-to-source voltage remains unchanged, the power dissipation of the MOSFET increases by a factor of S.

$$P' = I_D' \cdot V_{DS}' = \left(S \cdot I_D\right) \cdot V_{DS} = S \cdot P \qquad (3.74)$$

Finally, the power density (power dissipation per unit area) is found to increase by a factor of S^3 after constant-voltage scaling, with possible adverse effects on device reliability.

Quantity	Before Scaling	After Scaling
Oxide capacitance	C_{ox}	$C_{ox}' = S \cdot C_{ox}$
Drain current	I_D	$I_D' = S \cdot I_D$
Power dissipation	P	$P' = S \cdot P$
Power density	$P / Area$	$P'/Area' = S^3 \cdot (P/Area)$

Table 3.5.   Effects of constant-voltage scaling upon key device characteristics.

To summarize, constant-voltage scaling may be preferred over full (constant-field) scaling in many practical cases because of the external voltage-level constraints. It must be recognized, however, that constant-voltage scaling increases the drain current density and the power density by a factor of S^3. This large increase in current and power densities may eventually cause serious reliability problems for the scaled transistor, such as electromigration, hot-carrier degradation, oxide breakdown, and electrical over-stress.

As the device dimensions are systematically reduced through full scaling or constant-voltage scaling, various physical limitations become increasingly more prominent, and ultimately restrict the amount of feasible scaling for some device dimensions. Consequently, scaling may be carried out on a certain subset of MOSFET dimensions in many practical cases. Also, the simple gradual channel approximation (GCA) used for the derivation of current-voltage relationships does not accurately reflect the effects of scaling in smaller-size transistors. The current equations have to be modified accordingly. In the following, we will briefly investigate some of these small-geometry effects.

Short-Channel Effects

As a working definition, a MOS transistor is called a short-channel device if its channel length is the on same order of magnitude as the depletion region thicknesses of the source and drain junctions. Alternatively, a MOSFET can be defined as a short-channel device if the effective channel length L_{eff} is approximately equal to the source and drain junction depth x_j. The short-channel effects that arise in this case are attributed to two physical phenomena : (1) the limitations imposed on electron drift characteristics in the channel, and (2) the modification of the threshold voltage due to the shortening channel length.

Note that the lateral electric field E_y along the channel increases, as the effective channel length is decreased. While the electron drift velocity v_d in the channel is proportional to the electric field for lower field values, this drift velocity tends to saturate at high channel electric fields. For channel electric fields of $E_y = 10^5$ V/cm and higher, the electron drift velocity in the channel reaches a saturation value of about $v_d(sat) = 10^7$ cm/s. This velocity saturation has very significant implications upon the current-voltage characteristics of the short-channel MOSFET. Consider the saturation-mode drain current, under the assumption that carrier velocity in the channel has already reached its limit value. The effective channel length L_{eff} will be reduced due to channel-length shortening.

$$I_D(sat) = W \cdot v_d(sat) \cdot \int_0^{L_{eff}} q \cdot n(x)\ dx = W \cdot v_d(sat) \cdot |Q_I| \qquad (3.75)$$

Since the channel-end voltage is equal to V_{DSAT}, the saturation current can be found as follows.

$$I_D(sat) = W \cdot v_d(sat) \cdot C_{ox} \cdot V_{DSAT} \qquad (3.76)$$

Carrier velocity saturation actually reduces the saturation-mode current below the current value predicted by the conventional long-channel current equations. The current is no longer a quadratic function of the gate-to-source voltage V_{GS}, and it is virtually independent of the channel length. Also note that under these conditions, the device is defined to be in saturation when the carrier velocity in the channel approaches about 90% of its limit value.

In short-channel MOS transistors, the carrier velocity in the channel is also a function of the normal (vertical) electric-field component E_x. Since the vertical field influences the scattering of carriers (collisions suffered by the carriers) in the surface region, the surface mobility is reduced with respect to the bulk mobility. The dependence of the surface electron mobility on the vertical electric field can be expressed by the following empirical formula :

$$\mu_n(eff) = \frac{\mu_{no}}{1 + \Theta \cdot E_x} = \frac{\mu_{no}}{1 + \dfrac{\Theta}{t_{ox}} \dfrac{\varepsilon_{ox}}{\varepsilon_{Si}} \cdot (V_{GS} - V_c(y))} \tag{3.77}$$

where μ_{no} is the low-field surface electron mobility, and Θ is an empirical factor. For a simple estimation of field-related mobility reduction, (3.77) can be approximated by

$$\mu_n(eff) = \frac{\mu_{no}}{1 + \eta \cdot (V_{GS} - V_T)} \tag{3.78}$$

where η is also an empirical coefficient.

Next, we consider the modification of the threshold voltage due to short-channel effects. The threshold voltage expression (3.23) was derived for a long-channel MOSFET. Specifically, the channel depletion region was assumed to be created only by the applied gate voltage, and the depletion regions associated with the drain and source pn-junctions were neglected. The shape of this gate-induced bulk (channel) depletion region was assumed to be rectangular, extending from the source to the drain. In short-channel MOS transistors, however, the n^+ drain and source diffusion regions in the p-type substrate induce a significant amount of depletion charge; consequently, the long-channel threshold voltage expression derived earlier overestimates the depletion charge supported by the gate voltage. The threshold voltage value found by using (3.23) is therefore larger than the actual threshold voltage of the short-channel MOSFET.

Figure 3.25(a) shows the simplified geometry of the gate-induced bulk depletion region and the pn-junction depletion regions in a short-channel MOS transistor. Note that the bulk depletion region is assumed to have an asymmetric trapezoidal shape, instead of a rectangular shape, to represent accurately the gate-induced charge. The drain depletion region is expected to be larger than the source depletion region because the positive drain-to-source voltage reverse-biases the drain-substrate junction. We recognize that a significant portion of the total depletion region charge under the gate is actually due to the source and drain junction depletion, rather than the bulk depletion induced by the gate voltage. Since the bulk depletion charge in the short-channel device is smaller than expected, the threshold voltage expression must be modified to account for this reduction. Following the modification of the bulk charge term, the threshold voltage of the short-channel MOSFET can be written as

$$V_{T0}(short\ channel) = V_{T0} - \Delta V_{T0} \qquad (3.79)$$

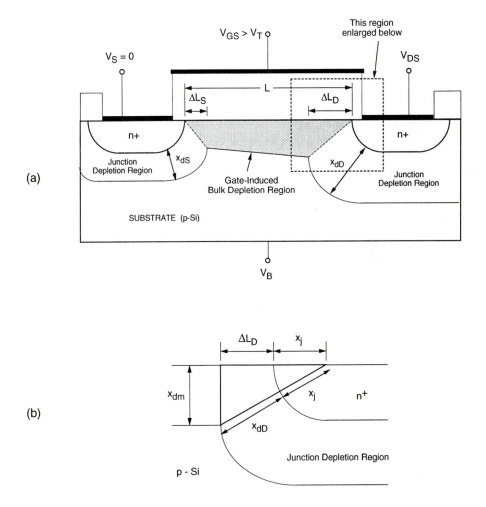

Figure 3.25. (a) Simplified geometry of the MOSFET channel region, with gate-induced bulk depletion region and the pn-junction depletion regions. (b) Close-up view of the drain diffusion edge.

where V_{T0} is the zero-bias threshold voltage calculated using the conventional long-channel formula (3.23), and ΔV_{T0} is the threshold voltage shift (reduction) due to the short-channel effect. The reduction term actually represents the amount of charge differential between a rectangular depletion region and a trapezoidal depletion region.

Let ΔL_S and ΔL_D represent the lateral extent of the depletion regions associated with the source junction and the drain junction, respectively. Then, the bulk depletion region charge contained within the trapezoidal region is

$$Q_{B0} = -\left(1 - \frac{\Delta L_S + \Delta L_D}{2L}\right) \cdot \sqrt{2 \cdot q \cdot \varepsilon_{Si} \cdot N_A \cdot |2\phi_F|} \qquad (3.80)$$

To calculate ΔL_S and ΔL_D, we will use the simplified geometry shown in Fig. 3.25(b). Here, x_{dS} and x_{dD} represent the depth of the pn-junction depletion regions associated with the source and the drain, respectively. The edges of the source and drain diffusion regions are represented by quarter-circular arcs, each with a radius equal to the junction depth, x_j. The vertical extent of the bulk depletion region into the substrate is represented by x_{dm}. The junction depletion region depths can be approximated by

$$x_{dS} = \sqrt{\frac{2 \cdot \varepsilon_{Si}}{q \cdot N_A} \cdot \phi_0} \qquad (3.81)$$

$$x_{dD} = \sqrt{\frac{2 \cdot \varepsilon_{Si}}{q \cdot N_A} \cdot (\phi_0 + V_{DS})} \qquad (3.82)$$

with the junction built-in voltage

$$\phi_0 = \frac{kT}{q} \cdot ln\left(\frac{N_D \cdot N_A}{n_i^2}\right) \qquad (3.83)$$

From Fig. 3.25(b), we find the following relationship between ΔL_D and the depletion region depths.

$$\left(x_j + x_{dD}\right)^2 = x_{dm}^2 + \left(x_j + \Delta L_D\right)^2 \qquad (3.84)$$

$$\Delta L_D^2 + 2 \cdot x_j \cdot \Delta L_D + x_{dm}^2 - x_{dD}^2 - 2 \cdot x_j \cdot x_{dD} = 0 \qquad (3.85)$$

Solving for ΔL_D, we obtain:

$$\Delta L_D = -x_j + \sqrt{x_j^2 - \left(x_{dm}^2 - x_{dD}^2\right) + 2x_j x_{dD}} \cong x_j \cdot \left(\sqrt{1 + \frac{2x_{dD}}{x_j}} - 1\right) \tag{3.86}$$

Similarly, the length ΔL_S can also be found as follows.

$$\Delta L_S \cong x_j \cdot \left(\sqrt{1 + \frac{2x_{dS}}{x_j}} - 1\right) \tag{3.87}$$

Now, the amount of threshold voltage reduction ΔV_{T0} due to short-channel effects can be found as:

$$\Delta V_{T0} = \frac{1}{C_{ox}} \cdot \sqrt{2q\,\varepsilon_{Si}\,N_A|2\phi_F|} \cdot \frac{x_j}{2L} \cdot \left[\left(\sqrt{1 + \frac{2x_{dS}}{x_j}} - 1\right) + \left(\sqrt{1 + \frac{2x_{dD}}{x_j}} - 1\right)\right] \tag{3.88}$$

The threshold voltage shift term is proportional to (x_j/L). As a result, this term becomes more prominent for MOS transistors with shorter channel lengths, and it approaches zero for long-channel MOSFETs where $L \gg x_j$. The following example illustrates the variation of the threshold voltage as a function of channel length in short-channel devices.

Example 3.6.

Consider an n-channel MOSFET process with the following parameters: substrate doping density $N_A = 10^{16}$ cm^{-3}, polysilicon gate doping density N_D (gate) $= 2 \times 10^{20}$ cm^{-3}, gate oxide thickness $t_{ox} = 50$ nm, oxide-interface fixed charge density $N_{ox} = 4 \times 10^{10}$ cm^{-2}, and source and drain diffusion doping density $N_D = 10^{17}$ cm^{-3}. In addition, the channel region is implanted with p-type impurities (impurity concentration $N_I = 2 \times 10^{11}$ cm^{-2}) to adjust the threshold voltage. The junction depth of the source and drain diffusion regions is $x_j = 1.0$ μm.

Plot the variation of the zero-bias threshold voltage V_{T0} as a function of the channel length (assume that $V_{DS} = V_{SB} = 0$). Also find V_{T0} for $L = 0.7$ μm, $V_{DS} = 5$ V, $V_{SB} = 0$.

First, we have to find the zero-bias threshold voltage using the conventional formula (3.23). The threshold voltage *without* the channel implant was already calculated for the same process parameters in Example 3.2, and was found to be $V_{T0} = 0.40$ V. The additional p-type channel implant will increase the threshold voltage by an amount of qN_I / C_{ox}. Thus, we find the long-channel zero-bias threshold voltage for the process described above as

$$V_{T0} = 0.34\, \text{V} + \frac{q \cdot N_I}{C_{ox}} = 0.40\, \text{V} + \frac{1.6 \times 10^{-19} \cdot 2 \times 10^{11}}{7.03 \times 10^{-8}} = 0.855\, \text{V}$$

Next, the amount of threshold voltage reduction due to short-channel effects must be calculated using (3.88). The source and drain junction built-in voltage is

$$\phi_0 = \frac{kT}{q} \cdot \ln\left(\frac{N_D \cdot N_A}{n_i^2}\right) = 0.026\ \text{V} \cdot \ln\left(\frac{10^{17} \cdot 10^{16}}{2.1 \times 10^{20}}\right) = 0.76\ \text{V}$$

For zero drain bias, the depth of source and drain junction depletion regions is found as

$$x_{dS} = x_{dD} = \sqrt{\frac{2 \cdot \varepsilon_{Si}}{q \cdot N_A} \cdot \phi_0} = \sqrt{\frac{2 \cdot 11.7 \cdot 8.85 \times 10^{-14}}{1.6 \times 10^{-19} \cdot 10^{16}} \cdot 0.76}$$

$$= 31.4 \times 10^{-6}\, \text{cm} = 0.314\ \mu\text{m}$$

Now, the threshold voltage shift ΔV_{T0} due to short-channel effects can be calculated as a function of the gate (channel) length L.

$$\Delta V_{T0} = \frac{1}{C_{ox}} \cdot \sqrt{2q\,\varepsilon_{Si}\, N_A |2\phi_F|} \cdot \frac{x_j}{2L} \cdot \left[\left(\sqrt{1 + \frac{2x_{dS}}{x_j}} - 1\right) + \left(\sqrt{1 + \frac{2x_{dD}}{x_j}} - 1\right)\right]$$

$$= \frac{4.82 \times 10^{-8}\ \text{C/cm}^2}{7.03 \times 10^{-8}\ \text{F/cm}^2} \cdot \frac{1.0\ \mu\text{m}}{L} \cdot \left(\sqrt{1 + \frac{2 \cdot 0.314\ \mu\text{m}}{1.0\ \mu\text{m}}} - 1\right)$$

Finally, the zero-bias threshold voltage is found as

$$V_{T0}(short\ channel) = 0.855\, \text{V} - 0.19\, \text{V} \cdot \frac{1}{L[\mu\text{m}]}$$

The following plot shows the variation of the threshold voltage with the channel length. The threshold voltage decreases by as much as 50% for channel lengths in the submicron range, while it approaches the value of 0.8 V for larger channel lengths.

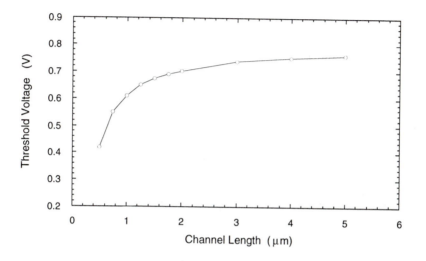

Since the conventional threshold voltage expression (3.23) is not capable of accounting for this drastic reduction of V_{T0} at smaller channel lengths, its application for short-channel MOSFETs must be carefully restricted.

Now, consider the variation of the threshold voltage with the applied drain-to-source voltage. Equation (3.82) shows that the depth of the drain junction depletion region increases with the voltage V_{DS}. For a drain-to-source voltage of $V_{DS} = 5$ V, the drain depletion depth is found as :

$$x_{dD} = \sqrt{\frac{2 \cdot \varepsilon_{Si}}{q \cdot N_A} \cdot (\phi_0 + V_{DS})} = \sqrt{\frac{2 \cdot 11.7 \cdot 8.85 \times 10^{-14}}{1.6 \times 10^{-19} \cdot 10^{16}} \cdot (0.76 + 5.0)} = 0.863 \ \mu m$$

The resulting threshold voltage shift can be calculated by substituting x_{dD} found above in (3.88).

$$\Delta V_{T0} = \frac{1}{C_{ox}} \cdot \sqrt{2 q \varepsilon_{Si} N_A |2\phi_F|} \cdot \frac{x_j}{2L} \cdot \left[\left(\sqrt{1 + \frac{2 x_{dS}}{x_j}} - 1 \right) + \left(\sqrt{1 + \frac{2 x_{dD}}{x_j}} - 1 \right) \right]$$

$$= \frac{4.82 \times 10^{-8}}{7.03 \times 10^{-8}} \cdot \frac{1.0}{2 \cdot 0.7} \cdot \left[\left(\sqrt{1 + \frac{2 \cdot 0.314}{1.0}} - 1 \right) + \left(\sqrt{1 + \frac{2 \cdot 0.863}{1.0}} - 1 \right) \right]$$

$$= 0.45 \ V$$

The threshold voltage of this short-channel MOS transistor is calculated as

$$V_{T0} = 0.855\,V - 0.45\,V = 0.405\,V$$

which is significantly lower than the threshold voltage predicted by the conventional long-channel formula (3.23).

Narrow-Channel Effects

MOS transistors that have channel widths W on the same order of magnitude as the maximum depletion region thickness x_{dm} are defined as narrow-channel devices. Similar to the short-channel effects examined earlier, the narrow-channel MOSFETs also exhibit typical characteristics which are not accounted for by the conventional GCA analysis. The most significant narrow-channel effect is that the actual threshold voltage of such a device is *larger* than that predicted by the conventional threshold voltage formula (3.23). In the following, we will briefly review the physical reasons that cause this discrepancy. A typical cross-sectional view of a narrow-channel device is shown in Fig. 3.26. The oxide thickness in the channel region is t_{ox}, while the regions around the channel are covered by a thick *field oxide* (FOX). Since the gate electrode also overlaps with the field oxide as shown in Fig. 3.26, a relatively shallow depletion region forms underneath this FOX-overlap area as well. Consequently, the gate voltage must also support this additional depletion charge in order to establish the conducting channel. The charge contribution of this fringe depletion region to the overall channel depletion charge is negligible in wider devices. For MOSFETs with small channel widths, however, the actual threshold voltage increases as a result of this extra depletion charge.

$$V_{T0}(narrow\ channel) = V_{T0} + \Delta V_{T0} \tag{3.89}$$

The additional contribution to the threshold voltage due to narrow-channel effects can be modeled as follows.

$$\Delta V_{T0} = \frac{1}{C_{ox}} \cdot \sqrt{2\,q\,\varepsilon_{Si}\,N_A\,|2\phi_F|} \cdot \frac{\kappa \cdot x_{dm}}{W} \tag{3.90}$$

where κ is an empirical parameter depending on the shape of the fringe depletion region. Assuming that the depletion region edges are modeled by quarter-circular arcs, for example, the parameter κ can be found as

$$\kappa = \frac{\pi}{2} \tag{3.91}$$

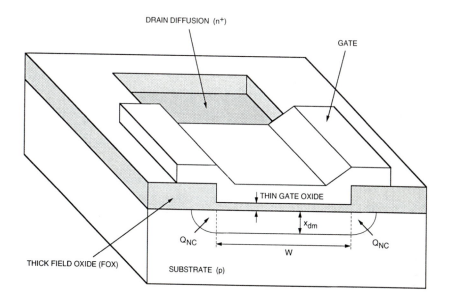

Figure 3.26. Cross-sectional view (across the channel) of a narrow-channel MOSFET. Note that Q_{NC} indicates the extra depletion charge due to narrow-channel effects.

The simple formula given in (3.90) can be modified for various device geometries and manufacturing processes, such as LOCOS, fully-recessed LOCOS, and thick-field-oxide MOSFET process. In all cases, we recognize that the additional contribution to V_{T0} is proportional to (x_{dm} / W). The amount of threshold voltage increase becomes significant only for devices which have a channel width W on the same order of magnitude as x_{dm}. Finally, note that for minimum-geometry MOSFETs which have a small channel length *and* a small channel width, the threshold voltage variations due to short- and narrow-channel effects may tend to cancel each other out.

Other Limitations Imposed by Small-Device Geometries

In small-geometry MOSFETs, the characteristics of current flow in the channel between the source and the drain can be explained as being controlled by the two-dimensional electric field vector $\vec{E}(x, y)$. The simple one-dimensional gradual channel approximation (GCA) assumes that the electric field components parallel to the surface and perpendicular to the surface are effectively decoupled and, therefore, cannot fully account for some of the observed device characteristics. These small-geometry device characteristics, however, may severely restrict the operating conditions of the transistor and impose limitations upon the practical utility of the device. Accurate identification and characterization of these small-geometry effects are crucial, especially for submicron MOSFETs.

One typical condition, which is due to the two-dimensional nature of channel current flow, is the *subthreshold conduction* in small-geometry MOS transistors. As already discussed in the previous sections, the current flow in the channel depends on creating and sustaining an inversion layer on the surface. If the gate bias voltage is not sufficient to invert the surface, i.e., $V_{GS} < V_{T0}$, the carriers (electrons) in the channel face a *potential barrier* that blocks the flow. Increasing the gate voltage reduces this potential barrier and, eventually, allows the flow of carriers under the influence of the channel electric field. This simple picture becomes more complicated in small-geometry MOSFETs, because the potential barrier is controlled by both the gate-to-source voltage V_{GS} *and* the drain-to-source voltage V_{DS}. If the drain voltage is increased, the potential barrier in the channel decreases, leading to *drain-induced barrier lowering* (DIBL). The reduction of the potential barrier eventually allows electron flow between the source and the drain, even if the gate-to-source voltage is lower than the threshold voltage. The channel current that flows under these conditions ($V_{GS} < V_{T0}$) is called the *subthreshold current*. Note that the GCA cannot account for any nonzero drain current I_D for $V_{GS} < V_{T0}$. Two-dimensional analysis of the small-geometry MOSFET yields the following approximate expression for the subthreshold current.

$$I_D(subthreshold) \cong \frac{qD_n W x_c n_0}{L_B} \cdot e^{\frac{q\phi_r}{kT}} \cdot e^{\frac{q}{kT}(A \cdot V_{GS} + B V_{DS})} \qquad (3.92)$$

Here, x_c is the subthreshold channel depth, D_n is the electron diffusion coefficient, L_B is the length of the barrier region in the channel, and ϕ_r is a reference potential. Note the exponential dependence of the subthreshold current on both the gate and the drain voltages. Identifying subthreshold conduction is very important for circuit applications where small amounts of current flow may significantly disturb the circuit operation.

We remember from the previous analysis that in small-geometry MOSFETs, the channel length is on the same order of magnitude as the source and drain depletion region thicknesses. For large drain-bias voltages, the depletion region surrounding the drain can extend farther toward the source, and the two depletion regions can eventually merge. This condition is termed *punch-through*; the gate voltage loses its control upon the drain current, and the current rises sharply once punch-through occurs. Being able to cause permanent damage to the transistor by localized melting of material, punch-through is obviously an undesirable condition, and should be prevented in normal circuit operation.

As some device dimensions, such as the channel length, are scaled down with each new generation, we find that some dimensions cannot be arbitrarily scaled because of physical limitations. One such dimension is the gate oxide thickness t_{ox}. The reduction of t_{ox} by a scaling factor of S, i.e., building a MOSFET with $t_{ox}' = t_{ox} / S$, is restricted by processing difficulties involved in growing very thin, uniform silicon-dioxide layers. Localized sites of nonuniform oxide growth, also called *pinholes*, may cause electrical shorts between the gate electrode and the substrate. Another limitation on the scaling of t_{ox} is the possibility of *oxide breakdown*. If the oxide electric field perpendicular to the surface is larger than a certain *breakdown field*, the silicon-dioxide layer may sustain permanent damage during operation, leading to device failure.

Finally, we will consider another reliability problem caused by high electric fields within the device. We have seen that advances in VLSI fabrication technologies are primarily based on the reduction of device dimensions, such as the channel length, the junction depth, and the gate oxide thickness, without proportional scaling of the power supply voltage (constant-voltage scaling). This decrease in critical device dimensions to submicron ranges, accompanied by increasing substrate doping densities, results in a significant increase of the horizontal and vertical electric fields in the channel region. Electrons and holes gaining high kinetic energies in the electric field (*hot carriers*) may, however, be injected into the gate oxide, and cause permanent

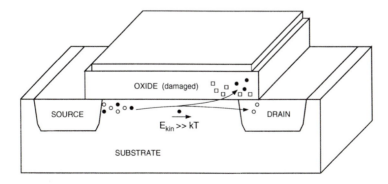

Figure 3.27.   Hot-carrier injection into the gate oxide and resulting oxide damage.

changes in the oxide-interface charge distribution, degrading the current-voltage characteristics of the MOSFET (Fig. 3.27). Since the likelihood of hot-carrier induced degradation increases with shrinking device dimensions, this problem was identified as one of the important factors that may impose strict limitations on maximum achievable device densities in VLSI circuits.

The channel hot-electron (CHE) effect is caused by electrons flowing in the channel region, from the source to the drain. This effect is more pronounced at large drain-to-source voltages, at which the lateral electric field in the drain end of the channel accelerates the electrons. The electrons arriving at the Si-SiO$_2$ interface with enough kinetic energy to surmount the surface potential barrier are injected into the oxide. Electrons and holes generated by impact ionization also contribute to the charge injection. Note that the channel hot-electron current and the subsequent damage in the gate oxide are localized near the drain junction (Fig. 3.27).

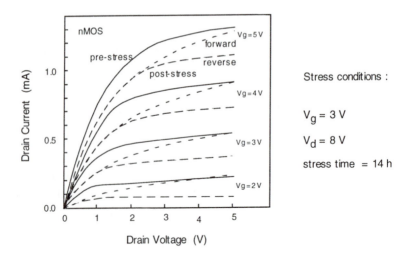

Figure 3.28.   Typical drain current vs. drain voltage characteristics of an n-channel MOS transistor before and after hot-carrier induced oxide damage.

The hot-carrier induced damage in nMOS transistors has been found to result in either trapping of carriers on defect sites in the oxide or the creation of interface states at the silicon-oxide interface, or both. The damage caused by hot-carrier injection affects the transistor characteristics by causing a degradation in transconductance, a shift in the threshold voltage, and a general decrease in the drain current capability (Fig. 3.28). This performance degradation in the devices leads to the degradation of circuit performance over time. Hence, new MOSFET technologies based on smaller device dimensions must carefully account for the hot-carrier effects and also ensure reliable long-term operation of the devices.

Other reliability concerns for small-geometry devices include interconnect damage through electromigration, electrostatic discharge (ESD) and electrical over-stress (EOS).

3.6. MOSFET Capacitances

The majority of the topics covered in this chapter has been related to the steady-state behavior of the MOS transistor. The current-voltage characteristics investigated here can be applied for investigating the DC response of MOS circuits under various operating conditions. In order to examine the transient (AC) response of MOSFETs and digital circuits consisting of MOSFETs, on the other hand, we have to determine the nature and the amount of parasitic capacitances associated with the MOS transistor.

The on-chip capacitances found in MOS circuits are in general complicated functions of the layout geometries and the manufacturing processes. Most of these capacitances are not lumped, but *distributed*, and their exact calculations would usually require complex, three-dimensional nonlinear charge-voltage models. In the following, we will develop simple approximations for the on-chip MOSFET capacitances that can be used in most hand calculations. These capacitance models are sufficiently accurate to represent the crucial characteristics of MOSFET charge-voltage behavior, and the equations are all based on fundamental semiconductor device theory, which should be familiar to most readers. We will also stress the distinction between the device-related capacitances and the interconnect capacitances. The capacitive contribution of metal interconnections between various devices is a very important component of the total parasitic capacitance observed in digital circuits. The estimation of this interconnect capacitance will be handled in Chapter 6.

Figure 3.29 shows the cross-sectional view and the top view (mask view) of a typical n-channel MOSFET. Until now, we concentrated on the cross-sectional view of the device, since we were primarily concerned with the flow of carriers within the MOSFET. As we study the parasitic device capacitances, we will have to become more familiar with the top view of the MOSFET. In this figure, the *mask length* (drawn length) of the gate is indicated by L_M, and the actual channel length is indicated by L. The extent of both the gate-source and the gate-drain overlap are L_D; thus, the channel length is given by

$$L = L_M - 2 \cdot L_D \qquad\qquad (3.93)$$

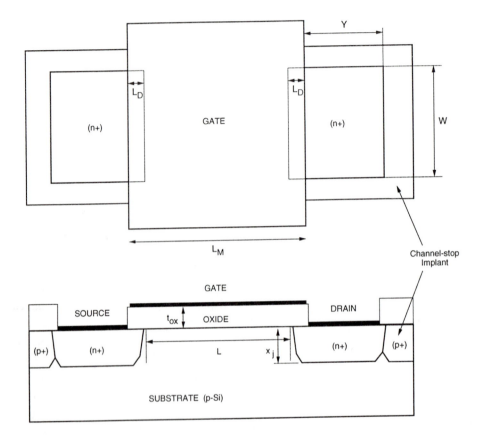

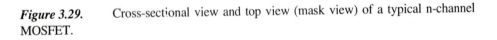

Figure 3.29.   Cross-sectional view and top view (mask view) of a typical n-channel MOSFET.

Note that the source and drain overlap region lengths are usually equal to each other because of the symmetry of the MOSFET structure. Typically, L_D is on the order of 0.1 μm. Both the source and the drain diffusion regions have the width of W. The typical diffusion region length is denoted by Y. Note that both the source diffusion region and the drain diffusion region are surrounded by a p^+ doped region, also called the channel-stop implant. As the name indicates, the purpose of this additional p^+ region is to prevent the formation of any unwanted (parasitic) channels between two neighboring n^+ diffusion regions, i.e., to ensure that the surface between two such regions cannot be inverted. Hence, the p^+ channel-stop implants act to electrically isolate neighboring devices built on the same substrate.

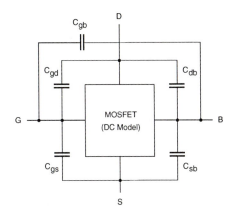

Figure 3.30. Lumped representation of the parasitic MOSFET capacitances.

We will identify the parasitic capacitances associated with this typical MOSFET structure as lumped equivalent capacitances *observed* between the device terminals (Fig. 3.30), since such a lumped representation can be easily used to analyze the dynamic transient behavior of the device. The reader must always be reminded, however, that in reality most parasitic device capacitances are due to three-dimensional, distributed charge-voltage relations within the device structure. Based on their physical origins, the parasitic device capacitances can be classified into two major groups: *oxide-related capacitances* and *junction capacitances*. First, the oxide-related capacitances will be considered.

Oxide-related Capacitances

It was shown earlier that the gate electrode overlaps both the source region and the drain region at the edges. The two overlap capacitances that arise as a result of this structural arrangement are called C_{GD} *(overlap)* and C_{GS} *(overlap)*, respectively. Assuming that both the source and the drain diffusion regions have the same width W, the overlap capacitances can be found as

$$C_{GS}(overlap) = C_{ox} \cdot W \cdot L_D$$
$$C_{GD}(overlap) = C_{ox} \cdot W \cdot L_D$$

(3.94)

with

$$C_{ox} = \frac{\varepsilon_{ox}}{t_{ox}}$$

(3.95)

Note that both of these overlap capacitances do not depend on the bias conditions, i.e., they are voltage-independent.

Now consider the capacitances which result due to the interaction between the gate voltage and the channel charge. Since the channel region is connected to the source, the drain, and the substrate, we can identify three capacitances between the gate and these regions: C_{gs}, C_{gd}, and C_{gb}. Notice that in reality, the gate-to-channel capacitance is distributed and voltage-dependent. Then, the gate-to-source capacitance C_{gs} is actually the gate-to-channel capacitance *seen* between the gate and the source terminals; the gate-to-drain capacitance C_{gd} is actually the gate-to-channel capacitance *seen* between the gate and the drain terminals. A simplified view of their bias-dependence can be obtained by observing the conditions in the channel region during cut-off, linear mode, and saturation.

In cut-off (Fig. 3.31(a)), the surface is not inverted. Consequently, there is no conducting channel that links the surface to the source and to the drain. Therefore, the gate-to-source and the gate-to-drain capacitances are both equal to zero: $C_{gs} = C_{gd} = 0$. The gate-to-substrate capacitance can be approximated by

$$C_{gb} = C_{ox} \cdot W \cdot L \qquad (3.96)$$

In linear-mode operation, the inverted channel extends across the MOSFET, between the source and the drain (Fig. 3.31(b)). This conducting inversion layer on the surface effectively shields the substrate from the gate electric field; thus, $C_{gb} = 0$. In this case, the distributed gate-to-channel capacitance may be viewed as being shared equally between the source and the drain, yielding

$$C_{gs} \cong C_{gd} \cong \frac{1}{2} \cdot C_{ox} \cdot W \cdot L \qquad (3.97)$$

When the MOSFET is operating in saturation mode, the inversion layer on the surface does not extend to the drain, but it is pinched off (Fig. 3.31(c)). The gate-to-drain capacitance component is therefore equal to zero ($C_{gd} = 0$). Since the source is still linked to the conducting channel, its shielding effect also forces the gate-to-substrate capacitance to be zero, $C_{gb} = 0$. Finally, the distributed gate-to-channel capacitance as seen between the gate and the source can be approximated by

$$C_{gs} \cong \frac{2}{3} \cdot C_{ox} \cdot W \cdot L \qquad (3.98)$$

Table 3.6 lists a summary of the approximate oxide capacitance values in three different operating modes of the MOSFET. The variation of the parasitic oxide capacitances as functions of the gate-to-source voltage V_{GS} is also shown in Fig. 3.32.

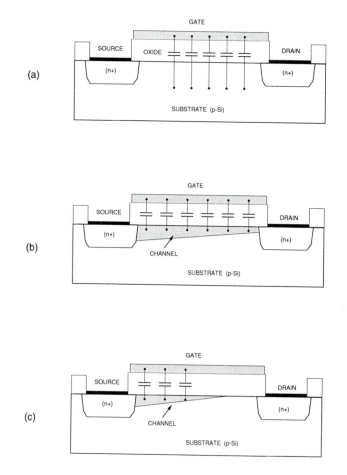

Figure 3.31.   Schematic representation of MOSFET oxide capacitances during (a) cut-off, (b) linear-mode operation, and (c) saturation.

Obviously, we have to combine C_{gs} and C_{gd} values found here with the relevant overlap capacitance values, in order to calculate the total capacitance between the external device terminals. It is also worth mentioning that the sum of all three voltage-dependent gate oxide capacitances $(C_{gb} + C_{gs} + C_{gd})$ has a minimum value of $0.66\,C_{ox}$ WL (in saturation mode) and a maximum value of $C_{ox}\,WL$ (in cut-off and in linear mode). For simple hand calculations where all three capacitances can be considered to be connected in parallel, a constant worst-case value of $C_{ox} WL$ can be used for the sum of MOSFET gate oxide capacitances.

Capacitance	Cut-off	Linear	Saturation
C_{gb}	$C_{ox}WL$	0	0
C_{gd}	0	$\frac{1}{2}C_{ox}WL$	0
C_{gs}	0	$\frac{1}{2}C_{ox}WL$	$\frac{2}{3}C_{ox}WL$

Table 3.6.   Approximate oxide capacitance values for three operating modes of the MOS transistor.

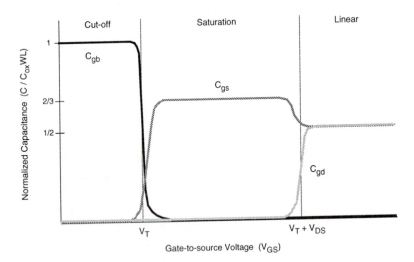

Figure 3.32.   Variation of oxide capacitances as functions of gate-to-source voltage V_{GS}.

Junction Capacitances

Now we consider the voltage-dependent source-substrate and drain-substrate junction capacitances, C_{sb} and C_{db}. Both of these capacitances are due to the depletion charge surrounding the respective source or drain diffusion regions embedded in the substrate. The calculation of the associated junction capacitances is complicated by the three-dimensional shape of the diffusion regions that form the source-substrate and the drain-substrate junctions. Note that both of these junctions are reverse-biased under normal operating conditions of the MOSFET and that the amount of junction capacitance is a function of the applied terminal voltages. Figure 3.33 shows the simplified, partial geometry of a typical n-channel enhancement MOSFET, focussing on the n-type diffusion region within the p-type substrate. The analysis to be carried out in the following will apply to both n-channel and p-channel MOS transistors.

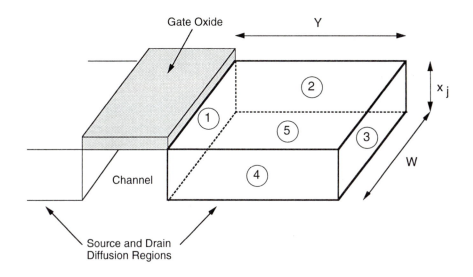

Figure 3.33.   Three-dimensional view of the n^+ diffusion region within the p-type substrate.

As seen in Fig. 3.33, the n^+ diffusion region forms a number of planar pn-junctions with the surrounding p-type substrate, indicated here with 1 through 5. The dimensions of the rectangular box representing the diffusion region are given as W, Y, and x_j. Abrupt (step) pn-junction profiles will be assumed for all junctions for simplicity. Also, comparing this three-dimensional view with Fig. 3.29, we recognize that three of the five planar junctions shown here (2, 3, and 4) are actually surrounded by the p^+ channel-stop implant. The junction labeled (1) is facing the channel, and the bottom junction (5) is facing the p-type substrate, which has a doping density of N_A. Since the p^+ channel-stop implant density is usually about $10N_A$, the junction capacitances associated with these *sidewalls* will be different from the other junction capacitances (see Table 3.7). Note that in general, the actual shape of the diffusion regions as well as the doping profiles are much more complicated. However, this simplified analysis provides sufficient insight for the first-order estimation of junction-related capacitances.

Junction	Area	Type
1	$W \cdot x_j$	n^+/p
2	$Y \cdot x_j$	n^+/p^+
3	$W \cdot x_j$	n^+/p^+
4	$Y \cdot x_j$	n^+/p^+
5	$W \cdot Y$	n^+/p

Table 3.7.   Types and areas of the pn-junctions shown in Figure 3.33.

To calculate the depletion capacitance of a reverse-biased abrupt pn-junction, consider first the depletion region thickness, x_d. Assuming that the n-type and p-type doping densities are given by N_D and N_A, respectively, and that the reverse bias voltage is given by V (negative), the depletion region thickness can be found as follows:

$$x_d = \sqrt{\frac{2 \cdot \varepsilon_{Si}}{q} \cdot \frac{N_A + N_D}{N_A \cdot N_D} \cdot (\phi_0 - V)} \qquad (3.99)$$

where the built-in junction potential is calculated as

$$\phi_0 = \frac{kT}{q} \cdot \ln\left(\frac{N_A \cdot N_D}{n_i^2}\right) \qquad (3.100)$$

Note that the junction is forward-biased for a *positive* bias voltage V, and reverse-biased for a *negative* bias voltage. The depletion-region charge stored in this area can be written in terms of the depletion region thickness, x_d.

$$Q_j = A \cdot q \cdot \left(\frac{N_A \cdot N_D}{N_A + N_D}\right) \cdot x_d = A \sqrt{2 \cdot \varepsilon_{Si} \cdot q \cdot \left(\frac{N_A \cdot N_D}{N_A + N_D}\right) \cdot (\phi_0 - V)} \qquad (3.101)$$

Here, A indicates the junction area. The junction capacitance associated with the depletion region is defined as

$$C_j = \left| \frac{dQ_j}{dV} \right| \tag{3.102}$$

By differentiating (3.101) with respect to the bias voltage V, we can now obtain the expression for the junction capacitance as follows.

$$C_j(V) = A \cdot \sqrt{\frac{\varepsilon_{Si} \cdot q}{2} \cdot \left(\frac{N_A \cdot N_D}{N_A + N_D} \right)} \cdot \frac{1}{\sqrt{\phi_0 - V}} \tag{3.103}$$

This expression can be rewritten in a more general form, to account for the junction grading.

$$C_j(V) = \frac{A \cdot C_{j0}}{\left(1 - \dfrac{V}{\phi_0} \right)^m} \tag{3.104}$$

The parameter m in (3.104) is called the *grading coefficient*. Its value is equal to 1/2 for an abrupt junction profile, and 1/3 for a linearly graded junction profile. Obviously, for an abrupt pn-junction profile, i.e., for $m = 1/2$, the equations (3.103) and (3.104) become identical. The zero-bias junction capacitance per unit area C_{j0} is defined as

$$C_{j0} = \sqrt{\frac{\varepsilon_{Si} \cdot q}{2} \cdot \left(\frac{N_A \cdot N_D}{N_A + N_D} \right)} \cdot \frac{1}{\phi_0} \tag{3.105}$$

Note that the value of the junction capacitance C_j given by (3.104) ultimately depends on the external bias voltage that is applied across the pn-junction. Since the terminal voltages of a MOSFET will change during dynamic operation, accurate estimation of the junction capacitances under transient conditions is quite complicated; the instantaneous values of all junction capacitances will also change accordingly. The problem of estimating capacitance values under changing bias conditions can be simplified, if we calculate a large-signal average (linear) junction capacitance instead, which, by definition, is independent of the bias potential. This *equivalent large-signal capacitance* can be defined as follows:

$$C_{eq} = \frac{\Delta Q}{\Delta V} = \frac{Q_j(V_2) - Q_j(V_1)}{V_2 - V_1} = \frac{1}{V_2 - V_1} \cdot \int_{V_1}^{V_2} C_j(V) dV \tag{3.106}$$

Here, the reverse bias voltage across the pn-junction is assumed to change from V_1 to V_2. Hence, the equivalent capacitance C_{eq} is always calculated for a *transition between two known voltage levels*. By substituting (3.104) into (3.106), we obtain

$$C_{eq} = -\frac{A \cdot C_{j0} \cdot \phi_0}{(V_2 - V_1) \cdot (1-m)} \cdot \left[\left(1 - \frac{V_2}{\phi_0}\right)^{1-m} - \left(1 - \frac{V_1}{\phi_0}\right)^{1-m} \right] \qquad (3.107)$$

For the special case of abrupt pn-junctions, equation (3.107) becomes

$$C_{eq} = -\frac{2 \cdot A \cdot C_{j0} \cdot \phi_0}{(V_2 - V_1)} \cdot \left[\sqrt{1 - \frac{V_2}{\phi_0}} - \sqrt{1 - \frac{V_1}{\phi_0}} \right] \qquad (3.108)$$

This equation can be rewritten in a simpler form by defining a dimensionless coefficient K_{eq}, as follows:

$$C_{eq} = A \cdot C_{j0} \cdot K_{eq} \qquad (3.109)$$

$$K_{eq} = -\frac{2\sqrt{\phi_0}}{V_2 - V_1} \cdot \left(\sqrt{\phi_0 - V_2} - \sqrt{\phi_0 - V_1} \right) \qquad (3.110)$$

where K_{eq} is the *voltage equivalence factor* (note that $0 < K_{eq} < 1$). Thus, the coefficient K_{eq} allows us to take into account the voltage-dependent variations of the junction capacitance. The accuracy of the large-signal equivalent junction capacitance C_{eq} found by using (3.109) and (3.110) is usually sufficient for most first-order hand calculations. Practical applications of the capacitance calculation methods discussed here will be illustrated in the following examples.

Example 3.7.

Consider a simple abrupt pn-junction, which is reverse-biased with a voltage V_{bias}. The doping density of the n-type region is $N_D = 10^{20}$ cm^{-3}, and the doping density of the p-type region is given as $N_A = 10^{16}$ cm^{-3}. The junction area is $A = 20$ μm x 20 μm.

First, we will calculate the zero-bias junction capacitance per unit area, C_{j0}, for this structure. The built-in junction potential is found as

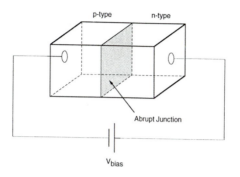

Abrupt Junction

$$\phi_0 = \frac{kT}{q} \cdot \ln\left(\frac{N_A \cdot N_D}{n_i^2}\right) = 0.026 \text{ V} \cdot \ln\left(\frac{10^{20} \cdot 10^{16}}{2.1 \times 10^{20}}\right) = 0.94 \text{ V}$$

Using (3.105), we can calculate the zero-bias junction capacitance :

$$C_{j0} = \sqrt{\frac{\varepsilon_{Si} \cdot q}{2} \cdot \left(\frac{N_A \cdot N_D}{N_A + N_D}\right) \cdot \frac{1}{\phi_0}}$$

$$= \sqrt{\frac{11.7 \cdot 8.85 \times 10^{-14} \text{F/cm} \cdot 1.6 \times 10^{-19} \text{C}}{2} \cdot \left(\frac{10^{20} \cdot 10^{16}}{10^{20} + 10^{16}}\right) \cdot \frac{1}{0.94 \text{ V}}}$$

$$= 2.9 \times 10^{-8} \text{ F/cm}^2$$

Next, find the equivalent large-signal junction capacitance assuming that the reverse bias voltage changes from $V_1 = 0$ to $V_2 = -5$ V. The voltage equivalence factor for this transition can be found as follows.

$$K_{eq} = -\frac{2\sqrt{\phi_0}}{V_2 - V_1} \cdot \left(\sqrt{\phi_0 - V_2} - \sqrt{\phi_0 - V_1}\right)$$

$$= -\frac{2\sqrt{0.94}}{-5} \cdot \left(\sqrt{0.94 - (-5)} - \sqrt{0.94}\right) = 0.57$$

Then, the average junction capacitance can be found simply by using (3.109).

$$C_{eq} = A \cdot C_{j0} \cdot K_{eq} = 400 \times 10^{-8} \text{ cm}^2 \cdot 2.98 \times 10^{-8} \text{ F/cm}^2 \cdot 0.57 = 68 \text{ fF}$$

It was shown in Fig. 3.29 and Fig. 3.33 that the sidewalls of a typical MOSFET source or drain diffusion region are surrounded by a p$^+$ channel-stop implant, with a higher doping density than the substrate doping density N_A. Consequently, the sidewall zero-bias capacitance C_{j0sw}, as well as the sidewall voltage equivalence factor $K_{eq}(sw)$ will be different from those of the bottom junction. Assuming that the sidewall doping density is given by $N_A(sw)$, the zero-bias capacitance per unit area can be found as follows.

$$C_{j0sw} = \sqrt{\frac{\varepsilon_{Si} \cdot q}{2} \cdot \left(\frac{N_A(sw) \cdot N_D}{N_A(sw) + N_D}\right) \cdot \frac{1}{\phi_{0sw}}} \qquad (3.111)$$

where ϕ_{0sw} is the built-in potential of the sidewall junctions. Since all sidewalls in a typical diffusion structure have approximately the same depth of x_j, we can define a zero-bias sidewall junction capacitance per unit length.

$$C_{jsw} = C_{j0sw} \cdot x_j \qquad (3.112)$$

The sidewall voltage equivalence factor $K_{eq}(sw)$ for a voltage swing between V_1 and V_2 is defined as follows:

$$K_{eq}(sw) = -\frac{2\sqrt{\phi_{0sw}}}{V_2 - V_1} \cdot \left(\sqrt{\phi_{0sw} - V_2} - \sqrt{\phi_{0sw} - V_1}\right) \qquad (3.113)$$

Combining the equations (3.111) through (3.113), the equivalent large-signal junction capacitance $C_{eq}(sw)$ for a sidewall of length (perimeter) P can be calculated as

$$C_{eq}(sw) = P \cdot C_{jsw} \cdot K_{eq}(sw) \qquad (3.114)$$

Example 3.8.

Consider the n-channel enhancement-type MOSFET shown below. The process parameters are given as follows:

Substrate doping $N_A = 2 \times 10^{15} \text{ cm}^{-3}$
Source / drain doping $N_D = 10^{20} \text{ cm}^{-3}$

Sidewall (p+) doping $N_A(sw) = 4 \times 10^{16}\,\text{cm}^{-3}$
Gate oxide thickness $t_{ox} = 45\,\text{nm}$
Junction depth $x_j = 1.0\,\mu\text{m}$

Note that both the source and the drain diffusion regions are surrounded by p$^+$ channel-stop diffusion. The substrate is biased at 0 V. Assuming that the drain voltage is changing from 0.5 V to 5 V, find the average drain-substrate junction capacitance C_{db}.

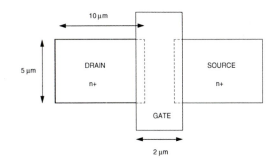

First, we recognize that three sidewalls of the rectangular drain diffusion structure form n$^+$/p$^+$ junctions with the p$^+$ channel-stop implant, while the bottom area and the sidewall facing the channel form n$^+$/p junctions. Start by calculating the built-in potentials for both types of junctions.

$$\phi_0 = \frac{kT}{q} \cdot \ln\left(\frac{N_A \cdot N_D}{n_i^2}\right) = 0.026\,\text{V} \cdot \ln\left(\frac{10^{20} \cdot 2 \times 10^{15}}{2.1 \times 10^{20}}\right) = 0.896\,\text{V}$$

$$\phi_{0sw} = \frac{kT}{q} \cdot \ln\left(\frac{N_A(sw) \cdot N_D}{n_i^2}\right) = 0.026\,\text{V} \cdot \ln\left(\frac{10^{20} \cdot 4 \times 10^{16}}{2.1 \times 10^{20}}\right) = 0.975\,\text{V}$$

Next, we calculate the zero-bias junction capacitances per unit area:

$$C_{j0} = \sqrt{\frac{\varepsilon_{Si} \cdot q}{2} \cdot \left(\frac{N_A \cdot N_D}{N_A + N_D}\right) \cdot \frac{1}{\phi_0}}$$

$$= \sqrt{\frac{11.7 \cdot 8.85 \times 10^{-14}\,\text{F/cm} \cdot 1.6 \times 10^{-19}\,\text{C}}{2} \cdot \left(\frac{10^{20} \cdot 2 \times 10^{15}}{10^{20} + 2 \times 10^{15}}\right) \cdot \frac{1}{0.896\,\text{V}}}$$

$$= 1.35 \times 10^{-8}\,\text{F/cm}^2$$

$$C_{j0sw} = \sqrt{\frac{\varepsilon_{Si} \cdot q}{2} \cdot \left(\frac{N_A(sw) \cdot N_D}{N_A(sw) + N_D}\right)} \cdot \frac{1}{\phi_{0sw}}$$

$$= \sqrt{\frac{11.7 \cdot 8.85 \times 10^{-14} \, F/cm \cdot 1.6 \times 10^{-19} \, C}{2} \cdot \left(\frac{10^{20} \cdot 4 \times 10^{16}}{10^{20} + 4 \times 10^{16}}\right)} \cdot \frac{1}{0.975 \, V}$$

$$= 5.83 \times 10^{-8} \, F/cm^2$$

The zero-bias sidewall junction capacitance per unit length can also be found as follows.

$$C_{jsw} = C_{j0sw} \cdot x_j = 5.83 \times 10^{-8} \, F/cm^2 \cdot 10^{-4} \, cm = 5.83 \, pF/cm$$

In order to take the given drain voltage variation into account, we must now calculate the voltage equivalence factors, K_{eq} and $K_{eq}(sw)$, for both types of junctions. This will allow us to find the average large-signal capacitance values.

$$K_{eq} = -\frac{2\sqrt{0.9}}{-5-(-0.5)} \cdot \left(\sqrt{0.9+5} - \sqrt{0.9+0.5}\right) = 0.52$$

$$K_{eq}(sw) = -\frac{2\sqrt{0.925}}{-5-(-0.5)} \cdot \left(\sqrt{0.925+5} - \sqrt{0.925+0.5}\right) = 0.53 \cong K_{eq}$$

The total area of the n^+/p junctions is calculated as the sum of the bottom area and the sidewall area facing the channel region.

$$A = (10 \times 5) \, \mu m^2 + (5 \times 1) \, \mu m^2 = 55 \, \mu m^2$$

The total length of the n^+/p^+ junction perimeter, on the other hand, is equal to the sum of three sides of the drain diffusion area. Thus, the combined equivalent (average) drain-substrate junction capacitance can be found as follows:

$$\langle C_{db} \rangle = A \cdot C_{j0} \cdot K_{eq} + P \cdot C_{j0sw} \cdot K_{eq}(sw)$$

$$= 55 \times 10^{-8} \, cm^2 \cdot 1.35 \times 10^{-8} \, F/cm^2 \cdot 0.52$$

$$+ 25 \times 10^{-4} \, cm \cdot 5.83 \times 10^{-12} \, F/cm \cdot 0.53 = 11.58 \times 10^{-15} \, F = 11.6 \, fF$$

References

1.   A.S. Grove, *Physics and Technology of Semiconductor Devices*, New York, NY: Wiley, 1967.

2.   R.S. Muller and T. Kamins, *Device Electronics for Integrated Circuits*, second edition, New York, NY: Wiley, 1986.

3.   Y.P. Tsividis, *Operation and Modeling of the MOS Transistor*, New York, NY: McGraw-Hill, 1987.

4.   C.-T. Sah, *Fundamentals of Solid-State Electronics*, River Ridge, NJ: World Scientific Publishing Co., 1991.

5.   S.M. Sze, *VLSI Technology*, New York, NY: McGraw-Hill, 1983.

6.   S.M. Sze, *Physics of Semiconductor Devices*, second edition, New York, NY: Wiley, 1981.

Exercise Problems

3.1   Consider a MOS system with the following parameters:

$$t_{ox} = 200 \text{ Å}$$
$$\phi_{GC} = -0.85 \text{ V}$$
$$N_A = 2 \cdot 10^{15} \text{ cm}^{-3}$$
$$Q_{ox} = q\, 2 \cdot 10^{11} \text{ C/cm}^2$$

(a)   Determine the threshold voltage V_{T0} under zero bias at room temperature (T = 300 K). Note that $\varepsilon_{ox} = 3.97\varepsilon_0$ and $\varepsilon_{si} = 11.7\varepsilon_0$.

(b)   Determine the type (p-type or n-type) and amount of channel implant (N_I/cm^2) required to change the threshold voltage to 0.8 V.

3.2   Consider a diffusion area which has the dimensions 10 µm x 5 µm, and the abrupt junction depth is 0.5 µm. Its n-type impurity doping level is $N_D = 1 \cdot 10^{20}$ cm^{-3} and the surrounding p-type substrate doping level is $N_A = 1 \cdot 10^{16}$ cm^{-3}. Determine the capacitance when the diffusion area is biased at 5 V and the substrate is biased at 0 V. In this problem, assume that there is no channel-stop implant.

3.3   Describe the relationship between the mask channel length, L_{mask}, and the electrical channel length, L. Are they identical? If not, how would you express L in terms of L_{mask} and other parameters?

3.4   How is the device junction temperature affected by the power dissipation of the chip and its package? Can you describe the relationship between the device junction temperature, ambient temperature, chip power dissipation, and packaging quality?

3.5   Describe the three main components of the load capacitance C_{load}, when a logic gate is driving other fanout gates.

3.6   Consider the layout of an nMOS transistor shown in Fig. P3.6. The process parameters are:

$$N_D = 2 \cdot 10^{20}\ cm^{-3}$$
$$N_A = 1 \cdot 10^{15}\ cm^{-3}$$
$$X_j = 0.5\ \mu m$$
$$L_D = 0.5\ \mu m$$
$$t_{ox} = 0.05\ \mu m$$
$$V_{T0} = 0.8\ V$$
Channel stop doping = 16.0 x (p-type substrate doping)

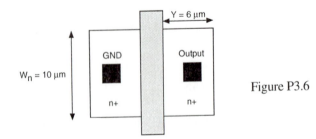

Figure P3.6

Find the effective drain parasitic capacitance when the drain node voltage changes from 5 V to 2.5 V.

3.7   A set of *I-V* characteristics for an nMOS transistor at room temperature is shown below for different biasing conditions. Figure P3.7 shows the measurement setup.

Using the data, find : (a) the threshold voltage V_{T0}, (b) electron mobility μ_n, and (c) body effect coefficient gamma (γ).

Some of the parameters are given as: $W/L = 1.0$, $t_{ox} = 345$ Å, $|2\phi_F| = 0.64$ V.

V_{GS}	V_{DS}	V_{SB}	I_D [μA]
4 V	4 V	0.0 V	256
5 V	5 V	0.0 V	441
4 V	4 V	2.6 V	144
5 V	5 V	2.6 V	256

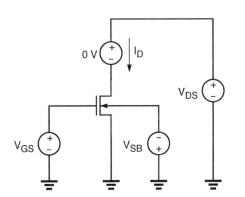

Figure P3.7

3.8   Compare the two technology scaling methods, namely, (i) the constant electric-field scaling and (ii) the constant power-supply voltage scaling. In particular, show analytically by using equations how the delay time, power dissipation, and power density are affected in terms of the scaling factor, S.

To be more specific, what would happen if the design rules change from, say, 1 μm to $1/S$ μm ($S > 1$) ?

3.9   A pMOS transistor was fabricated on an n-type substrate with a bulk doping density of $N_D = 10^{16}$ cm^{-3}, gate doping density (n-type poly) of $N_D = 10^{20}$ cm^{-3}, $Q_{ox}/q = 4 \cdot 10^{10}$ cm^{-2}, and gate oxide thickness of $t_{ox} = 0.1$ μm. Calculate the threshold voltage at room temperature for $V_{SB} = 0$. Use $\varepsilon_{si} = 11.7\varepsilon_0$.

3.10   A depletion-type nMOS transistor has the following device parameters:

$$\mu_n = 500 \text{ cm}^2/\text{V}\cdot\text{s}$$
$$t_{ox} = 345 \text{ Å}$$
$$|2\phi_F| = 0.84 \text{ V}$$
$$W/L = 1.0$$

Some laboratory measurement results of the terminal behavior of this device are shown in the table below. Using the data in the table, find the missing value of the gate voltage in the last entry. Show all of the details of your calculation.

I_{DS} [μA]	V_D	V_S	V_B	V_G
50.0	3 V	0	0	0
40.0	5 V	3 V	0	?

3.11   Using the parameters given below, calculate the current through two nMOS transistors in series (see Figure P3.11), when the drain of the top transistor is tied to V_{DD}, the source of the bottom transistor is tied to $V_{SS}=0$ and their gates are tied to V_{DD}. The substrate is also tied to $V_{SS}=0$ V. Assume that $W/L = 10$ for both transistors.

$$k' = 25 \ \mu A/V^2$$
$$V_{T0} = 1.0 \ V$$
$$\gamma = 0.39 \ V^{1/2}$$
$$|2\phi_F| = 0.6 \ V$$

Hint : The solution requires some iterations and the body effect on threshold voltage has to be taken into account. Start with the KCL equation.

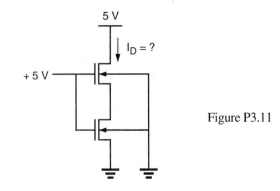

Figure P3.11

3.12   The following parameters are given for an nMOS process:

$$t_{ox} = 500 \ \text{Å}$$
substrate doping $N_A = 1 \cdot 10^{16} \ \text{cm}^{-3}$
polysilicon gate doping $N_D = 1 \cdot 10^{20} \ \text{cm}^{-3}$
oxide-interface fixed-charge density $N_{ox} = 2 \cdot 10^{10} \ \text{cm}^{-3}$

(a)   Calculate V_T for an unimplanted transistor.

(b)   What type and what concentration of impurities must be implanted to achieve $V_T = +2$ V and $V_T = -2$ V ?

3.13   Using the measured data given below, determine the device parameters V_{T0}, k, γ, and λ, assuming $2\phi_F = -0.6$ V.

V_{GS}	V_{DS}	V_{BS}	$I_D[\mu A]$
2	5	0	10
5	5	0	400
5	5	-3	280
5	8	0	480

3.14   Using the design rules specified in Chapter 2, sketch a simple layout of an nMOS transistor on grid paper. Use a minimum feature size of 3 μm. Neglect the substrate connection. After you complete the layout, calculate approximate values for C_g, C_{sb}, and C_{db}. The following parameters are given.

Substrate doping $N_A = 10^{16}$ cm^{-3}
Drain/source doping $N_D = 10^{19}$ cm^{-3}
$W = 15$ μm
$L = 3$ μm
$t_{ox} = 0.05$ μm

3.15   Derive the current equation for a p-channel MOS transistor operating in the linear region, i.e., for $V_{SG} + V_{TP} > V_{SD}$.

3.16   An enhancement-type nMOS transistor has the following parameters:

$V_{T0} = 0.8$ V
$\gamma = 0.2$ V$^{1/2}$
$\lambda = 0.05$ V^{-1}
$|2\phi_F| = 0.58$ V
$k' = 20$ μA/V^2

(a)   When the transistor is biased with $V_G = 2.8$ V, $V_D = 5$ V, $V_S = 1$ V, and $V_B = 0$ V, the drain current is $I_D = 0.24$ mA. Determine W/L.

(b)   Calculate I_D for $V_G = 5$ V, $V_D = 4$ V, $V_S = 2$ V, and $V_B = 0$ V.

(c)   If $\mu_n = 500$ cm^2/V·s and $C_g = C_{ox}·W·L = 1.0 \times 10^{-15}$ F, find W and L.

3.17   An nMOS transistor is fabricated with the following physical parameters:

$$N_D = 10^{20} \text{ cm}^{-3}$$
$$N_A(\text{substrate}) = 10^{16} \text{ cm}^{-3}$$
$$N_A^+(\text{chan. stop}) = 10^{19} \text{ cm}^{-3}$$
$$W = 10 \text{ } \mu m$$
$$Y = 5 \text{ } \mu m$$
$$L = 1.5 \text{ } \mu m$$
$$L_D = 0.25 \text{ } \mu m$$
$$X_j = 0.4 \text{ } \mu m$$

(a)   Determine the drain diffusion capacitance for $V_{DB} = 5$ V and 2.5 V.

(b)   Calculate the overlap capacitance between gate and drain for an oxide thickness of $t_{ox} = 200$ Å.

CHAPTER 4

MODELING OF MOS TRANSISTORS USING SPICE

SPICE (Simulation Program with Integrated Circuit Emphasis) is a general purpose circuit simulator which is used very widely both in the microelectronics industry and in educational institutions as an essential computer-aided design (CAD) tool for circuit design. After almost two decades of running on various platforms around the world, it can be regarded as the de facto standard in circuit simulation. Most engineers and circuit designers using SPICE acknowledge how critical the input *models* for transistors are, to obtain simulation outputs matching the experimental data. In the fast-advancing field of VLSI design, a sound knowledge of physical models used to describe the behavior of transistors and knowledge of various device parameters are essential for performing detailed circuit simulations and for optimizing design. This chapter will describe the physical aspects of various MOSFET models built-in in SPICE, and discuss the model equations as well as the model parameters. Also, practical comparisons among the different MOSFET models available in SPICE will be offered, which may help the users to select the most appropriate device model for a given simulation task. It will be assumed that the reader already has a working knowledge of SPICE, the structure of circuit input files, and the use of .MODEL description statements.

SPICE has three built-in MOSFET models: LEVEL 1 (MOS1) is described by a square-law current-voltage characteristic, LEVEL 2 (MOS2) is a detailed analytical MOSFET model, and LEVEL 3 (MOS3) is a semi-empirical model. Both MOS2 and MOS3 include second-order effects such as the short-channel threshold voltage, subthreshold conduction, scattering-limited velocity saturation, and charge-controlled capacitances. The level (type) of the MOSFET model to be used in a particular simulation task is declared on the .MODEL statement. In addition, the user can describe a large number of model parameters on this statement. Information about the geometry of a particular device, such as the channel length, channel width, and source and drain areas is usually given on the element description line of that device. Some typical MOSFET element description lines and .MODEL statements are given below.

```
M1      3    1    0    0    NMOD    L=1U     W=10U     AD=120P    PD=42U

MDEV32    14    9    12    5    PMOD    L=1.2U   W=20U

.MODEL NMOD NMOS (LEVEL=1 VTO=1.4 KP=4.5E-5 CBD=5PF CBS=2PF)

.MODEL PMOD PMOS     (VTO=-2    KP=3.0E-5    LAMBDA=0.02    GAMMA=0.4
+                    CBD=4PF    CBS=2PF    RD=5    RS=3    CGDO=1PF
+                    CGSO=1PF   CGBO=1PF)
```

4.1. Basic Concepts

In the following, we will examine the various model equations and the model parameters associated with the built-in MOSFET models. We will also discuss the usual (normal) range for the parameter values and the default values of the model parameters which are already incorporated into SPICE. A list of all MOSFET model parameters is given in Table 4.1 on pages 114-115.

The equivalent circuit structure of the NMOS LEVEL 1 model, which is the default MOSFET model in SPICE, is shown in Fig. 4.1. This basic structure is also typical for the LEVEL 2 and LEVEL 3 models. Note that the voltage-controlled current source I_D determines the steady-state current-voltage behavior of the device, while the voltage-controlled (nonlinear) capacitors connected between the terminals represent the parasitic oxide-related and junction capacitances. The source-substrate and the drain-substrate junctions, which are reverse-biased under normal operating conditions, are represented by ideal diodes in this equivalent circuit. Finally, the parasitic source and drain resistances are represented by the resistors R_D and R_S, connected between the drain current source and the respective terminals.

The basic geometry of an MOS transistor can be described by specifying the nominal channel (gate) length L and the channel width W, both of which are indicated on the element description line. The channel width W is, by definition, the *width* of the area covered by the thin gate oxide. Note that the effective channel length L_{eff} is defined as the distance on the surface between the two (source and drain) diffusion regions. Thus, in order to find the effective channel length, the gate-source overlap distance and the gate-drain overlap distance must be subtracted from the nominal (mask) gate length specified on the device description line. The amount of gate overlap over the source and the drain can be specified by using the *lateral diffusion* coefficient L_D in SPICE.

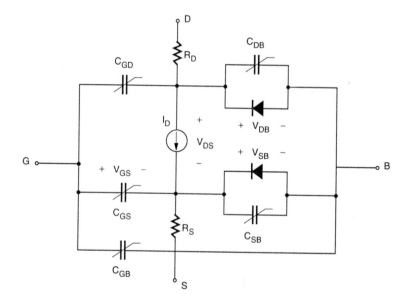

Figure 4.1.   Equivalent circuit structure of the LEVEL 1 MOSFET model in SPICE.

For modeling p-channel MOS transistors, the direction of the dependent current source, the polarities of the terminal voltages, and the directions of the two diodes representing the source-substrate and the drain-substrate junctions must be reversed. Otherwise, the equations to be presented in the following sections apply to p-channel MOSFETs as well.

4.2. The LEVEL 1 Model Equations

The LEVEL 1 model is the simplest current-voltage description of the MOSFET, which is basically the GCA-based quadratic model originally proposed by Shichman

TABLE 4-1 Model Parameters for the MOST (Copyright © 1988 by McGraw-Hill, Inc.).

Symbol	SPICE keyword	LEVEL	Parameter description	Default value	Typical value	Units
			Parameters of the MOST			
V_{T0}	VTO	1–3	Zero-bias threshold voltage	1.0	1.0	V
KP	KP	1–3	Transconductance parameter	2×10^{-5}	3×10^{-5}	A/V^2
γ	GAMMA	1–3	Body-effect parameter	0.0	0.35	$V^{1/2}$
$2\phi_F$	PHI	1–3	Surface inversion potential	0.6	0.65	V
λ	LAMBDA	1, 2	Channel-length modulation	0.0	0.02	V^{-1}
t_{ox}	TOX	1–3	Thin oxide thickness	1×10^{-7}	1×10^{-7}	m
N_b	NSUB	1–3	Substrate doping	0.0	1×10^{15}	cm^{-3}
N_{SS}	NSS	2, 3	Surface state density	0.0	1×10^{10}	cm^{-2}
N_{FS}	NFS	2, 3	Surface-fast state density	0.0	1×10^{10}	cm^{-2}
N_{eff}	NEFF	2	Total channel charge coefficient	1	5	
X_j	XJ	2, 3	Metallurgical junction depth	0.0	1×10^{-6}	m
X_{jl}	LD	1–3	Lateral diffusion	0.0	0.8×10^{-6}	m
T_{PG}	TPG	2, 3	Type of gate material	1	1	
μ_0	UO	1–3	Surface mobility	600	700	$cm^2/(V \cdot s)$
U_c	UCRIT	2	Critical electric field for mobility	1×10^4	1×10^4	V/cm
U_e	UEXP	2	Exponential coefficient for mobility	0.0	0.1	
U_t	UTRA	2	Transverse field coefficient	0.0	0.5	
v_{max}	VMAX	2, 3	Maximum drift velocity of carriers	0.0	5×10^4	m/sec
X_{QC}	XQC	2, 3	Coefficient of channel charge share	0.0	0.4	
δ	DELTA	2, 3	Width effect on threshold voltage	0.0	1.0	

Symbol	SPICE name	Range	Description			Units
η	ETA	3	Static feedback on threshold voltage	0.0	1.0	
θ	THETA	3	Mobility modulation	0.0	0.05	V^{-1}
A_F	AF	1–3	Flicker-noise exponent	1.0	1.2	
K_F	KF	1–3	Flicker-noise coefficient	0.0	1×10^{-26}	

Parameters of parasitic effects

Symbol	SPICE name	Range	Description			Units
I_s	IS	1–3	Bulk-junction saturation current	1×10^{-14}	1×10^{-15}	A
J_s	JS	1–3	Bulk-junction saturation current per square meter	0.0	1×10^{-8}	A
ϕ_j	PB	1–3	Bulk-junction potential	0.80	0.75	V
C_j	CJ	1–3	Zero-bias bulk capacitance per square meter	0.0	2×10^{-4}	F/m^2
M_j	MJ	1–3	Bulk-junction grading coefficient	0.5	0.5	
C_{jsw}	CJSW	1–3	Zero-bias perimeter capacitance per meter	0.0	1×10^{-9}	F/m
M_{jsw}	MJSW	1–3	Perimeter capacitance grading coefficient	0.33	0.33	
FC	FC	1–3	Bulk-junction forward-bias coefficient	0.5	0.5	
C_{GBO}	CGBO	1–3	Gate-bulk overlap capacitance per meter	0.0	2×10^{-10}	F/m
C_{GDO}	CGDO	1–3	Gate-drain overlap capacitance per meter	0.0	4×10^{-11}	F/m
C_{GSO}	CGSO	1–3	Gate-source overlap capacitance per meter	0.0	4×10^{-11}	F/m
R_D	RD	1–3	Drain ohmic resistance	0.0	10.	Ω
R_S	RS	1–3	Source ohmic resistance	0.0	10.	Ω
R_{sh}	RSH	1–3	Source and drain sheet resistance	0.0	30.	Ω

and Hodges. The equations used for the LEVEL 1 n-channel MOSFET model in SPICE are as follows.

Linear Region

$$I_D = \frac{k'}{2} \cdot \frac{W}{L_{eff}} \cdot \left[2 \cdot (V_{GS} - V_T)V_{DS} - V_{DS}^2\right] \cdot (1 + \lambda V_{DS}) \quad for \quad V_{GS} \geq V_T$$

$$and \quad V_{DS} < V_{GS} - V_T \tag{4.1}$$

Saturation Region

$$I_D = \frac{k'}{2} \cdot \frac{W}{L_{eff}} \cdot (V_{GS} - V_T)^2 \cdot (1 + \lambda \cdot V_{DS}) \quad for \quad V_{GS} \geq V_T$$

$$and \quad V_{DS} \geq V_{GS} - V_T \tag{4.2}$$

where the threshold voltage V_T is calculated as

$$V_T = V_{T0} + \gamma \cdot \left(\sqrt{|2\phi_F| + V_{SB}} - \sqrt{|2\phi_F|}\right) \tag{4.3}$$

Note that the effective channel length L_{eff} used in these equations is found as follows.

$$L_{eff} = L - 2 \cdot L_D \tag{4.4}$$

The empirical channel length modulation term $(1 + \lambda V_{DS})$ appears both in the linear region and in the saturation region equations, although the physical channel length shortening effect is observed only in saturation. This term is included in the linear region equations to ensure the continuity of the first-order derivatives at the linear-saturation region boundary.

Five electrical parameters completely characterize this model: k', V_{T0}, γ, $|2\phi_F|$, and λ. These parameters (KP, VTO, GAMMA, PHI, and LAMBDA, respectively) can be specified directly in the .MODEL statement, or some of them can be calculated from the physical parameters, as follows.

$$k' = \mu \cdot C_{ox}, \quad where \quad C_{ox} = \frac{\varepsilon_{ox}}{t_{ox}} \tag{4.5}$$

$$\gamma = \frac{\sqrt{2 \cdot \varepsilon_{Si} \cdot q \cdot N_A}}{C_{ox}} \tag{4.6}$$

$$2\phi_F = 2\frac{kT}{q} \cdot \ln\left(\frac{n_i}{N_A}\right) \tag{4.7}$$

Thus, it is also possible to specify the physical parameters $\mu, t_{ox},$ and N_A in the .MODEL statement instead of the electrical parameters, or a combination of both types of parameters. If a conflict occurs (for example, if both the electron mobility μ and the electrical parameter k' are specified in the .MODEL statement), the given value of the electrical parameter (in this case k') overrides the physical parameter. Typical variations of the drain current with the significant electrical model parameters KP, VTO, GAMMA, and LAMBDA are shown in Figs. 4.2 through 4.6. Nominal parameter values used in the simulations are:

$k' = 27.6\ \mu\text{A/V}^2$	KP = 27.6U
$V_{TO} = 1.0\ \text{V}$	VTO = 1
$\gamma = 0.53\ \text{V}^{1/2}$	GAMMA = 0.53
$2\phi_F = -0.58$	PHI = 0.58
$\lambda = 0$	LAMBDA = 0

The corresponding physical parameter values (which may be overridden by the electrical parameters listed above) are as follows:

$\mu_n = 800\ \text{cm}^2/\text{V}\cdot\text{s}$	UO = 800
$t_{ox} = 100\ \text{nm}$	TOX = 100E-9
$N_A = 10^{15}\ \text{cm}^{-3}$	NSUB = 1E15
$L_D = 0.8\ \mu\text{m}$	LD = 0.8E-6

In summary, for simple simulation problems the LEVEL 1 model offers a useful estimate of the circuit performance without using a large number of device model parameters. In the following two sections, we present basic equations for LEVEL 2 and LEVEL 3 models, although they are not identical to the SPICE-implemented versions.

4.3. The LEVEL 2 Model Equations

To obtain a more accurate model for the drain current, it is necessary to eliminate some of the simplifying assumptions made in the original GCA analysis. Specifically, the bulk depletion charge must be calculated by taking into account its dependence on the

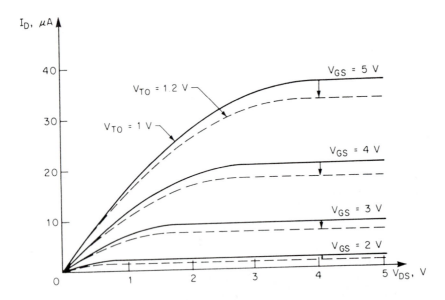

Figure 4.2.   Variation of the drain current with model parameter VTO, for the LEVEL1 model (Copyright © 1988 by McGraw-Hill, Inc.).

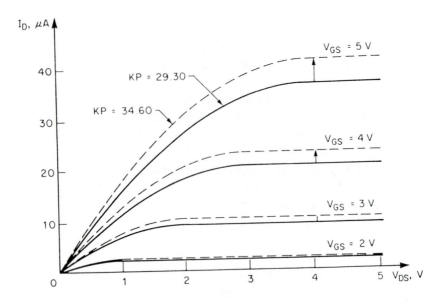

Figure 4.3.   Variation of the drain current with model parameter KP, for the LEVEL1 model (Copyright © 1988 by McGraw-Hill, Inc.).

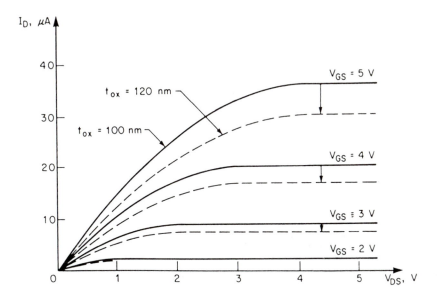

Figure 4.4. Variation of the drain current with model parameter TOX, for the LEVEL1 model (Copyright © 1988 by McGraw-Hill, Inc.).

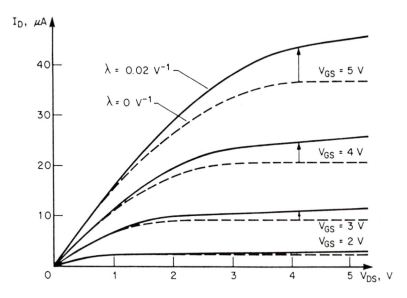

Figure 4.5. Variation of the drain current with parameter LAMBDA, for the LEVEL1 model (Copyright © 1988 by McGraw-Hill, Inc.).

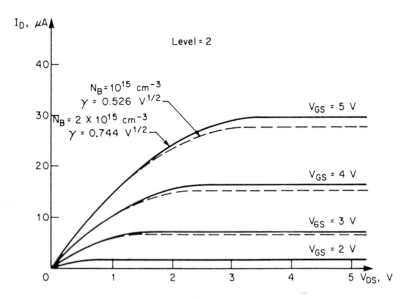

Figure 4.6. Variation of the drain current with parameter GAMMA, for the LEVEL2 model (Copyright © 1988 by McGraw-Hill, Inc.).

channel voltage. Solving the drain current equation using the voltage-dependent bulk charge term, the following current-voltage characteristics can be obtained.

$$
I_D = \frac{k'}{\left(1 - \lambda \cdot V_{DS}\right)} \cdot \frac{W}{L_{eff}} \cdot \left\{ \left(V_{GS} - V_{FB} - |2\phi_F| - \frac{V_{DS}}{2} \right) \cdot V_{DS} \right.
$$
$$
\left. - \frac{2}{3} \cdot \gamma \cdot \left[\left(V_{DS} - V_{BS} + |2\phi_F| \right)^{3/2} - \left(-V_{BS} + |2\phi_F| \right)^{3/2} \right] \right\}
\tag{4.8}
$$

Here, V_{FB} denotes the *flat-band* voltage of the MOSFET. Note that the corrective term for channel length modulation is present in the denominator in this current equation. The model equation (4.8) also includes a variation of the drain current with the parameter γ, even if the substrate-to-source voltage V_{BS} is equal to zero. The saturation condition is reached when the channel (inversion) charge at the drain end becomes equal to zero. From this definition, the saturation voltage V_{DSAT} can be calculated as

$$
V_{DSAT} = V_{GS} - V_{FB} - |2\phi_F| + \gamma^2 \cdot \left(1 - \sqrt{1 + \frac{2}{\gamma^2} \cdot \left(V_{GS} - V_{FB} \right)} \right)
\tag{4.9}
$$

The saturation mode current is

$$I_D = I_{Dsat} \cdot \frac{1}{(1 - \lambda \cdot V_{DS})} \tag{4.10}$$

where I_{Dsat} is calculated from (4.9) using $V_{DS} = V_{DSAT}$. The zero-bias threshold voltage V_{T0} corresponding to the LEVEL 2 model can be calculated from (4.8) as follows.

$$V_{T0} = \Phi_{GC} - \frac{q \cdot N_{SS}}{C_{ox}} + |2\phi_F| + \gamma \cdot \sqrt{|2\phi_F|} \tag{4.11}$$

where Φ_{GC} denotes the gate-to-channel work function difference, and N_{SS} denotes the fixed surface (interface) charge density. The LEVEL 2 model yields more accurate results than the simple LEVEL 1 model, but its accuracy is still not sufficient to achieve good agreement with experimental data, especially for short- and narrow-channel MOSFETs. Consequently, a number of semi-empirical corrections have been added to the basic equations, in order to improve their precision. Some of these enhancements will be discussed in the following.

Variation of Mobility with Electric Field

In the current equations presented above, the surface carrier mobility has been assumed constant, and its variation with applied terminal voltages has been neglected. This approximation simplifies the calculation of the drain current integral; but in reality, the surface mobility *decreases* with the increasing gate voltage. In order to simulate this mobility variation, which is primarily due to the scattering of carriers in the channel, the parameter k' is modified as follows.

$$k'(new) = k' \cdot \left(\frac{\varepsilon_{Si}}{\varepsilon_{ox}} \cdot \frac{t_{oc} \cdot U_c}{(V_{GS} - V_T - U_t \cdot V_{DS})} \right)^{U_e} \tag{4.12}$$

Here, the parameter U_c represents the gate-to-channel critical field, the parameter U_t represents the contribution of the drain voltage to the gate-to-channel field, and U_e is an exponential fitting parameter. U_t is usually chosen between 0 and 0.5. This formula allows a good agreement between SPICE simulation results and experimental data, for long-channel MOSFETs.

Variation of Channel Length in Saturation Mode

Both the LEVEL 1 and the LEVEL 2 model equations use the empirical parameter λ to account for the channel length modulation effect in the saturation region. The LEVEL 2 model also offers the possibility of using a physical expression for calculating the channel length in saturation mode:

$$L'_{eff} = L_{eff} - \Delta L \tag{4.13}$$

where

$$\Delta L = \sqrt{\frac{2 \cdot \varepsilon_{Si}}{q \cdot N_A}} \cdot \left[\frac{V_{DS} - V_{DSAT}}{4} + \sqrt{1 + \left(\frac{V_{DS} - V_{DSAT}}{4} \right)^2} \right] \tag{4.14}$$

Thus, if the empirical channel length shortening (modulation) coefficient λ is not specified in the .MODEL statement, its value can be found as

$$\lambda = \frac{\Delta L}{L_{eff} \cdot V_{DS}} \tag{4.15}$$

The slope of the I_D-V_{DS} curve in saturation can be adjusted and fitted to experimental data by changing the substrate doping parameter N_A. In this case, however, other N_A-dependent electrical parameters such as $2\phi_F$ and γ must be specified separately in the .MODEL statement, since N_A is being used as a fitting parameter for the saturation mode slope.

Saturation of Carrier Velocity

The calculation of the saturation voltage V_{DSAT} in (4.9) is based on the assumption that the channel charge near the drain becomes equal to zero when the device enters saturation. This hypothesis is actually incorrect, since a minimum charge concentration greater than zero must exist in the channel, due to the carriers that sustain the saturation current. This minimum concentration depends on the speed of the carriers. Moreover, carriers in the channel usually reach the maximum speed limit, i.e., their velocity saturates, before the channel charge approaches zero. This maximum carrier speed in the channel is denoted as v_{max}. The inversion layer charge at the channel-end (for $V_{DS} = V_{DSAT}$) is found as

$$Q_{inv} = \frac{I_{Dsat}}{W \cdot v_{max}} \tag{4.16}$$

The value of the saturation voltage V_{DSAT} can also be found from this expression. If the parameter is specified in the .MODEL statement, the amount of channel length shortening (ΔL) in saturation is calculated by the following formula, instead of (4.14).

$$\Delta L = X_D \cdot \sqrt{\left(\frac{X_D \cdot v_{max}}{2 \cdot \mu}\right)^2 + V_{DS} - V_{DSAT}} - \frac{X_D^2 \cdot v_{max}}{2 \cdot \mu} \qquad (4.17)$$

where

$$X_D = \sqrt{\frac{2 \cdot \varepsilon_{Si}}{q \cdot N_A \cdot N_{eff}}} \qquad (4.18)$$

Here, the parameter N_{eff} is used as a fitting parameter. This model provides good agreement with experimental data for MOSFETs with long channel lengths. On the other hand, its implementation is complicated because of a discontinuity of the first derivative at the saturation region boundary. This difficulty may sometimes cause convergence problems in the Newton-Raphson algorithm.

Subthreshold Conduction

The basic model implemented in SPICE calculates the drift current in the channel when the surface potential is equal to or larger than $2\phi_F$, i.e., in strong surface inversion. In reality, as already explained in Chapter 3, a significant concentration of electrons exists near the surface for $V_{GS} < V_T$; therefore, there is a channel current even when the surface is not in strong inversion. This *subthreshold current* is due mainly to diffusion between the source and the channel. The model implemented in SPICE introduces an exponential, semi-empirical dependence of the drain current on V_{GS} in the *weak inversion region*. A voltage V_{on} is defined as the boundary between the regions of weak and strong inversion (Fig. 4.7).

$$I_D(\text{weak inversion}) = I_{on} \cdot e^{(V_{GS} - V_{on}) \cdot \left(\frac{q}{nkT}\right)} \qquad (4.19)$$

Here, I_{on} is the current in strong inversion, for $V_{GS} = V_{on}$, and the voltage V_{on} is found as

$$V_{on} = V_T + \frac{nkT}{q} \qquad (4.20)$$

where

$$n = 1 + \frac{q \cdot N_{FS}}{C_{ox}} + \frac{C_d}{C_{ox}} \qquad (4.21)$$

The parameter N_{FS} is defined as the number of fast superficial states, and is used as a fitting parameter that determines the slope of the subthreshold current-voltage characteristics. C_d is the capacitance associated with the depletion region. It is clear that the model introduces a discontinuity for $V_{GS} = V_{on}$; therefore, the simulation of the transition region between weak and strong inversion is not very precise.

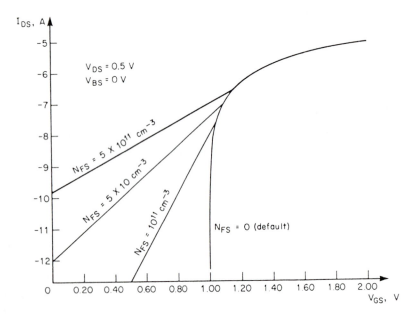

Figure 4.7. Variation of the drain current in the weak inversion region, as a function of the gate voltage and for different values of the parameter N_{FS}, in the LEVEL 2 model (Copyright © 1988 by McGraw-Hill, Inc.).

Other Small-Geometry Corrections

The current-voltage equations (4.8)–(4.10) have been obtained from a theory that does not take two-dimensional effects and small-geometry effects into account. Therefore, these equations do not include a link between the threshold voltage and the channel dimensions W and L. As already discussed in Chapter 3, however, the threshold voltage is a function of both device dimensions in small-geometry MOSFETs.

The model used in SPICE (LEVEL 2) incorporates essentially the same equations presented in Chapter 3 for short- and narrow-channel effects. The model parameters x_j and N_A can be used as fitting parameters for the short-channel effect, but it is difficult to obtain satisfactory results over a large range of channel lengths. Moreover, this model does not adequately explain the dependence of V_T on the drain voltage. For narrow-channel effects, the empirical parameter δ is used for fitting the experimental data. The calculation of narrow-channel threshold voltage variations can be disabled by specifying $\delta = 0$ in the .MODEL statement.

4.4. The LEVEL 3 Model Equations

The LEVEL 3 model has been developed for simulating short-channel MOS transistors; it can represent the characteristics of MOSFETs quite precisely for channel lengths down to 2 μm. The current-voltage equations are formulated in the same way as for the LEVEL 2 model. However, the current equation in the linear region has been simplified with a Taylor series expansion of (4.8). This approximation allows the development of more manageable basic current equations compared to those for the LEVEL 2 model. The short-channel and other small-geometry effects are introduced in the threshold voltage and mobility calculations.

The majority of the LEVEL 3 model equations are empirical. The aims of using empirical equations instead of analytical models are both to improve the accuracy of the model and to limit the complexity of the calculations and, hence, the amount of required simulation time. The drain current in the linear region is given by the following equation.

$$I_D = \mu_s \cdot C_{ox} \cdot \frac{W}{L_{eff}} \cdot \left(V_{GS} - V_T - \frac{1 + F_B}{2} \cdot V_{DS} \right) \cdot V_{DS} \qquad (4.22)$$

where

$$F_B = \frac{\gamma \cdot F_s}{4 \cdot \sqrt{|2\phi_F| + V_{SB}}} + F_n \qquad (4.23)$$

The empirical parameter F_B expresses the dependence of the bulk depletion charge on the three-dimensional geometry of the MOSFET. Here, the parameters V_T, F_s, and μ_s are influenced by the short-channel effects, while the parameter F_n is influenced by the narrow-channel effects. The dependence of the surface mobility on the gate electric field is simulated as follows.

$$\mu_s = \frac{\mu}{1 + \theta \cdot (V_{GS} - V_T)} \qquad (4.24)$$

The LEVEL 3 model also includes a simple equation that accounts for the decrease in the effective mobility with the average lateral electric field:

$$\mu_{eff} = \frac{\mu_s}{1 + \mu_s \cdot \dfrac{V_{DS}}{v_{max} \cdot L_{eff}}} \tag{4.25}$$

where μ_s is the surface mobility calculated from (4.24). The model for the weak inversion region is the same as that for the LEVEL 2 model.

4.5. Capacitance Models

The SPICE MOSFET models account for the parasitic device capacitances, by using separate sets of equations in cut-off, linear mode, and saturation mode. Oxide capacitances and depletion capacitances are calculated as nonlinear functions of the bias voltages, using the basic parasitic capacitance information (such as the zero-bias capacitance values) and the device geometry (such as the junction areas and perimeters) supplied by the user.

Gate Oxide Capacitances

SPICE uses a simple gate oxide capacitance model that represents the charge storage effect by three nonlinear two-terminal capacitors : C_{GB}, C_{GS}, and C_{GD}. The voltage dependence of these capacitances (cf. Meyer's capacitance model) is quite similar to that shown in Fig. 3.32 in Chapter 3. The geometry information required for the calculation of gate oxide capacitances are: gate oxide thickness TOX, channel width W, channel length L, and the lateral diffusion LD. This information is to be provided by the user in the respective device description line. The capacitances CGBO, CGSO, and CGDO, which are specified in the .MODEL statement, are the overlap capacitances between the gate and the other terminals *outside* the channel region.

If the parameter XQC is specified in the .MODEL statement, then SPICE uses a simplified version of the charge-controlled capacitance model proposed by Ward, instead of Meyer's model. Ward's model analytically calculates the charge in the gate and in the substrate. While this model avoids errors caused in the simulation in which some nodes in the network cannot change their charge, it may occasionally lead to convergence problems during simulation. The variation of oxide capacitances as functions of the gate voltage according to Ward's model are shown in Fig. 4.8.

Junction Capacitances

The parasitic capacitances of the source and drain diffusion regions are simulated in SPICE with the simple pn-junction model. Since both the source and the

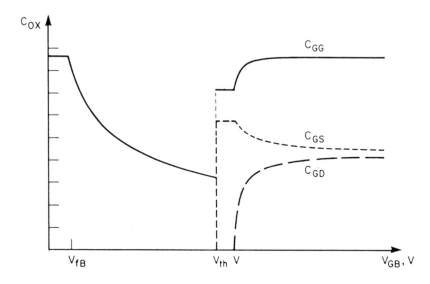

Figure 4.8.   Oxide capacitances as functions of the gate-to-substrate voltage, according to Ward's capacitance model (Copyright © 1988 by McGraw-Hill, Inc.).

drain diffusion regions are surrounded by a p^+ doped sidewall (channel-stop implant), two separate models are used for the bottom area depletion capacitance and for the sidewall area depletion capacitance.

$$C_{SB} = \frac{C_j \cdot AS}{\left(1 - \dfrac{V_{BS}}{\phi_0}\right)^{M_j}} + \frac{C_{jsw} \cdot PS}{\left(1 - \dfrac{V_{BS}}{\phi_0}\right)^{M_{jsw}}} \tag{4.26}$$

$$C_{DB} = \frac{C_j \cdot AD}{\left(1 - \dfrac{V_{BD}}{\phi_0}\right)^{M_j}} + \frac{C_{jsw} \cdot PD}{\left(1 - \dfrac{V_{BD}}{\phi_0}\right)^{M_{jsw}}} \tag{4.27}$$

Here, C_j is the zero-bias depletion capacitance per unit area at the bottom junction of the drain or the source diffusion region, and C_{jsw} is the zero-bias depletion capacitance per unit length at the sidewall junctions. For typical sidewall doping, which is about 10 times larger than the substrate doping concentration, C_{jsw} can be approximated by

$$C_{jsw} \cong \sqrt{10} \cdot C_j \cdot x_j \qquad (4.28)$$

AS and *AD* are the source and the drain areas; *PS* and *PD* are the source and the drain perimeters, respectively. This geometry information must be specified in the corresponding device description line. Note that although in reality only three sides of a rectangular diffusion region are surrounded by the p^+ channel-stop implant, entire source and drain perimeters are usually specified as *PS* and *PD*. This tends to overestimate the total diffusion capacitance, but the difference is not very significant. Also, the built-in junction potential ϕ_0, which is actually a function of the doping densities, is assumed to be equal for both the bottom junction and the sidewall junction.

Finally, the parameters M_j and M_{jsw} denote the junction grading coefficients for the bottom and the sidewall junctions, respectively. Default values are $M_j = 0.5$ (assuming abrupt junction profile for the bottom area) and $M_{jsw} = 0.33$ (assuming linearly graded junction profile for the sidewalls).

Example 4.1.

The top view of an n-channel MOSFET is shown in the figure below. The process parameters for this device are :

$$N_A = 10^{15} \text{ cm}^{-3}$$
$$N_A \text{ (sidewall)} = 2.1 \times 10^{16} \text{ cm}^{-3}$$
$$N_D = 10^{20} \text{ cm}^{-3}$$
$$x_j = 0.8 \text{ μm}$$
$$t_{ox} = 300 \text{ Å}$$
$$L_D = 0.5 \text{ μm}$$

The zero-bias threshold voltage is measured as 0.85 V, and k' is determined to be 45 $\mu A/V^2$. The channel length modulation coefficient is $\lambda = 0.05$. The source, drain, gate, and substrate nodes of the device are labeled by node numbers 4, 6, 12, and 7, respectively. Prepare the device description line and the .MODEL line for SPICE simulation. Use the LEVEL 1 model, and avoid conflicting parameter definitions.

The gate oxide capacitance per unit area is

$$C_{ox} = \frac{\varepsilon_{ox}}{t_{ox}} = \frac{3.9 \cdot 8.85 \times 10^{-14}}{600 \times 10^{-8}} = 5.75 \times 10^{-8} \text{ F/cm}^2$$

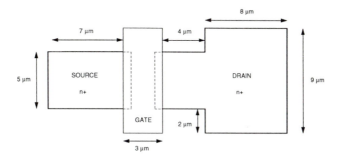

The substrate bias coefficient (GAMMA) and the surface inversion potential (PHI) are found as follows.

$$\gamma = \frac{\sqrt{2 \cdot q \cdot N_A \cdot \varepsilon_{Si}}}{C_{ox}} = \frac{\sqrt{2 \cdot 1.6 \cdot 10^{-19} \cdot 10^{15} \cdot 11.7 \cdot 8.85 \cdot 10^{-14}}}{5.75 \times 10^{-8}} = 0.32 \ \mathrm{V}^{\frac{1}{2}}$$

$$|2 \cdot \phi_F(substrate)| = \left| 2 \cdot \frac{kT}{q} \ln\left(\frac{n_i}{N_A}\right) \right| = \left| 0.026 \ \mathrm{V} \cdot \ln\left(\frac{1.45 \cdot 10^{10}}{10^{15}}\right) \right| = 0.58 \ \mathrm{V}$$

Now we start to calculate the parameters needed for the parasitic capacitance descriptions. The built-in junction potential (PB) for the bottom diffusion area is

$$\phi_0 = \frac{kT}{q} \cdot \ln\left(\frac{N_A \cdot N_D}{n_i^2}\right) = 0.026 \ \mathrm{V} \cdot \ln\left(\frac{10^{20} \cdot 10^{15}}{2.1 \times 10^{20}}\right) = 0.878 \ \mathrm{V}$$

Note that this junction potential is used for calculating *all* junction capacitances, which results in an *overestimation* of the sidewall junction capacitance. The zero-bias depletion capacitances associated with the bottom junction (CJ) and the sidewall junctions (CJSW) are found as

$$C_{j0} = \sqrt{\frac{\varepsilon_{Si} \cdot q}{2} \cdot \left(\frac{N_A \cdot N_D}{N_A + N_D}\right) \cdot \frac{1}{\phi_0}}$$

$$= \sqrt{\frac{11.7 \cdot 8.85 \cdot 10^{-14} \, \mathrm{F/cm} \cdot 1.6 \cdot 10^{-19} \, \mathrm{C}}{2} \cdot \left(\frac{10^{20} \cdot 10^{15}}{10^{20} + 10^{15}}\right) \cdot \frac{1}{0.878}}$$

$$= 9.7 \times 10^{-9} \ \mathrm{F/cm}^2$$

$$C_{j0sw} = x_j \cdot \sqrt{\frac{\varepsilon_{Si} \cdot q}{2} \cdot \left(\frac{N_A(sw) \cdot N_D}{N_A(sw) + N_D}\right) \cdot \frac{1}{\phi_0}}$$

$$= 0.8 \times 10^{-4} \cdot \sqrt{\frac{11.7 \cdot 8.85 \cdot 10^{-14} \cdot 1.6 \cdot 10^{-19}}{2} \cdot \left(\frac{10^{20} \cdot 2.1 \cdot 10^{16}}{10^{20} + 2.1 \cdot 10^{16}}\right) \cdot \frac{1}{0.878}}$$

$$= 3.56 \times 10^{-12} \text{ F/cm}$$

We will assume abrupt junction profiles for both the bottom junctions and the sidewall junctions, thus, MJ = 0.5 and MJSW = 0.5. The gate overlap capacitances per unit length (CGSO and CGDO) are calculated as:

$$C_{GSO} = C_{GDO} = C_{ox} \cdot L_D = 5.75 \times 10^{-8} \cdot 0.5 \times 10^{-4} = 2.87 \text{ pF/cm}$$

Finally, we calculate the area and the perimeter (in m^2 and m, respectively) of the source and the drain diffusion regions. Now, we can write the device description line and the .MODEL line that correspond to this device, as follows:

```
M1   6   12   4   7   NM1        W=5U L=3U LD=0.5U AS=37.5P PS=25U
+                                AD=94.5P PD=43U

.MODEL   NM1   NMOS   (VTO=0.85 KP=45U LAMBDA=0.05 GAMMA=0.32
+                     PHI=0.58 PB=0.878 CJ=9.7E-5 CJSW=3.56E-10
+                     CGSO=2.87E-10  CGDO=2.87E-10  MJ=0.5
+                     MJSW=0.5)
```

The junction capacitances are very important not only for the correct description of the circuit, but also because the convergence of the variable time-step integration algorithms used in SPICE actually improves with the presence of node capacitances. Therefore, it is useful to define the areas of the source and the drain diffusions early in the design process, even if the actual areas and the perimeters in the SPICE circuit description file must be updated following the completion of the layout.

4.6. Comparison of the SPICE MOSFET Models

Now we can briefly reexamine the main differences between the three MOSFET models presented in this chapter. This comparison may be useful for choosing the best model for a given circuit simulation task.

The LEVEL 1 model is usually not very precise because the GCA used in the derivation of model equations is too approximate, and the number of fitting parameters is too small. The LEVEL 1 model is useful for a quick and rough estimate of the circuit performance without much accuracy.

The LEVEL 2 model can be used with differing complexities by adding the parameters relating to the various effects that this model supports. However, if all model parameters are specified by the user, i.e., if the greatest level of complexity is obtained, this model requires a large amount of CPU time for the calculations. Moreover, the LEVEL 2 model may occasionally cause convergence problems in the Newton-Raphson algorithm used in SPICE.

A comparison between the LEVEL 2 and the LEVEL 3 models is interesting (Fig. 4.9). The LEVEL 3 model usually achieves about the same level of accuracy as the LEVEL 2 model, but the CPU time needed for model evaluation is less and the number of iterations are significantly fewer for the LEVEL 3 model. Consequently, the only disadvantage of the LEVEL 3 model is the complexity of calculating some of its parameters. It is best to use the LEVEL 3 model, if possible, and the LEVEL 1 model, if large precision is not required.

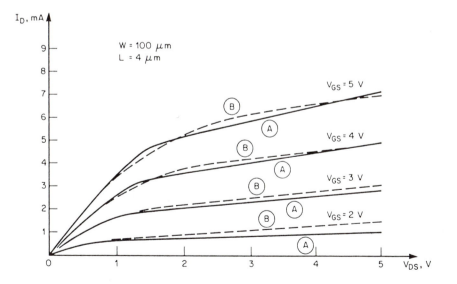

Figure 4.9.   Drain current vs. drain voltage characteristics of an n-channel MOSFET calculated with the LEVEL 2 model (A) and the LEVEL 3 model (B) (Copyright © 1988 by McGraw-Hill, Inc.).

The parameters common for both models are : VTO = 1, XJ = 1.0E-6, LD = 0.8E-6.
The parameters of the LEVEL 2 model are : UO = 800, UCRIT = 5.0E4, UEXP = 0.15.
The parameters of the LEVEL 3 model are : UO = 850, THETA = 0.04.

References

1.   L.W. Nagel, *SPICE2: A Computer Program to Simulate Semiconductor Circuits*, Memo ERL-M520, Berkeley, CA: University of California, 1975.

2.   G. Massobrio and P. Antognetti, *Semiconductor Device Modeling with SPICE*, second edition, New York, NY: McGraw-Hill, 1993.

3.   H. Schichman and D. A. Hodges, "Modeling and simulation of insulated-gate field-effect transistors," *IEEE Journal of Solid-State Circuits*, vol. SC-3, no. 5, pp. 285-289, September 1968.

4.   J. E. Meyer, "MOS models and circuit simulation," *RCA Review*, 32, pp. 42-63, March 1971.

5.   M. C. Jeng, P. M. Lee, M. M. Kuo, P. K. Ko, and C. Hu, *Theory, Algorithms, and User's Guide for BSIM and SCALP*, Electronic Research Laboratory Memorandum, UCB/ERL M87/35, Berkeley, CA: University of California, 1983.

6.   B. J. Sheu, D. L. Scharfetter, P. K. Ko, and M. C. Jeng, "BSIM, Berkeley short-channel IGFET model," *IEEE Journal of Solid-State Circuits*, vol. SC-22, pp. 558-566, 1987.

7.   J. R. Brews, "A charge-sheet model of the MOSFET," *Solid-State Electronics*, vol. 21, pp. 345-355, 1978.

8.   P. Yang and P. K. Chatterjee, "SPICE modeling for small geometry MOSFET circuits," *IEEE Transactions on Computer-Aided Design*, vol. CAD-1, no. 4, pp. 169-182, 1982.

9.   S.-W. Lee and R. C. Rennick, "A compact IGFET model-ASIM," *IEEE Transactions on Computer-Aided Design*, vol. 7, no. 9, pp. 952-975, September 1988.

10.   K. Lee, M. Shur, T. A. Fjeldly, and Y. Ytterdal, *Semiconductor Device Modeling for VLSI*, Englewood Cliffs, NJ: Prentice-Hall, Inc., 1993.

Exercise Problems

4.1 Rewrite the SPICE code for the nMOS model in Example 4.1 for a graded junction using the following parameters:

- $N_A = 2 \cdot 10^{15}$ cm^{-3}
- N_A (*sidewall*) $= 2.1 \cdot 10^{16}$ cm^{-3}
- $N_D = 10^{19}$ cm^{-3}
- $X_j = 0.5$ μm
- $t_{ox} = 200$ Å
- $L_D = 0.2$ μm

4.2 A layout of nMOS transistors in the NAND2 gate is shown below. Write a SPICE description corresponding to the layout. The diffusion region between two polysilicon gates can be split equally between the two transistors.

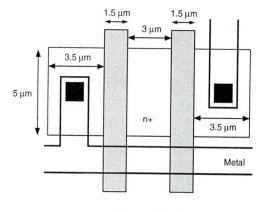

Figure P4.2

4.3 Using the SPICE LEVEL 1 MOSFET model equations, derive an expression for the sensitivity of the drain current I_D with respect to temperature. Calculate the sensitivity at room temperature $T = 300$ K by using the values in Example 4.1. For simplicity, assume that n_i is independent of temperature. Also verify your solution numerically by calculating I_D at 310 K and then $\Delta I_D/\Delta T$.

4.4 Explain why several versions of nMOS transistor models and pMOS transistor models coexist despite the fact that some models are more accurate than others.

CHAPTER 5

MOS INVERTERS:
STATIC CHARACTERISTICS

The inverter is the most fundamental logic gate that performs a Boolean operation on a single input variable. In this chapter, we will examine the DC (static) characteristics of various MOS inverter circuits. It will be seen later that many of the basic principles employed in the design and analysis of MOS inverters can be directly applied to more complex logic circuits, such as NAND and NOR gates, as well. The inverter design therefore forms a significant basis for digital circuit design. Consequently, the DC analysis of MOS inverters will be carried out rigorously, and in detail. While analytic techniques constitute most of the material presented in this chapter, numerical comparisons with circuit simulation using SPICE will also be emphasized, since computer-aided techniques are integral components of the analysis and design processes for digital circuits.

5.1. Introduction

The logic symbol and the truth table of the ideal inverter are shown in Fig. 5.1. In MOS inverter circuits, both the input variable A and the output variable B are represented by node voltages, referenced to the ground potential. Using *positive logic convention*, the Boolean (or logic) value of "1" can be represented by a high voltage of V_{DD}, and the Boolean (or logic) value of "0" can be represented by a low voltage of 0. The DC voltage transfer characteristic (VTC) of the ideal inverter circuit is shown in Fig. 5.2.

The voltage V_{th} is called the *inverter threshold voltage*. Note that for any input voltage between 0 and $V_{th} = V_{DD}/2$, the output voltage is equal to V_{DD} (logic "1"). The output switches from V_{DD} to 0 when the input is equal to V_{th}. For any input voltage between V_{th} and V_{DD}, the output voltage assumes a value of 0 (logic "0"). Thus, an input voltage $0 < V_{in} < V_{th}$ is interpreted by this ideal inverter as a logic "0," while an input voltage $V_{th} < V_{in} < V_{DD}$ is interpreted as a logic "1." The DC characteristics of actual inverter circuits will obviously differ in various degrees from the ideal characteristic shown in Fig. 5.2. The accurate estimation and the manipulation of the shape of VTC for various inverter types are actually important parts of the design process.

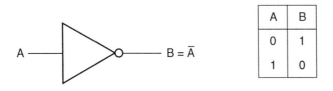

A	B
0	1
1	0

Symbol   Truth Table

Figure 5.1.   Logic symbol and truth table of the inverter.

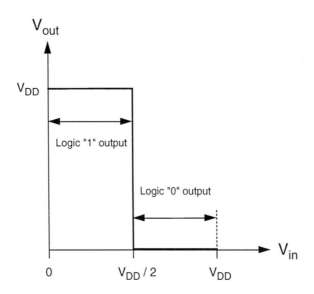

Figure 5.2.   Voltage transfer characteristic (VTC) of the ideal inverter.

Figure 5.3 shows the generalized circuit structure of an nMOS inverter. The input voltage of the inverter circuit is also the gate-to-source voltage of the nMOS transistor ($V_{in} = V_{GS}$), while the output voltage of the circuit is equal to the drain-to-source voltage ($V_{out} = V_{DS}$). The source and the substrate terminals of the nMOS transistor, also called the *driver* transistor, are connected to ground potential; hence, the source-to-substrate voltage is $V_{SB} = 0$. In this generalized representation, the *load* device is represented as a two-terminal circuit element with terminal current I_L and terminal voltage $V_L(I_L)$. One terminal of the load device is connected to the drain of the n-channel MOSFET, while the other terminal is connected to V_{DD}, the power supply voltage. We will see shortly that the characteristics of the inverter circuit actually depend very strongly upon the type and the characteristics of the load device.

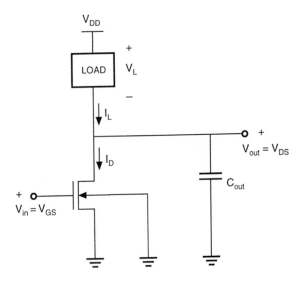

Figure 5.3.   General circuit structure of an nMOS inverter.

The output terminal of the inverter shown in Fig. 5.3 is connected to the input of another MOS inverter. Consequently, the next circuit *seen* by the output node can be represented as a lumped capacitance, C_{out}. Since the DC gate current of an MOS transistor is negligible for all practical purposes, there will be no current flow into or out of the input and output terminals of the inverter in DC steady state.

Voltage Transfer Characteristic (VTC)

Applying Kirchhoff's Current Law (KCL) to this simple circuit, we see that the load current is always equal to the nMOS drain current.

$$I_D(V_{in}, V_{out}) = I_L(V_L) \qquad\qquad (5.1)$$

The voltage transfer characteristic describing V_{out} as a function of V_{in} under DC conditions can then be found by analytically solving (5.1) for various input voltage values. The typical VTC of a realistic nMOS inverter is shown in Fig. 5.4. Upon examination, we can identify a number of important properties of this DC transfer characteristic.

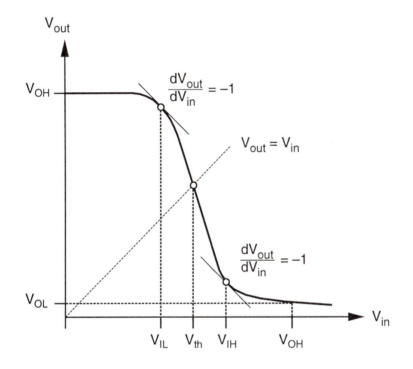

Figure 5.4.   Typical voltage transfer characteristic (VTC) of a realistic nMOS inverter.

The general shape of the VTC in Fig. 5.4 is qualitatively similar to that of the ideal inverter transfer characteristic shown in Fig. 5.2. There are, however, several significant differences that deserve special attention. For very low input voltage levels, the output voltage V_{out} is equal to the high value of V_{OH} (*output high voltage*). In this case, the driver nMOS transistor is in cut-off, and hence, does not conduct any current. Consequently, the voltage drop across the load device is very small in magnitude, and the output voltage level is high. As the input voltage V_{in} increases, the driver transistor starts conducting a certain drain current, and the output voltage eventually starts to decrease. Notice that this drop in the output voltage level does not occur abruptly, such as the vertical drop assumed for the ideal inverter VTC, but rather gradually and with a finite slope. We identify two *critical voltage points* on this curve, where the slope of the $V_{out}(V_{in})$ characteristic becomes equal to -1, i.e.,

$$\frac{dV_{out}}{dV_{in}} = -1 \qquad\qquad (5.2)$$

The smaller input voltage value satisfying this condition is called the *input low voltage* V_{IL}, and the larger input voltage satisfying this condition is called the *input high voltage* V_{IH}. Both of these voltages play significant roles in determining the noise margins of the inverter circuit, as we will discuss in the following sections. The physical justification for selecting these voltage points will also be examined in the context of noise immunity.

   As the input voltage is further increased, the output voltage continues to drop and reaches a value of V_{OL} (*output low voltage*) when the input voltage is equal to V_{OH}. The *inverter threshold voltage* V_{th}, which is considered as the transition voltage, is defined as the point where $V_{in} = V_{out}$ on the VTC. Thus, a total of five critical voltages, V_{OL}, V_{OH}, V_{IL}, V_{IH}, and V_{th}, characterize the DC input-output voltage behavior of the inverter circuit. The functional definitions for the first four of these critical voltages are given below.

V_{OH} :   Maximum output voltage when the output level is logic "1."
V_{OL} :   Minimum output voltage when the output level is logic "0."
V_{IL} :   Maximum input voltage which can be *interpreted* as logic "0."
V_{IH} :   Minimum input voltage which can be *interpreted* as logic "1."

   These definitions obviously imply that logic levels in digital circuits are not represented by quantized voltage values, but rather by voltage ranges corresponding to these logic levels. According to the definitions above, any input voltage level between the lowest available voltage in the system (usually the ground potential) and V_{IL} is interpreted as a logic "0" input, while any input voltage level between the highest available voltage in the system (usually the power supply voltage) and V_{IH} is interpreted as a logic "1" input. Any output voltage level between the lowest available voltage in the system and V_{OL} is interpreted as a logic "0" output, while any output voltage level between the highest available voltage in the system and V_{OH} is interpreted as a logic "1" output. These voltage ranges are also shown on the inverter voltage transfer characteristic shown in Fig. 5.4.

   The ability of an inverter to interpret an input signal within a voltage range as either a logic "0" or as a logic "1" allows digital circuits to operate with a certain *tolerance* to external signal perturbations. This tolerance to variations in the signal level is especially valuable in environments in which circuit noise can significantly corrupt the signals. Here the *circuit noise* corresponds to unwanted signals that are coupled to some part of the circuit from neighboring lines (usually interconnection lines) by capacitive or inductive coupling, or from outside of the system. The result of this interference is that the signal level at one end of an interconnection line may be significantly different from the signal level at the other end.

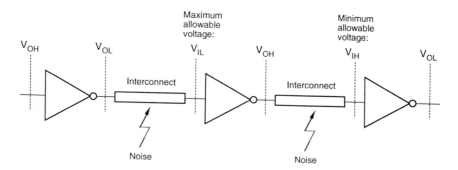

Figure 5.5.   Propagation of digital signals under the influence of noise.

Noise Immunity and Noise Margins

To illustrate the effect of noise on the circuit reliability, we will consider the circuit consisting of three cascaded inverters, as shown in Fig. 5.5. Assume that all inverters are identical, and that the input voltage of the first inverter is equal to V_{OH}, i.e., a logic "1." By definition, the output voltage of the first inverter will be equal to V_{OL}, corresponding to a logic "0" level. Now, this output signal is being transmitted to the next inverter input via an interconnect, which could be a metal or polysilicon line connecting the two gates. Since on-chip interconnects are generally prone to signal noise, the output signal of the first inverter will be perturbed during transmission. Consequently, the voltage level at the input of the second inverter will be different from V_{OL}, i.e., either larger or smaller. If the input voltage of the second inverter is smaller than V_{OL}, this signal will be interpreted correctly as a logic "0" input by the second inverter. On the other hand, if the input voltage becomes larger than V_{IL} as a result of noise, then it may not be interpreted correctly by the inverter. Thus, we conclude that V_{IL} is the maximum allowable voltage at the input of the second inverter, which is low enough to ensure a logic "1" output.

Now consider the signal transmission from the output of the second inverter to the input of the third inverter, assuming that the second inverter produces an output voltage level of V_{OH}. As in the previous case, this output signal will be perturbed because of noise interference, and the voltage level at the input of the third inverter will be different from V_{OH}. If the input voltage of the third inverter is larger than V_{OH}, this signal will be interpreted correctly as a logic "1" input by the third inverter. If the voltage level drops below V_{IH} due to noise, however, the input cannot be interpreted as a logic "1." Consequently, V_{IH} is the minimum allowable voltage at the input of the third inverter which is high enough to ensure a logic "0" output.

These observations lead us to the definition of noise tolerances for digital circuits, called *noise margins* and denoted by *NM*. The noise immunity of the circuit increases with *NM*. Two noise margins will be defined: the noise margin for low signal levels (NM_L) and the noise margin for high signal levels (NM_H).

$$NM_L = V_{IL} - V_{OL} \qquad (5.3)$$

$$NM_H = V_{OH} - V_{IH} \qquad (5.4)$$

Figure 5.6 shows a graphical illustration of the noise margins. Here, the shaded areas indicate the valid regions of the input and output voltages, and the noise margins are shown as the amount of variation in the signal levels that can be allowed while the signal is transmitted from the output of one gate to the input of the next gate.

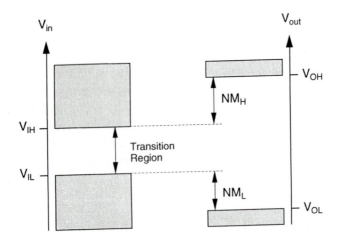

Figure 5.6.   Definition of noise margins NM_L and NM_H. Note that the shaded regions indicate valid high and low levels for the input and output signals.

The selection of the two critical voltage points, V_{IL} and V_{IH}, using the slope condition (5.2) can now be justified based on noise considerations. We know that the output voltage V_{out} of an inverter under noise-free, steady-state conditions is a nonlinear function of the input voltage, V_{in}. The shape of this function is described by the voltage transfer characteristic (VTC).

$$V_{out} = f(V_{in})$$

(5.5)

If the input signal is perturbed from its nominal value because of external influences, such as noise, the output voltage will also deviate from its nominal level. Here, ΔV_{noise} represents the voltage perturbation that is assumed to affect the input voltage of the inverter.

$$V_{out}' = f(V_{in} + \Delta V_{noise})$$

(5.6)

By using a simple first-order Taylor series expansion and by neglecting higher-order terms, we can express the perturbed output voltage V_{out}' as follows.

$$V_{out}' = f(V_{in}) + \frac{dV_{out}}{dV_{in}} \cdot \Delta V_{noise} + higher\, order\, terms\, (neglected)$$

(5.7)

Notice that here, $f(V_{in})$ represents the nominal (non-perturbed) output signal, and the term $\left(dV_{out}/dV_{in} \right)$ represents the *voltage gain* of the inverter at the nominal input voltage value V_{in}. Thus, the equation (5.7) can also be expressed as:

$$Perturbed\, Output = Nominal\, Output + Gain \times External\, Perturbation$$

(5.8)

If the magnitude of the voltage gain at the nominal input voltage V_{in} is smaller than unity, then the input perturbation is not amplified and, consequently, the output perturbation remains relatively small. Otherwise, with the voltage gain larger than unity, a small perturbation in the input voltage level will cause a rather large perturbation in the output voltage. Hence, we define the boundaries of the valid input signal regions as the voltage points where the magnitude of the inverter voltage gain is equal to unity.

Finally, note that there is a voltage range between V_{IL} and V_{IH} (Fig. 5.6), corresponding to input voltage values that may not be processed correctly either as a logic "0" input or as a logic "1" input by the inverter. This region is called the *uncertain region* or, alternatively, the *transition region*. Ideally, the slope of the voltage transfer characteristic should be very large between V_{IL} and V_{IH}, because a narrow uncertain (transition) region obviously allows for larger noise margins. We will see that reducing the width of the uncertain region is one of the most important design objectives.

The preceding discussion of inverter static (DC) characteristics shows that the shape of the VTC in general and the noise immunity properties, in particular, are very significant criteria that ultimately dictate the design priorities. For any inverter circuit, the five critical voltage points V_{OL}, V_{OH}, V_{IL}, V_{IH}, and V_{th} fully determine the DC input-output voltage behavior, the noise margins, and the width and location of the transition region. Accurate estimation of these critical voltage points for different inverter designs will be an important task throughout this chapter.

Power and Area Considerations

In addition to these concerns, we can identify two other issues that play significant roles in inverter design: *power* consumption and the *chip area* occupied by the inverter circuit. About one million logic gates can be accommodated on a very-large-scale integrated (VLSI) chip using 0.5 μm MOS technology, and the circuit density is expected to increase even further in future-generation chips. Since each gate on the chip dissipates power and thus generates heat, the removal of this thermal energy, i.e., cooling of the chip, becomes an essential and usually very expensive task (note that the junction temperature is given as $T_j = T_a + \Theta P$, where T_a is the ambient temperature, Θ is the thermal resistance, and P is the amount of power dissipated). Also, most portable systems, such as cellular communication devices and laptop and palmtop computers, operate from a limited power supply, and the extension of battery-based operation time is a significant design goal. Therefore, it is very important to reduce the amount of power dissipated by the circuit in both DC and dynamic operation.

The DC power dissipation of an inverter circuit can be calculated as the product of its power supply voltage and the amount of current drawn from the power supply during steady state.

$$P_{DC} = V_{DD} \cdot I_{DC} \tag{5.9}$$

Notice that the DC current drawn by the inverter circuit may vary depending on the input and output voltage levels. Assuming that the input voltage level corresponds to logic "0" during 50% of the operation time and to logic "1" during the other 50%, the overall DC power consumption of the circuit can be estimated as follows.

$$P_{DC} = \frac{V_{DD}}{2} \cdot \left[I_{DC}\left(V_{in} = low\right) + I_{DC}\left(V_{in} = high\right) \right] \tag{5.10}$$

It will be seen in the following sections that the DC power dissipation of different inverter designs varies significantly from each other, and that these differences may become a very important factor in the selection of a particular circuit type, e.g., CMOS vs. depletion-load NMOS, for a given design task.

To reduce the chip area occupied by the inverter circuit, it is necessary to reduce the area of the MOS transistors used in the circuit. As a practical measure, we use the gate area of the MOS transistor, i.e., the product of W and L. Thus, an MOS transistor has minimum area when both of the gate (channel) dimensions are made as small as possible within the constraints of the particular technology. It follows that the ratio of the gate width to gate length (W/L) should also be as close to unity as possible, in order to achieve minimum transistor area. This requirement, however, usually contradicts other design criteria, such as the noise margins, the output current driving capability, and the dynamic switching speed. We will observe in the following sections that the design of inverter circuits involves a detailed consideration and a trade-off of these criteria.

Our examination of different MOS inverter structures will begin with the resistive-load MOS inverter. The analysis of this simple circuit will help illustrate some of the basic aspects encountered in inverter design. We will briefly examine the enhancement-load inverter, and then devote most of our attention throughout the remainder of this chapter to two very important inverter structures: the depletion-load NMOS inverter and the CMOS inverter.

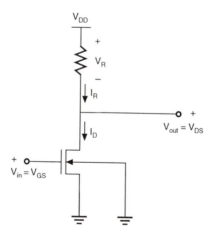

Figure 5.7.   Resistive-load inverter circuit.

5.2. Resistive-Load Inverter

The basic structure of the resistive-load inverter circuit is shown in Fig. 5.7. As in the general inverter circuit already examined in Fig. 5.3, an enhancement-type nMOS transistor acts as the driver device. The load consists of a simple linear resistor, R_L. The power supply voltage of this circuit is V_{DD}. Since the following analysis concentrates on the static behavior of the circuit, the output load capacitance is not shown in this figure.

As already noted in Section 5.1, the drain current I_D of the driver MOSFET is equal to the load current I_R, in DC steady-state operation. To simplify the calculations, the channel-length modulation effect will be neglected in the following, i.e., $\lambda = 0$. Also, note that the source and the substrate terminals of the driver transistor are both connected to the ground; hence, $V_{SB} = 0$. Consequently, the threshold voltage of the driver transistor is always equal to V_{T0}. We start our analysis by identifying the various operating regions of the driver transistor under steady-state conditions.

For input voltages smaller than the threshold voltage V_{T0}, the transistor is in cut-off, and does not conduct any drain current. Since the voltage drop across the load resistor is equal to zero, the output voltage must be equal to the power supply voltage, V_{DD}. As the input voltage is increased beyond V_{T0}, the driver transistor starts conducting a nonzero drain current. Note that the driver MOSFET is initially in saturation, since its drain-to-source voltage $(V_{DS} = V_{out})$ is larger than $(V_{in} - V_{T0})$. Thus,

$$I_R = \frac{k_n}{2} \cdot \left(V_{in} - V_{T0}\right)^2 \tag{5.11}$$

With increasing input voltage, the drain current of the driver also increases, and the output voltage V_{out} starts to drop. Eventually, for input voltages larger than $V_{out} + V_{T0}$, the driver transistor enters the linear operation region. At larger input voltages, the transistor remains in linear mode, as the output voltage continues to decrease.

$$I_R = \frac{k_n}{2} \cdot \left[2 \cdot \left(V_{in} - V_{T0}\right) \cdot V_{out} - V_{out}^2\right] \tag{5.12}$$

The various operating regions of the driver transistor and the corresponding input-output conditions are listed in the following table.

Input Voltage Range	Operating Mode
$V_{in} < V_{T0}$	cut-off
$V_{T0} \leq V_{in} < V_{out} + V_{T0}$	saturation
$V_{in} \geq V_{out} + V_{T0}$	linear

Figure 5.8 shows the voltage transfer characteristic of a typical resistive-load inverter circuit, indicating the operating modes of the driver transistor and the critical voltage points on the VTC. Now, we start with the calculation of the five critical voltage points, which determine the steady-state input-output behavior of the inverter.

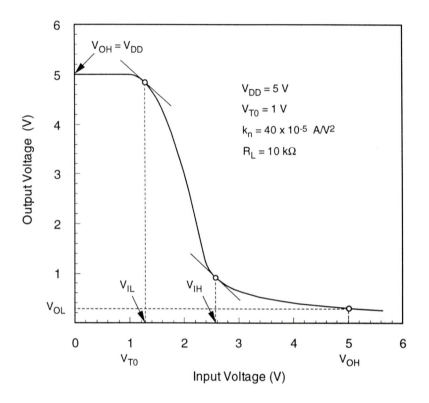

Figure 5.8. Typical VTC of a resistive-load inverter circuit. Important design parameters of the circuit are shown in the inset.

Calculation of V_{OH}

First, we note that the output voltage V_{out} is given by

$$V_{out} = V_{DD} - R_L \cdot I_R \qquad (5.13)$$

When the input voltage V_{in} is low, i.e., smaller than the threshold voltage of the driver MOSFET, the driver transistor is cut-off. Since the drain current of the driver transistor is equal to the load current, $I_R = I_D = 0$. It follows that the output voltage of the inverter under these conditions is:

$$V_{OH} = V_{DD} \tag{5.14}$$

Calculation of V_{OL}

To calculate the output low voltage V_{OL}, we assume that the input voltage is equal to V_{OH}, i.e., $V_{in} = V_{OH} = V_{DD}$. Since $V_{in} - V_{T0} > V_{out}$ in this case, the driver transistor operates in the linear region. Also note that the load current I_R is

$$I_R = \frac{V_{DD} - V_{out}}{R_L} \tag{5.15}$$

Using KCL for the output node, i.e., $I_R = I_D$, we can write the following equation.

$$\frac{V_{DD} - V_{OL}}{R_L} = \frac{k_n}{2} \cdot \left[2 \cdot (V_{DD} - V_{T0}) \cdot V_{OL} - V_{OL}{}^2 \right] \tag{5.16}$$

This equation yields a simple quadratic in V_{OL}, which is solved to find the value of the output low voltage.

$$V_{OL}{}^2 - 2 \cdot \left(V_{DD} - V_{T0} + \frac{1}{k_n R_L} \right) \cdot V_{OL} + \frac{2}{k_n R_L} \cdot V_{DD} = 0 \tag{5.17}$$

Note that of the two possible solutions of (5.17), we must choose the one that is physically correct, i.e., the value of the output low voltage must be between 0 and V_{DD}. The solution of (5.17) is given below. It can be seen that the product $(k_n R_L)$ is one of the important design parameters that determine the value of V_{OL}.

$$V_{OL} = V_{DD} - V_{T0} + \frac{1}{k_n R_L} - \sqrt{\left(V_{DD} - V_{T0} + \frac{1}{k_n R_L} \right)^2 - \frac{2 V_{DD}}{k_n R_L}} \tag{5.18}$$

Calculation of V_{IL}

By definition, V_{IL} is the smaller of the two input voltage values at which the slope of the VTC becomes equal to (-1), i.e., $dV_{out}/dV_{in} = -1$. Simple inspection of Fig. 5.8 shows that when the input is equal to V_{IL}, the output voltage (V_{out}) is only slightly smaller than V_{OH}. Consequently, $V_{out} > V_{in} - V_{T0}$, and the driver transistor operates in saturation. We start our analysis by writing the KCL for the output node.

$$\frac{V_{DD} - V_{out}}{R_L} = \frac{k_n}{2} \cdot \left(V_{in} - V_{T0} \right)^2 \tag{5.19}$$

To satisfy the derivative condition, we differentiate both sides of (5.19) with respect to V_{in}, which results in the following equation.

$$-\frac{1}{R_L} \cdot \frac{dV_{out}}{dV_{in}} = k_n \cdot \left(V_{in} - V_{T0} \right) \tag{5.20}$$

Since the derivative of the output voltage with respect to the input voltage is equal to (-1) at V_{IL}, we can substitute $dV_{out} / dV_{in} = -1$ in (5.20).

$$-\frac{1}{R_L} \cdot (-1) = k_n \cdot \left(V_{IL} - V_{T0} \right) \tag{5.21}$$

Solving (5.21) for V_{IL}, we obtain

$$V_{IL} = V_{T0} + \frac{1}{k_n R_L} \tag{5.22}$$

The value of the output voltage when the input is equal to V_{IL} can also be found by substituting (5.22) into (5.19), as follows:

$$V_{out}\left(V_{in} = V_{IL} \right) = V_{DD} - \frac{k_n R_L}{2} \cdot \left(V_{T0} + \frac{1}{k_n R_L} - V_{T0} \right)^2$$

$$= V_{DD} - \frac{1}{2 k_n R_L} \tag{5.23}$$

Calculation of V_{IH}

V_{IH} is the larger of the two voltage points on VTC at which the slope is equal to (-1). It can be seen from Fig. 5.8 that when the input voltage is equal to V_{IH}, the output voltage V_{out} is only slightly larger than the output low voltage V_{OL}. Hence, in this case,

$V_{out} < V_{in} - V_{T0}$, and the driver transistor operates in the linear region. The KCL equation for the output node is given below.

$$\frac{V_{DD} - V_{out}}{R_L} = \frac{k_n}{2} \cdot \left[2 \cdot (V_{in} - V_{T0}) \cdot V_{out} - V_{out}^2 \right] \tag{5.24}$$

Differentiating both sides of (5.24) with respect to V_{in}, we obtain

$$-\frac{1}{R_L} \cdot \frac{dV_{out}}{dV_{in}} = \frac{k_n}{2} \cdot \left[2 \cdot (V_{in} - V_{T0}) \cdot \frac{dV_{out}}{dV_{in}} + 2V_{out} - 2V_{out} \cdot \frac{dV_{out}}{dV_{in}} \right] \tag{5.25}$$

Next, we can substitute $dV_{out} / dV_{in} = -1$ into (5.25), since the slope of the VTC is equal to (−1) also at $V_{in} = V_{IH}$.

$$-\frac{1}{R_L} \cdot (-1) = k_n \cdot \left[(V_{IH} - V_{T0}) \cdot (-1) + 2V_{out} \right] \tag{5.26}$$

Solving (5.26) for V_{IH} yields the following expression.

$$V_{IH} = V_{T0} + 2V_{out} - \frac{1}{k_n R_L} \tag{5.27}$$

Thus, we obtain two linear equations, (5.24) and (5.27), for two unknowns, V_{IH} and V_{out}. To determine the unknown variables, we substitute (5.26) into the current equation given by (5.24) above.

$$\frac{V_{DD} - V_{out}}{R_L} = \frac{k_n}{2} \cdot \left[2 \cdot \left(V_{T0} + 2V_{out} - \frac{1}{k_n R_L} - V_{T0} \right) \cdot V_{out} - V_{out}^2 \right] \tag{5.28}$$

The positive solution of this second-order equation gives the output voltage V_{out} when the input is equal to V_{IH}.

$$V_{out}(V_{in} = V_{IH}) = \sqrt{\frac{2}{3} \cdot \frac{V_{DD}}{k_n R_L}} \tag{5.29}$$

Finally, V_{IH} can be found by substituting (5.29) into (5.27), as follows.

$$V_{IH} = V_{T0} + \sqrt{\frac{8}{3} \cdot \frac{V_{DD}}{k_n R_L}} - \frac{1}{k_n R_L} \qquad (5.30)$$

The four critical voltage points V_{OL}, V_{OH}, V_{IL}, V_{IH} can now be used to determine the noise margins, NM_L and NM_H, of the resistive-load inverter circuit. In addition to these voltage points, which characterize the static input-output behavior, the inverter threshold voltage V_{th} may also be calculated in a straightforward manner. Note that the driver transistor operates in saturation mode at this point. Thus, the inverter threshold voltage can be found simply by substituting $V_{in} = V_{out} = V_{th}$ into (5.19), and by solving the resulting quadratic for V_{th}.

It can be seen from the preceding discussion that the term $(k_n R_L)$ plays an important role in determining the shape of the voltage transfer characteristic, and that it appears as a critical parameter in expressions for V_{OL} (5.18), V_{IL} (5.22), and V_{IH} (5.30). Assuming that parameters such as the power supply voltage V_{DD} and the driver MOSFET threshold voltage V_{T0} are dictated by system- and processing-related constraints, the term $(k_n R_L)$ remains as the only design parameter which can be adjusted by the circuit designer to achieve certain design goals.

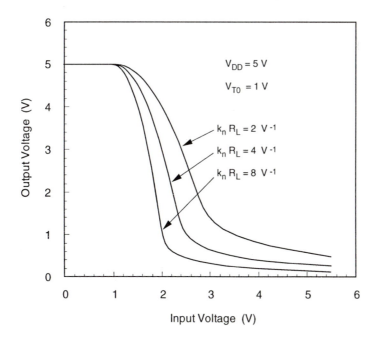

Figure 5.9.   Voltage transfer characteristics of the resistive-load inverter, for different values of the parameter $(k_n R_L)$.

The output high voltage V_{OH} is determined primarily by the power supply voltage, V_{DD}. Among the other three critical voltage points, the adjustment of V_{OL} receives primary attention, while V_{IL} and V_{IH} are usually treated as secondary design variables. Figure 5.9 shows the VTC of a resistive-load inverter for different values of $(k_n R_L)$. Note that for larger $(k_n R_L)$ values, the output low voltage V_{OL} becomes smaller and that the shape of the VTC approaches that of the ideal inverter, with very large transition slope. Achieving larger $(k_n R_L)$ values in a design, however, may involve other trade-offs with the area and the power consumption of the circuit.

Power Consumption and Chip Area

The average DC power consumption of the resistive-load inverter circuit is found by considering two cases, $V_{in} = V_{OL}$ (low) and $V_{in} = V_{OH}$ (high). When the input voltage is equal to V_{OL}, the driver transistor is in cut-off. Consequently, there is no steady-state current flow in the circuit ($I_D = I_R = 0$), and the DC power dissipation is equal to zero. When the input voltage is equal to V_{OH}, on the other hand, both the driver MOSFET and the load resistor conduct a nonzero current. Since the output voltage in this case is equal to V_{OL}, the current drawn from the power supply can be found as

$$I_D = I_R = \frac{V_{DD} - V_{OL}}{R_L} \qquad (5.31)$$

Assuming that the input voltage is "low" during 50% of the operation time, and "high" during the remaining 50%, the average DC power consumption of the inverter can be estimated as follows:

$$P_{DC}(average) = \frac{V_{DD}}{2} \cdot \frac{V_{DD} - V_{OL}}{R_L} \qquad (5.32)$$

Example 5.1

Consider the following inverter design problem: Given $V_{DD} = 5$ V, $k_n' = 30\,\mu\text{A/V}^2$, and $V_{T0} = 1$ V, design a resistive-load inverter circuit with $V_{OL} = 0.2$ V. Specifically, determine the (W/L) ratio of the driver transistor and the value of the load resistor R_L that achieve the required V_{OL}.

In order to satisfy the design specification on the output low voltage V_{OL}, we start our design by writing the relevant current equation. Note that the driver transistor is operating in the linear region when the output voltage is equal to V_{OL} and the input voltage is equal to $V_{OH} = V_{DD}$.

$$\frac{V_{DD} - V_{OL}}{R_L} = \frac{k_n'}{2} \cdot \frac{W}{L} \cdot \left[2 \cdot (V_{OH} - V_{T0}) \cdot V_{OL} - V_{OL}^2 \right]$$

Assuming $V_{OL} = 0.2$ V and using the given values for the power supply voltage, the driver threshold voltage and the driver transconductance k_n', we obtain the following equation:

$$\frac{5 - 0.20}{R_L} = \frac{30 \times 10^{-6}}{2} \cdot \frac{W}{L} \cdot \left(2 \cdot 4 \cdot 0.20 - 0.20^2 \right)$$

This equation can be rewritten as:

$$\frac{W}{L} \cdot R_L = 2.05 \times 10^5 \ \Omega$$

At this point, we recognize that the designer has a choice of different (W/L) and R_L values, all of which satisfy the given design specification, $V_{OL} = 0.2$ V. The selection of the pair of values to use for (W/L) and R_L in the final design ultimately depends on other considerations, such as the power consumption of the circuit and the silicon area. The table below lists some of the design possibilities, along with the average DC power consumption estimated for each design.

$\left(\dfrac{W}{L}\right)$ – Ratio	Load Resistor R_L [kΩ]	DC Power Consumption $P_{DC, average}$ [μW]
1	205.0	58.5
2	102.5	117.1
3	68.4	175.4
4	51.3	233.9
5	41.0	292.7
6	34.2	350.8

It is seen that the power consumption increases significantly as the value of the load resistor R_L is decreased, and the (W/L) ratio is increased. If lowering the DC power consumption is the overriding concern, we may choose a small (*W/L*) ratio and a large load resistor. On the other hand, if the fabrication of the large load resistor requires a large silicon area, a clear trade-off exists between the DC power dissipation and the area occupied by the inverter circuit.

The chip area occupied by the resistive-load inverter circuit depends on two parameters, the (*W/L*) ratio of the driver transistor and the value of the resistor R_L. The area of the driver transistor can be approximated by the gate area, (*W* x *L*). Assuming that the gate length *L* is kept at its smallest possible value for the given technology, the gate area will be proportional to the (*W/L*) ratio of the transistor. The resistor area, on the other hand, depends very strongly on the technology which is used to fabricate the resistor on chip.

We will briefly consider two possibilities for fabricating resistors using the standard MOS process: diffused resistor and polysilicon (undoped) resistor. The diffused resistor is fabricated, as the name implies, as an isolated n-type (or p-type) diffusion region with one contact on each end. The resistance is determined by the doping density of the diffusion region and the dimensions, i.e., the length-to-width ratio, of the resistor. Practical values of the diffusion-region sheet resistance range between 20 to 100 Ω/square. Consequently, very large length-to-width ratios would be required to achieve resistor values on the order of tens to hundreds of kΩ.

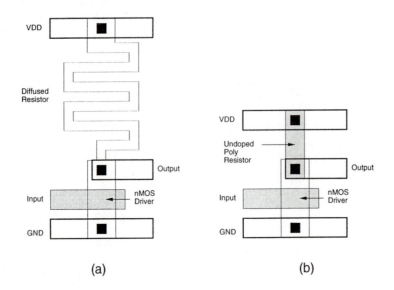

(a)   (b)

Figure 5.10.   Sample layout of resistive-load inverter circuits with (a) diffused resistor and (b) undoped polysilicon resistor.

The placement of these resistor structures on chip, commonly in a serpentine shape for compactness, requires significantly more area than the driver MOSFET, as illustrated in Fig. 5.10(a). A resistive inverter with a large diffused load resistor is, therefore, not a practical component for VLSI applications.

An alternative approach to save silicon area is to fabricate the load resistor using undoped polysilicon. In conventional poly-gate MOS technology, the polysilicon structures forming the gates of transistors and the interconnect lines are heavily doped in order to reduce resistivity. The sheet resistivity of doped polysilicon interconnects and gates is about 20 to 40 Ω/square. If sections of the polysilicon are masked off from this doping step, on the other hand, the resulting *undoped* polysilicon layer has a very high sheet resistivity, in the order of 10 MΩ/square. Thus, very compact and very high-valued resistors can be fabricated using undoped poly layers (Fig. 5.10(b)). One drawback of this approach is that the resistance value can not be controlled very accurately, which results in large variations of the VTC. Consequently, resistive-load inverters with undoped poly resistors are not commonly used in logic gate circuits where certain design criteria, such as noise margins, are expected to be met. Simple inverter structures with large, undoped poly load resistors are used primarily in *low-power* Static Random Access Memory (SRAM) cells, where the primary emphasis is on the reduction of steady-state (DC) power consumption, and the operation of the memory circuit is not significantly affected by the variations of the inverter VTCs. This issue will be discussed in further detail in Chapter 10.

Example 5.2

Consider a resistive-load inverter circuit with $V_{DD} = 5$ V, $k_n' = 20\,\mu$A/V^2, $V_{T0} = 0.8$ V, $R_L = 200$ kΩ, and $W/L = 2$. Calculate the critical voltages (V_{OL}, V_{OH}, V_{IL}, V_{IH}) on the VTC and find the noise margins of the circuit.

When the input voltage is low, i.e., when the driver nMOS transistor is cut-off, the output high voltage can be found as

$$V_{OH} = V_{DD} = 5 \text{ V}$$

Note that in this resistive-load inverter example, the transconductance of the driver transistor is $k_n = k_n'(W/L) = 40\,\mu$A/V^2 and, hence, $(k_n R_L) = 8$ V^{-1}.

The output low voltage V_{OL} is calculated by using (5.18):

$$V_{OL} = V_{DD} - V_{T0} + \frac{1}{k_n R_L} - \sqrt{\left(V_{DD} - V_{T0} + \frac{1}{k_n R_L}\right)^2 - \frac{2 V_{DD}}{k_n R_L}}$$

$$= 5 - 0.8 + \frac{1}{8} - \sqrt{\left(5 - 0.8 + \frac{1}{8}\right)^2 - \frac{2 \cdot 5}{8}}$$

$$= 0.147 \text{ V}$$

The critical voltage V_{IL} is found using (5.22), as follows.

$$V_{IL} = V_{T0} + \frac{1}{k_n R_L} = 0.8 + \frac{1}{8} = 0.925 \text{ V}$$

Finally, the critical voltage V_{IH} can be calculated by using (5.30).

$$V_{IH} = V_{T0} + \sqrt{\frac{8}{3} \cdot \frac{V_{DD}}{k_n R_L}} - \frac{1}{k_n R_L} = 0.8 + \sqrt{\frac{8}{3} \cdot \frac{5}{8}} - \frac{1}{8} = 1.97 \text{ V}$$

Now, the noise margins can be found, according to (5.3) and (5.4).

$$NM_L = V_{IL} - V_{OL} = 0.93 - 0.15 = 0.78 \text{ V}$$

$$NM_H = V_{OH} - V_{IH} = 5.0 - 1.97 = 3.03 \text{ V}$$

At this point, we can comment on the quality of this particular inverter design for DC operation. Notice that the noise margin NM_L found here is quite low, and it may eventually lead to misinterpretation of input signal levels. For better noise immunity, the noise margin for "low" signals should be at least about 25% of the power supply voltage V_{DD}, i.e., about 1.25 V.

5.3. Inverter with Enhancement-Type MOSFET Load

The simple resistive-load inverter circuit examined in the previous section is not a suitable candidate for most digital VLSI system applications, primarily because of the large area occupied by the load resistor. In this section, we will introduce inverter circuits, which use an enhancement-type nMOS transistor as the *active load* device, instead of the linear load resistor. The main advantage of using a MOSFET as the load device is that the silicon area occupied by the transistor is usually smaller than that occupied by a comparable resistive load. Moreover, inverter circuits with active loads can be designed to have better overall performance compared to that of passive-load inverters. In a chronological view, the development of inverters with an enhancement-type MOSFET load actually precedes other active-load inverter types, since the enhancement-type MOSFET fabrication process was perfected earlier than others.

Depending on the bias voltage applied to its gate terminal, the load transistor can be operated either in the saturation region or in the linear region. The ability to set the operation mode of the load device, regardless of the input and output voltage levels, allows us to design two different types of inverters with an enhancement MOSFET load. Both types of inverters have some distinct advantages and disadvantages from the circuit design point of view. In the following, we will consider the basic DC characteristics of the saturated enhancement load inverter circuit.

Saturated Enhancement-Load Inverter

The circuit configuration of an inverter with a saturated enhancement-type load is shown in Fig. 5.11. Note that both the driver and the load nMOS transistors are fabricated using the same process; hence, we can assume that the device parameters k_n' and V_{T0} are identical for both devices. As in the previous case, the channel-length modulation effect is being neglected for simplicity, i.e., $\lambda = 0$. The substrate terminals

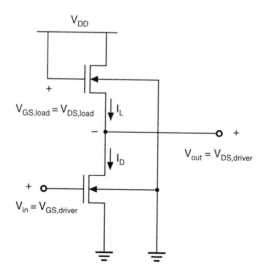

Figure 5.11.   Inverter circuit with saturated enhancement-type MOSFET load.

of the driver and the load transistor are both connected to ground. Consequently, the driver transistor, with $V_{SB,driver}=0$, will not be subject to the substrate-bias effect; thus, $V_{T,driver} = V_{T0}$. The source terminal of the load transistor, however, is at a higher potential, with $V_{SB,load} > 0$. As a result, the threshold voltage of the load device is a function of the source-to-substrate voltage, because of the substrate-bias effect.

First, consider the operating region of the load device shown in Fig. 5.11. Since both the gate and the drain terminals of the load transistor are connected to V_{DD}, we have $V_{GS,load} = V_{DS,load}$. The condition $V_{DS,load} > V_{GS,load} - V_{T,load}$ is always satisfied; hence, the load transistor operates in the saturation region as long as it is on, i.e., $V_{GS,load} > V_{T,load}$. The load current can be expressed as

$$I_{D,load} = \frac{k_n'}{2} \cdot \left(\frac{W}{L}\right)_{load} \left(V_{GS,load} - V_{T,load}\right)^2 \tag{5.33}$$

For input voltages smaller than V_{T0}, the driver transistor is cut-off and does not conduct a drain current. In this case, the output (high) voltage level is determined by the load transistor. As the input voltage is increased beyond V_{T0}, the driver transistor starts conducting a nonzero drain current, initially in saturation mode. Since the driver current is equal to the load current in steady-state, i.e., $I_{D,driver} = I_{D,load}$,

$$\frac{k_n'}{2} \cdot \left(\frac{W}{L}\right)_{load} \left(V_{DD} - V_{out} - V_{T,load}\right)^2 = \frac{k_n'}{2} \cdot \left(\frac{W}{L}\right)_{driver} \left(V_{in} - V_{T0}\right)^2 \tag{5.34}$$

Note that the gate-to-source voltage of the load device is $V_{GS,load} = V_{DD} - V_{out}$. Also, it can be seen from Fig. 5.11 that $V_{GS,driver} = V_{in}$ and $V_{DS,driver} = V_{out}$. The output voltage level starts to decrease with increasing input voltage, and the driver transistor enters the linear operating region.

$$\frac{k_n'}{2} \cdot \left(\frac{W}{L}\right)_{load} \left(V_{DD} - V_{out} - V_{T,load}\right)^2$$
$$= \frac{k_n'}{2} \cdot \left(\frac{W}{L}\right)_{driver} \left[2 \cdot \left(V_{in} - V_{T0}\right) \cdot V_{out} - V_{out}^2\right] \tag{5.35}$$

The current equations (5.34) and (5.35) describe the DC input-output characteristics of the saturated enhancement-load inverter circuit. In the following, we will briefly examine the critical voltage points which characterize the VTC of this inverter.

Calculation of V_{OH}

When the input voltage V_{in} is smaller than the threshold voltage V_{T0}, the driver transistor is cut-off. Since $I_{D,driver} = I_{D,load} = 0$ in this case, we can solve (5.32) to find the gate-to-source voltage of the load device as $V_{GS,load} = V_{DD} - V_{out} = V_{T,load}$. This equation yields the output high voltage as follows:

$$V_{OH} = V_{DD} - V_{T,load}(V_{OH})$$

$$= V_{DD} - \left[V_{T0} + \gamma \left(\sqrt{|2\,\phi_F| + V_{OH}} - \sqrt{|2\,\phi_F|} \right) \right] \qquad (5.36)$$

Note that the threshold voltage of the load transistor $V_{T,load}$ is actually a function of the output voltage because of the substrate-bias effect. To find V_{OH}, we reorganize (5.36), and square both sides.

$$V_{OH}^2 - \left[2\left(V_{DD} - V_{T0} + \gamma \sqrt{|2\,\phi_F|} \right) + \gamma^2 \right] V_{OH}$$

$$+ \left[\left(V_{DD} - V_{T0} + \gamma \sqrt{|2\,\phi_F|} \right)^2 - \gamma^2 |2\,\phi_F| \right] = 0 \qquad (5.37)$$

The solution of this quadratic equation yields the output high voltage V_{OH}. The important result here is that the output high voltage V_{OH} cannot exceed $V_{DD} - V_{T,load}$, a limitation which significantly reduces the noise margin NM_H for high signal levels.

Calculation of V_{OL}

For calculating the output low level V_{OL}, we assume that $V_{in} = V_{OH}$. In this case, the driver transistor is operating in the linear region, and the load transistor is operating in the saturation region. Using KCL for the output node, we obtain

$$\frac{k_n'}{2} \cdot \left(\frac{W}{L} \right)_{load} \left[V_{DD} - V_{OL} - V_{T,load}(V_{OL}) \right]^2$$

$$= \frac{k_n'}{2} \cdot \left(\frac{W}{L} \right)_{driver} \left[2 \cdot (V_{OH} - V_{T0}) \cdot V_{OL} - V_{OL}^2 \right] \qquad (5.38)$$

Here, the value of V_{OH} must be found from (5.37) first. Also, we note that the threshold voltage of the load transistor $V_{T,load}$ is a function of the output low voltage, as follows.

$$V_{T,load} = V_{T0} + \gamma\left(\sqrt{|2\,\phi_F| + V_{OL}} - \sqrt{|2\,\phi_F|}\right) \qquad (5.39)$$

The numerical value of the output low voltage V_{OL} can be found by simultaneously solving (5.38) and (5.39). Alternatively, V_{OL} can be calculated through a simple technique based on numerical iteration, as follows. We start our calculation by assuming that V_{OL} is a small number; hence, we initially assume $V_{T,load} = V_{T0}$. Solving (5.38) with this assumption yields the first estimate for V_{OL}, which can now be substituted into (5.39) to find a better estimate for $V_{T,load}$. The threshold voltage value found in this step is in turn used to solve (5.38) again, to obtain a more accurate estimate for V_{OL}. This cycle can be repeated several times, until subsequent iterations do not cause a significant change in the V_{OL} value. Figure 5.12 illustrates the iterative procedure used to calculate V_{OL}. Since the actual numerical value of V_{OL} is relatively small, only a few iterations are usually sufficient for an accurate result.

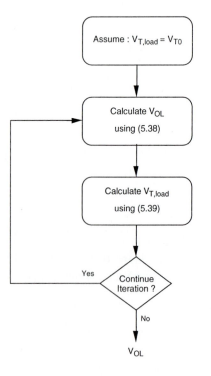

Figure 5.12.   Calculation of V_{OL} through numerical iterations.

This simple iterative method can also be implemented as a short program on a computer or on a programmable calculator. The same approach will be used on several occasions, where the iterative technique may offer a more convenient alternative to tedious analytical calculations.

Calculation of V_{IL}

At $V_{in} = V_{IL}$, we expect that the slope of the inverter VTC is equal to (-1), i.e., $dV_{out}/dV_{in} = -1$. Since the corresponding output voltage is expected to be very close to V_{OH}, we can assume that the driver transistor is turned on, and that it is operating in the saturation region. Hence, writing KCL for the output node at steady-state yields the following current equation.

$$\frac{k_{load}}{2} \cdot \left(V_{DD} - V_{out} - V_{T,load}\right)^2 = \frac{k_{driver}}{2} \cdot \left(V_{in} - V_{T0}\right)^2 \tag{5.40}$$

Differentiating both sides of (5.40) with respect to V_{in}, we obtain :

$$k_{load} \cdot \left(V_{DD} - V_{out} - V_{T,load}\right) \cdot \left(-\frac{dV_{out}}{dV_{in}} - \frac{dV_{T,load}}{dV_{out}} \cdot \frac{dV_{out}}{dV_{in}}\right)$$
$$= k_{driver} \cdot \left(V_{in} - V_{T0}\right) \cdot \left(\frac{dV_{in}}{dV_{in}}\right) \tag{5.41}$$

This equation can be rewritten as follows.

$$\frac{dV_{out}}{dV_{in}} = -\frac{k_{driver} \cdot \left(V_{in} - V_{T0}\right)}{\left(1 + \frac{dV_{T,load}}{dV_{out}}\right) \cdot k_{load} \cdot \left(V_{DD} - V_{out} - V_{T,load}\right)} \tag{5.42}$$

Assuming that the magnitude of $(dV_{T,load}/dV_{out})$ is negligible compared to 1, we obtain the following relationship.

$$\frac{dV_{out}}{dV_{in}} = -\frac{\sqrt{2 \cdot k_{driver} \cdot I_{D,driver}}}{\sqrt{2 \cdot k_{load} \cdot I_{D,load}}} \tag{5.43}$$

Since $I_{D,load} = I_{D,driver}$, this equation yields

$$\frac{dV_{out}}{dV_{in}} = -\sqrt{k_R} \qquad (5.44)$$

with

$$k_R = \frac{k_{driver}}{k_{load}} = \frac{\left(\dfrac{W}{L}\right)_{driver}}{\left(\dfrac{W}{L}\right)_{load}} \qquad if \ k'_{n,driver} = k'_{n,load} \qquad (5.45)$$

This result implies that the *slope* of the VTC will be equal to a negative constant, determined by the transconductances of the driver and the load devices, after the driver transistor is turned on and as long as both devices are in saturation. Hence, in this case, the critical voltage point V_{IL} is not defined as the point at which the slope of the VTC is equal to zero. For practical purposes, the voltage V_{IL} is taken as the point at which the driver transistor starts conducting, i.e., $V_{IL} = V_{T0}$.

Calculation of V_{IH}

When the input voltage is equal to V_{IH}, the output voltage V_{out} is expected to be slightly larger than V_{OL}. Assuming that $V_{out} < V_{IH} - V_{T0}$, the driver transistor operates in the linear region, while the load transistor is still in saturation.

$$\frac{k_{load}}{2} \cdot \left(V_{DD} - V_{out} - V_{T,load}\right)^2 = \frac{k_{driver}}{2} \cdot \left[2 \cdot \left(V_{in} - V_{T0}\right) \cdot V_{out} - V_{out}^2\right] \quad (5.46)$$

Here, we will define V_{IH} as the voltage point at which the slope of the VTC becomes equal to (−1). We differentiate both sides of (5.46) with respect to V_{in}, substitute V_{IH} for the input voltage, and let $dV_{out}/dV_{in} = -1$ to obtain the following expression.

$$V_{IH} = V_{T0} - \frac{k_{load}}{k_{driver}} \cdot \left(V_{DD} - V_{T,load}\right) + \left(2 + \frac{k_{load}}{k_{driver}}\right) \cdot V_{out} \qquad (5.47)$$

We can now substitute this expression into the KCL equation given by (5.46), and solve for the output voltage V_{out}.

$$V_{out} = \sqrt{\frac{k_{load}}{k_{load} + 3 \cdot k_{driver}}} \cdot \left(V_{DD} - V_{T,load}\right) \qquad (5.48)$$

Note that the threshold voltage of the load device is also a function of the output voltage. Therefore, as in the case of V_{OL}, equations (5.47) and (5.48) must be solved together with equation (5.39), using numerical iterations. Since the output voltage is relatively small, a few iterations usually produce a sufficiently accurate result.

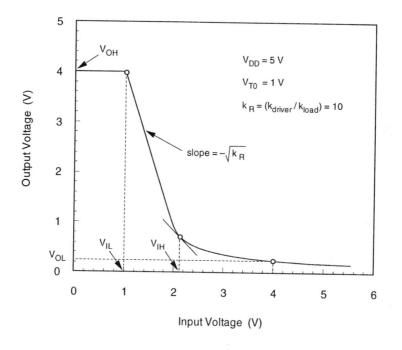

Figure 5.13.   Typical VTC of a saturated enhancement-load inverter circuit.

The typical DC voltage transfer characteristic (VTC) of the saturated enhancement-load inverter circuit is shown in Fig. 5.13. Obviously, the most significant performance deficiency of this circuit is that the output high voltage V_{OH} is lower than the power supply voltage V_{DD}, i.e., $V_{OH} = V_{DD} - V_{T,load}$. Thus, the noise margin for high signal levels (NM_H) is greatly reduced for this inverter.

Linear Enhancement-Load Inverter

One possible solution to this problem is to apply the gate bias of the enhancement-type load device from another voltage source V_{GG}, so that the load device operates in the linear region instead of the saturation region. This circuit configuration, illustrated in Fig. 5.14, is called the linear enhancement-load inverter. To ensure that the load transistor operates in the linear region regardless of input and output voltage levels, we must set the gate bias voltage as $V_{GG} > V_{DD} + V_{T,load}$. Upon examination, we can see that the output voltage reaches a high level of $V_{OH}=V_{DD}$ when the input is low, i.e., when the driver transistor is cut off. Hence, this circuit does not suffer from the limitations imposed on the noise margin NM_H by the lower V_{OH} value, as in the saturated enhancement-load inverter case. The price for this improvement is paid, however, by using two separate power supplies, V_{DD} and V_{GG}. To supply each circuit block designed in this fashion with three power connections (two for supply voltages, one ground) is a costly and complicated proposition in terms of area and routing requirements; this circuit configuration is not widely used in large-scale applications.

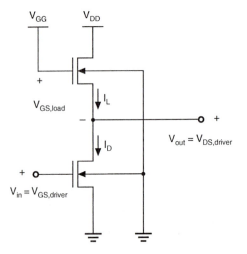

Figure 5.14.   Inverter circuit with linear enhancement-type MOSFET load.

In addition, both the saturated enhancement-load inverter and the linear enhancement-load inverter circuits require a relatively large driver-to-load ratio k_R to achieve a sharp VTC transition, lower V_{OL}, and larger noise margins. A large driver-to-load ratio, on the other hand, increases the total area occupied by the inverter circuit and, hence, is not well-suited for large-scale integration.

5.4. Inverter with Depletion-Type MOSFET Load

Several of the disadvantages of the enhancement-type load inverter can be avoided by using a depletion-type nMOS transistor as the load device. Obviously, the fabrication process for producing an inverter with an enhancement-type nMOS driver and a depletion-type nMOS load is slightly more complicated and requires additional processing steps, especially for the channel implant to adjust the threshold voltage of the load device. The resulting improvement of circuit performance and integration possibilities, however, easily justify the additional processing effort required for the fabrication of depletion-load inverters. The immediate advantages of implementing this circuit configuration are: (i) sharp VTC transition and better noise margins, (ii) single power supply, and (iii) smaller overall layout area.

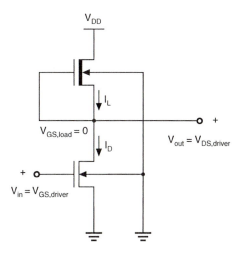

Figure 5.15.   Inverter circuit with depletion-type nMOS load.

The circuit diagram of the depletion-load inverter circuit is shown in Fig. 5.15. The driver device is an enhancement-type nMOS transistor, with $V_{T0,driver} > 0$, whereas the load is a depletion-type nMOS transistor, with $V_{T0,load} < 0$. The current-voltage equations to be used for the depletion-type load transistor are identical to those of the enhancement-type device, with the exception of the negative threshold voltage (cf. Section 3.3). The gate and the source nodes of the load transistor are connected, hence, $V_{GS,load} = 0$ always. Since the threshold voltage of the depletion-type load is negative, the condition $V_{GS,load} > V_{T,load}$ is satisfied, and the load device always has a conducting channel regardless of the input and output voltage levels. Also note that both the driver transistor and the load transistor are built on the same p-type substrate, which is connected to the ground. Consequently, the load device is subject to the substrate-bias effect, so that its threshold voltage is a function of its source-to-substrate voltage, $V_{SB,load} = V_{out}$.

$$V_{T,load} = V_{T0,load} + \gamma \left(\sqrt{|2\,\phi_F| + V_{out}} - \sqrt{|2\,\phi_F|} \right) \qquad (5.49)$$

The operating mode of the load transistor is determined by the output voltage level. When the output voltage is small, i.e., when $V_{out} < V_{DD} + V_{T,load}$, the load transistor is in saturation. Note that this condition corresponds to $V_{DS,load} > V_{GS,load} - V_{T,load}$. Then, the load current is given by the following equation.

$$I_{D,load} = \frac{k_{n,load}}{2} \cdot \left[-V_{T,load}(V_{out}) \right]^2 \tag{5.50}$$

For larger output voltage levels, i.e., for $V_{out} > V_{DD} + V_{T,load}$, the depletion-type load transistor operates in the linear region. The load current in this case is

$$I_{D,load} = \frac{k_{n,load}}{2} \cdot \left[2|V_{T,load}(V_{out})| \cdot (V_{DD} - V_{out}) - (V_{DD} - V_{out})^2 \right] \tag{5.51}$$

The voltage transfer characteristic (VTC) of this inverter can be constructed by setting $I_{D,driver} = I_{D,load}$, $V_{GS,driver} = V_{in}$, and $V_{DS,driver} = V_{out}$, and by solving the corresponding current equations for $V_{out} = f(V_{in})$. Figure 5.16 shows the VTC of a typical depletion-load inverter, with $k_{n,driver}' = k_{n,load}'$.

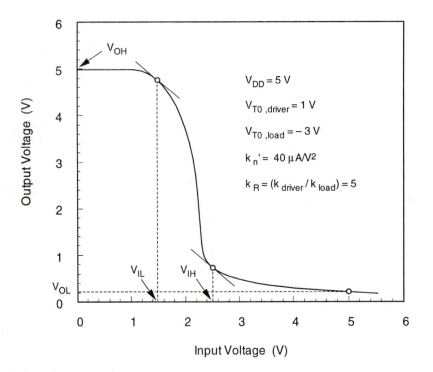

Figure 5.16.   Typical VTC of a depletion-load inverter circuit.

Next, we will consider the critical voltage points V_{OH}, V_{OL}, V_{IL}, and V_{IH} for this inverter circuit. The operating regions and the voltage levels of the driver and the load transistors at these critical points are listed below.

V_{in}	V_{out}	Driver operating region	Load operating region
V_{OL}	V_{OH}	cut-off	linear
V_{IL}	$\approx V_{OH}$	saturation	linear
V_{IH}	small	linear	saturation
V_{OH}	V_{OL}	linear	saturation

Calculation of V_{OH}

When the input voltage V_{in} is smaller than the driver threshold voltage V_{T0}, the driver transistor is turned off and does not conduct any drain current. Consequently, the load device, which operates in the linear region, also has zero drain current. Substituting V_{OH} for V_{out} in (5.51), and letting the load current $I_{D,load} = 0$, we obtain

$$I_{D,load} = \frac{k_{n,load}}{2} \cdot \left[2 \left| V_{T,load}(V_{OH}) \right| \cdot (V_{DD} - V_{OH}) - (V_{DD} - V_{OH})^2 \right] = 0 \qquad (5.52)$$

The solution of this equation gives $V_{OH} = V_{DD}$.

Calculation of V_{OL}

To calculate the output low voltage V_{OL}, we assume that the input voltage V_{in} of the inverter is equal to $V_{OH} = V_{DD}$. Note that in this case, the driver transistor operates in the linear region while the depletion-type load is in saturation.

$$\frac{k_{driver}}{2} \cdot \left[2 \cdot (V_{OH} - V_{T0}) \cdot V_{OL} - V_{OL}^2 \right] = \frac{k_{load}}{2} \cdot \left[-V_{T,load}(V_{OL}) \right]^2 \qquad (5.53)$$

This second-order equation in V_{OL} can be solved by temporarily neglecting the dependence of $V_{T,load}$ on V_{OL}, as follows.

$$V_{OL} = V_{OH} - V_{T0} - \sqrt{(V_{OH} - V_{T0})^2 - \left(\frac{k_{load}}{k_{driver}} \right) \cdot \left| V_{T,load}(V_{OL}) \right|^2} \qquad (5.54)$$

The actual value of the output low voltage can be found by solving the two equations (5.54) and (5.49) using numerical iterations, as illustrated in Fig. 5.12. The iterative method converges rapidly because the actual value of V_{OL} is relatively small.

Calculation of V_{IL}

By definition, the slope of the VTC is equal to (-1), i.e., $dV_{out}/dV_{in} = -1$ when the input voltage is $V_{in} = V_{IL}$. Note that in this case, the driver transistor operates in saturation while the load transistor operates in the linear region. Applying KCL for the output node, we obtain the following current equation.

$$\frac{k_{driver}}{2} \cdot \left(V_{in} - V_{T0}\right)^2 = \frac{k_{load}}{2} \cdot \left[2\left|V_{T,load}\left(V_{out}\right)\right| \cdot \left(V_{DD} - V_{out}\right) - \left(V_{DD} - V_{out}\right)^2\right] \quad (5.55)$$

To satisfy the derivative condition at V_{IL}, we differentiate both sides of (5.55) with respect to V_{in}.

$$k_{driver} \cdot \left(V_{in} - V_{T0}\right) = \frac{k_{load}}{2} \cdot \left[2\left|V_{T,load}\left(V_{out}\right)\right|\left(-\frac{dV_{out}}{dV_{in}}\right)\right.$$

$$\left. + 2\left(V_{DD} - V_{out}\right)\left(-\frac{dV_{T,load}}{dV_{in}}\right) - 2\left(V_{DD} - V_{out}\right)\left(-\frac{dV_{out}}{dV_{in}}\right)\right] \quad (5.56)$$

In general, we can assume that the term $(dV_{T,load}/dV_{in})$ is negligible with respect to the others. Substituting V_{IL} for V_{in}, and letting $dV_{out}/dV_{in} = -1$, we obtain V_{IL} as a function of the output voltage V_{out}.

$$V_{IL} = V_{T0} + \left(\frac{k_{load}}{k_{driver}}\right) \cdot \left[V_{out} - V_{DD} + \left|V_{T,load}\left(V_{out}\right)\right|\right] \quad (5.57)$$

To account for the substrate-bias effect on $V_{T,load}$, this equation must be solved together with (5.49) and (5.55) using successive iterations.

Calculation of V_{IH}

V_{IH} is the larger of the two voltage points on the VTC at which the slope is equal to (-1). Since the output voltage corresponding to this operating point is relatively small, the driver transistor is in the linear region and the load transistor is in saturation.

$$\frac{k_{driver}}{2} \cdot \left[2 \cdot \left(V_{in} - V_{T0} \right) \cdot V_{out} - V_{out}^2 \right] = \frac{k_{load}}{2} \cdot \left[-V_{T,load} \left(V_{out} \right) \right]^2 \qquad (5.58)$$

Differentiating both sides of (5.58) with respect to V_{in}, we obtain :

$$k_{driver} \cdot \left[V_{out} + \left(V_{in} - V_{T0} \right) \left(\frac{dV_{out}}{dV_{in}} \right) - V_{out} \left(\frac{dV_{out}}{dV_{in}} \right) \right]$$

$$= k_{load} \cdot \left[-V_{T,load} \left(V_{out} \right) \right] \cdot \left(\frac{dV_{T,load}}{dV_{out}} \right) \cdot \left(\frac{dV_{out}}{dV_{in}} \right) \qquad (5.59)$$

Now, substitute $dV_{out} / dV_{in} = -1$ into (5.59), and solve for $V_{in} = V_{IH}$.

$$V_{IH} = V_{T0} + 2 V_{out} + \left(\frac{k_{load}}{k_{driver}} \right) \cdot \left[-V_{T,load} \left(V_{out} \right) \right] \cdot \left(\frac{dV_{T,load}}{dV_{out}} \right) \qquad (5.60)$$

Note that the derivative of the load threshold voltage with respect to the output voltage cannot be neglected in this case.

$$\frac{dV_{T,load}}{dV_{out}} = \frac{\gamma}{2\sqrt{|2\,\phi_F| + V_{out}}} \qquad (5.61)$$

The actual values of V_{IH} and the corresponding output voltage V_{out} are determined by solving (5.61) together with the current equation (5.58) and with the threshold voltage expression (5.49), using numerical iterations. A relatively accurate first-order estimate for V_{IH} can be obtained by assuming that the output voltage in this case is approximately equal to V_{OL}.

It is seen from this discussion that the critical voltage points, the general shape of the inverter VTC, and ultimately, the noise margins, are determined essentially by the threshold voltages of the driver and the load devices, and by the driver-to-load ratio (k_{driver}/k_{load}). Since the threshold voltages are usually set by the fabrication process, the driver-to-load ratio emerges as a primary design parameter which can be adjusted to achieve the desired VTC shape. Notice that with $k'_{n,driver} = k'_{n,load}$, the driver-to-load ratio is solely determined by the (W/L) ratios of the driver and the load transistors, i.e., by the device geometries. Figure 5.17 shows the VTCs of depletion-load inverter circuits with different driver-to-load ratios $k_R = (k_{driver}/k_{load})$.

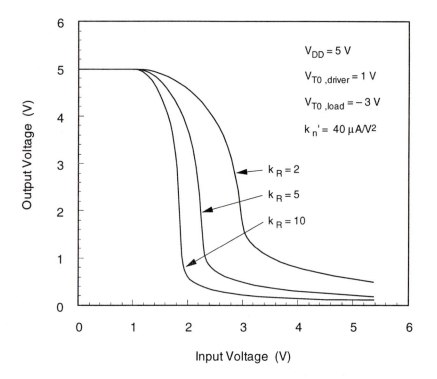

Figure 5.17. Voltage transfer characteristics of depletion-load inverters, with different driver-to-load ratios.

One important observation is that, unlike in the enhancement-load inverter case, a sharp VTC transition and larger noise margins can be obtained with relatively small driver-to-load ratios. Thus, the total area occupied by a depletion-load inverter circuit with an acceptable circuit performance is expected to be much smaller than the area occupied by a comparable resistive-load or enhancement-load inverter.

Design of Depletion-Load Inverters

Based on the VTC analysis given in the previous section, we can now consider the *design* of depletion-load inverters to satisfy certain DC performance criteria. In the broadest sense, the designable parameters in an inverter circuit are: (i) the power supply voltage V_{DD}, (ii) the threshold voltages of the driver and the load transistors, and (iii) the (W/L) ratios of the driver and the load transistors. In most practical cases, however, the power supply voltage and the device threshold voltages are dictated by other external constraints and by the fabrication process; thus, they cannot be adjusted for every individual inverter circuit to satisfy performance requirements. This leaves the (W/L) ratio of the transistors, and more specifically, the driver-to-load ratio k_R, as the primary design parameter.

Note that the power supply voltage V_{DD} of the inverter circuit also determines the level of the output high voltage V_{OH}, since $V_{OH} = V_{DD}$. Of the remaining three critical voltages on the VTC, the output low voltage V_{OL} is usually the most significant design constraint. Designing the inverter to achieve a certain V_{OL} value will automatically set the other two critical voltages, V_{IL} and V_{IH}, as well. Equation (5.54) can be rearranged to calculate the driver-to-load ratio that achieves a target V_{OL} value, assuming that the power supply voltage and the threshold voltage values are set previously by independent design and processing constraints.

$$k_R = \frac{k_{driver}}{k_{load}} = \frac{\left|V_{T,load}(V_{OL})\right|^2}{2(V_{OH} - V_{T0})V_{OL} - V_{OL}^2} \tag{5.62}$$

Here, the driver-to-load ratio is given by

$$k_R = \frac{k'_{n,driver} \cdot \left(\dfrac{W}{L}\right)_{driver}}{k'_{n,load} \cdot \left(\dfrac{W}{L}\right)_{load}} \tag{5.63}$$

Since the channel doping densities and, consequently, the channel electron mobilities of the enhancement-type driver transistor and the depletion-type load transistor are not equal, we shall expect that $k_{n,driver}' \neq k_{n,load}'$, in general. Only if $k_{n,driver}' \approx k_{n,load}'$, can the driver-to-load ratio be reduced to

$$k_R = \frac{\left(\dfrac{W}{L}\right)_{driver}}{\left(\dfrac{W}{L}\right)_{load}} \tag{5.64}$$

Finally, note that the design procedure summarized above determines the *ratio* of the driver and the load transconductances, but not the specific (W/L) ratio of each transistor. As a result, one can propose a number of designs with different (W/L) ratios for the driver (and for the load) device, each of which satisfies the driver-to-load ratio condition stated above. The actual sizes of the driver and the load transistors are usually determined by other design constraints, such as the current-drive capability, the steady-state power dissipation, and the transient switching speed.

Power and Area Considerations

The steady-state DC power consumption of the depletion-load inverter circuit can be easily found by calculating the amount of current being drawn from the power supply, during the input-low state and the input-high state. When the input voltage is low, i.e., when the driver transistor is cut-off and $V_{out} = V_{OH} = V_{DD}$, there is no significant current flow through the driver and the load transistors. Consequently, the inverter does not dissipate DC power under this condition. When the input is at the high state with $V_{in} \approx V_{DD}$ and $V_{out} = V_{OL}$, on the other hand, both the driver and the load transistors conduct a significant current, given by

$$I_{DC}(V_{in} = V_{DD}) = \frac{k_{load}}{2} \cdot \left[-V_{T,load}(V_{OL}) \right]^2$$
$$= \frac{k_{driver}}{2} \cdot \left[2 \cdot (V_{OH} - V_{T0}) \cdot V_{OL} - V_{OL}^2 \right] \tag{5.65}$$

Assuming that the input voltage level is low during 50% of the operation time and high during the other 50%, the overall average DC power consumption of this circuit can be estimated as follows.

$$P_{DC} = \frac{V_{DD}}{2} \cdot \frac{k_{load}}{2} \cdot \left[-V_{T,load}(V_{OL}) \right]^2 \tag{5.66}$$

Figure 5.18(a) shows a simplified layout of the depletion-load inverter circuit. Note that the drain of the enhancement-type driver and the source of the depletion-type load transistor share a common n^+ diffusion region, which saves silicon area compared to two separate diffusion regions. The threshold voltage of the depletion-type load device is adjusted by a donor implant into the channel region. The width-to-length ratio of the driver transistor is seen to be larger than the width-to-length ratio of the load, which yields a driver-to-load ratio of about 4. Overall, this inverter circuit configuration is relativ ly compact, and it occupies a significantly smaller area compared to those for resistive-load or enhancement-load inverters with similar performance.

The area requirements of the depletion-load inverter circuit can be reduced even further by using a *buried contact* for connecting the gate and the source of the

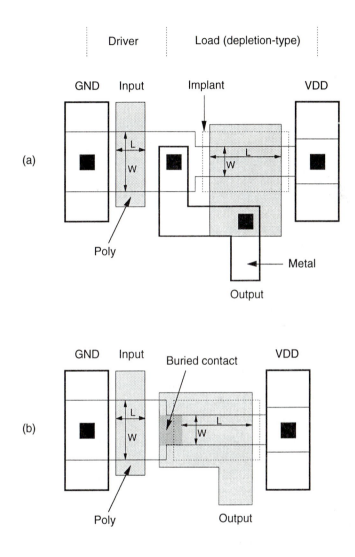

Figure 5.18. Sample layout of depletion-load inverter circuits (a) with output contact on diffusion, and (b) with buried contact.

load transistor, as shown in Figure 5.18(b). In this case, the polysilicon gate of the depletion-mode transistor makes a direct ohmic contact with the n^+ source diffusion. The contact window on the intermediate diffusion area can be omitted, which results in a further reduction of the total area occupied by the inverter.

Example 5.3

Consider a depletion-load inverter circuit with the following parameters:

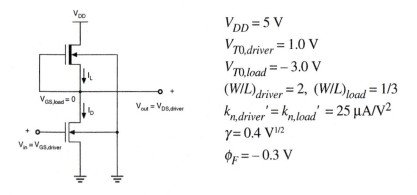

$$V_{DD} = 5 \text{ V}$$
$$V_{T0,driver} = 1.0 \text{ V}$$
$$V_{T0,load} = -3.0 \text{ V}$$
$$(W/L)_{driver} = 2, \ (W/L)_{load} = 1/3$$
$$k_{n,driver}' = k_{n,load}' = 25 \ \mu\text{A/V}^2$$
$$\gamma = 0.4 \text{ V}^{1/2}$$
$$\phi_F = -0.3 \text{ V}$$

Calculate the critical voltages (V_{OL}, V_{OH}, V_{IL}, V_{IH}) on the VTC and find the noise margins of the circuit.

First, the output high voltage is simply found according to (5.52) as $V_{OH} = V_{DD} = 5$ V.

To calculate the output low voltage V_{OL}, we must solve (5.49) and (5.54) simultaneously, using numerical iterations, since the threshold voltage of the load transistor is also a function of the inverter output voltage. We start the iterations by assuming that the output voltage is equal to zero; thus, letting $V_{T,load} = V_{T0,load} = -3$ V. Solving (5.54) with this assumption yields a first-order estimate for V_{OL}.

$$V_{OL} = V_{OH} - V_{T0} - \sqrt{\left(V_{OH} - V_{T0}\right)^2 - \left(\frac{k_{load}}{k_{driver}}\right) \cdot \left|V_{T,load}\left(V_{OL}\right)\right|^2}$$

$$= 5 - 1 - \sqrt{\left(5-1\right)^2 - \left(\frac{1}{6}\right)|3|^2} = 0.192 \text{ V}$$

Now, the threshold voltage of the depletion-load device can be updated by substituting this output voltage into (5.49).

$$V_{T,load} = V_{T0,load} + \gamma\left(\sqrt{|2\phi_F| + V_{OL}} - \sqrt{|2\phi_F|}\right)$$

$$= -3 + 0.4\left(\sqrt{0.6 + 0.2} - \sqrt{0.6}\right) = -2.95 \text{ V}$$

Using this new value for $V_{T,load}$, we now recalculate V_{OL}, again according to (5.54).

$$V_{OL} = 0.186 \text{ V}$$
$$V_{T,load} = -2.95 \text{ V}$$

At this point, we can stop the iteration process since the threshold voltage of the load device has not changed in the two significant digits after the decimal point. Continuing the iteration would not produce a perceptible improvement of V_{OL}.

The calculation of V_{IL} involves simultaneous solution of (5.49), (5.55), and (5.57), using numerical iterations. When the input voltage is equal to V_{IL}, we expect that the output voltage is slightly lower than the output high voltage, V_{OH}. As a first-order approximation, assume that $V_{out} = V_{OH} = 5$ V for $V_{in} = V_{IL}$. Then, the threshold voltage of the load device can be estimated as $V_{T,load}(V_{out} = 5 \text{ V}) = -2.36$ V. Substituting this value into (5.57) gives V_{IL} as a function of the output voltage V_{out}.

$$V_{IL}(V_{out}) = V_{T0} + \frac{k_{load}}{k_{driver}} \cdot \left[V_{out} - V_{DD} + \left| V_{T,load}(V_{out}) \right| \right]$$

$$= 1 + \left(\frac{1}{6} \right)(V_{out} - 5 + 2.36) = 0.167 V_{out} + 0.56$$

This expression can be rearranged as

$$V_{out} = 6 V_{IL} - 3.35$$

Now, substitute this into the KCL equation (5.55) to obtain the following quadratic equation for V_{IL}:

$$\frac{k_{driver}}{2} \cdot (V_{IL} - V_{T0})^2 = \frac{k_{load}}{2} \cdot \left[2 \left| V_{T,load}(V_{out}) \right| \cdot (V_{DD} - 6 V_{IL} + 3.35) \right.$$
$$\left. - (V_{DD} - 6 V_{IL} + 3.35)^2 \right]$$

$$2 \cdot (V_{IL} - 1)^2 = \frac{1}{3} \cdot \left[2 \cdot 2.36 \cdot (5 - 6 V_{IL} + 3.35) - (5 - 6 V_{IL} + 3.35)^2 \right]$$

The solution of this second-order equation yields two possible values for V_{IL}.

$$V_{IL} = \begin{cases} 0.98 \text{ V} \\ \underline{1.36 \text{ V}} \end{cases}$$

Note that V_{IL} must be larger than the threshold voltage V_{T0} of the driver transistor, hence, $V_{IL} = 1.36$ V is the physically correct solution. The output voltage level at this point can also be found as

$$V_{out} = 6 \cdot 1.36 - 3.35 = 4.81 \text{ V}$$

which significantly improves our initial assumption of $V_{out} = 5$ V. At this point, the threshold voltage of the load transistor must be recalculated, in order to update its value. Substituting $V_{out} = 4.81$ V into (5.49) yields $V_{T,load} = -2.38$ V. We observe that this value is only slightly higher (by 20 mV) than the threshold voltage value used in the previous calculations. For practical purposes, we can terminate the numerical iteration at this stage and accept $V_{IL} = 1.36$ V as a fairly accurate estimate.

To calculate V_{IH}, we have to find the numerical value of $(dV_{T,load} / V_{out})$ first, according to (5.61). When the input voltage is equal to V_{IH}, the output voltage is expected to be relatively low. As a first-order approximation, assume that the output voltage level is $V_{out} = V_{OL} \approx 0.2$ V when $V_{in} = V_{IH}$. The threshold voltage of the load device can also be estimated as $V_{T,load}(V_{out} = 0.2 \text{ V}) = -2.95$ V. Thus,

$$\frac{dV_{T,load}}{dV_{out}} = \frac{\gamma}{2\sqrt{|2\phi_F| + V_{out}}} = \frac{0.4}{2\sqrt{0.6 + 0.2}} = 0.22$$

This value can now be used in (5.60) to find V_{IH} as a function of the output voltage.

$$V_{IH}(V_{out}) = V_{T0} + 2 V_{out} + \frac{k_{load}}{k_{driver}} \cdot \left[-V_{T,load}(V_{out}) \right] \cdot \left(\frac{dV_{T,load}}{dV_{out}} \right)$$

$$= 1 + 2 V_{out} + \left(\frac{1}{6} \right) \cdot 2.95 \cdot 0.22 = 2 V_{out} + 1.1$$

This expression is rearranged as:

$$V_{out} = 0.5 V_{IH} - 0.55$$

Next, substitute V_{out} in the KCL equation (5.58), to obtain

$$2 \cdot \left[2 \cdot (V_{IH} - 1) \cdot (0.5 V_{IH} - 0.55) - (0.5 V_{IH} - 0.55)^2 \right] = \frac{1}{3} \cdot (2.95)^2$$

The solution of this simple quadratic equation yields two values for V_{IH}:

$$V_{IH} = \begin{cases} 0.10 \text{ V} \\ \underline{1.96 \text{ V}} \end{cases}$$

where $V_{IH} = 1.96$ V is the physically correct solution, since V_{IH} must be larger than the driver threshold voltage. The output voltage level at this point is calculated as

$$V_{out} = 0.5 \cdot 1.96 - 0.55 = 0.43 \text{ V}$$

With this updated output voltage value, we can now reevaluate the load threshold voltage as $V_{T,load}(V_{out} = 0.43 \text{ V}) = -2.9$ V, and the $(dV_{T,load} / V_{out})$ value as

$$\frac{dV_{T,load}}{dV_{out}} = 0.19$$

Note both of these values are fairly close to those used at the beginning of this iteration process. Repeating the iterative calculation will provide only a marginal improvement of accuracy, thus, we may accept $V_{IH} = 1.96$ V as a good estimate.

In conclusion, the noise margins for high signal levels and for low signal levels can be found as follows:

$$NM_H = V_{OH} - V_{IH} = 3.04 \text{ V}$$
$$NM_L = V_{IL} - V_{OL} = 1.17 \text{ V}$$

5.5. CMOS Inverter

All of the inverter circuits considered so far had the general circuit structure shown in Fig. 5.3, consisting of an enhancement-type nMOS driver transistor, and a load device which can be a resistor, an enhancement-type nMOS transistor, or a depletion-type nMOS transistor acting as a nonlinear resistor. In this general configuration, the input signal is always applied to the gate of the driver transistor, and the operation of the inverter is controlled primarily by switching the driver. Now, we will turn our attention to a radically different inverter structure, which consists of an enhancement-type nMOS transistor and an enhancement-type pMOS transistor, operating in complementary mode (Fig. 5.19). This configuration is called Complementary MOS, or CMOS. The circuit topology is complementary push-pull in the sense that for high input, the nMOS transistor drives (pulls down) the output node while the pMOS transistor acts as the load, and for low input the pMOS transistor drives (pulls up) the output node while the nMOS transistor acts as the load. Consequently, both devices contribute equally to the circuit operation characteristics.

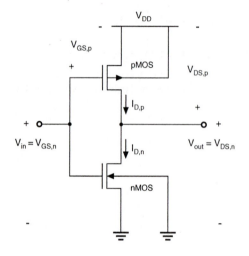

Figure 5.19.   CMOS inverter circuit.

The CMOS inverter has two important advantages over the other inverter configurations. The first and perhaps the most important advantage is that the steady-state power dissipation of the CMOS inverter circuit is virtually negligible, except for small power dissipation due to leakage currents. In all other inverter structures examined so far, a nonzero steady-state current is drawn from the power source when the driver transistor is turned on, which results in a significant DC power consumption. The other advantages of the CMOS configuration are that the voltage transfer characteristic (VTC) exhibits a full output voltage swing between 0 V and V_{DD}, and that the VTC transition is usually very sharp. Thus, the VTC of the CMOS inverter resembles that of an ideal inverter.

Since nMOS and pMOS transistors must be fabricated on the same chip side-by-side, the CMOS process is more complex than the standard nMOS-only process. In particular, the CMOS process must provide an n-type substrate for the pMOS transistors and a p-type substrate for the nMOS transistors. This can be achieved by building either n-type *tubs* (wells) on a p-type wafer, or by building p-type tubs on an n-type wafer (cf. Chapter 2). In addition, the close proximity of an nMOS and a pMOS transistor may lead to the formation of two parasitic bipolar transistors, causing a *latch-up* condition. In order to prevent this undesirable effect, additional *guard rings* must be built around the nMOS and the pMOS transistors as well. The increased process complexity of CMOS fabrication may be considered as the price being paid for the improvements achieved in power consumption and noise margins.

Circuit Operation

In Fig. 5.19, note that the input voltage is connected to the gate terminals of both the nMOS and the pMOS transistors. Thus, both transistors are driven directly by the input signal, V_{in}. The substrate of the nMOS transistor is connected to the ground, while the substrate of the pMOS transistor is connected to the power supply voltage, V_{DD}, in order to reverse-bias the source and drain junctions. Since $V_{SB} = 0$ for both devices, there will be no substrate-bias effect for either device. It can be seen from the circuit diagram in Fig. 5.19 that

$$V_{GS,n} = V_{in}$$
$$V_{DS,n} = V_{out} \tag{5.67}$$

and also,

$$V_{GS,p} = -\left(V_{DD} - V_{in}\right)$$
$$V_{DS,p} = -\left(V_{DD} - V_{out}\right) \tag{5.68}$$

We will start our analysis by considering two simple cases. When the input voltage is smaller than the nMOS threshold voltage, i.e., when $V_{in} < V_{T0,n}$, the nMOS transistor is cut-off. At the same time, the pMOS transistor is on, operating in the linear region. Since the drain currents of both transistors are approximately equal to zero (except for small leakage currents), i.e.,

$$I_{D,n} = I_{D,p} = 0 \tag{5.69}$$

the drain-to-source voltage of the pMOS transistor is also equal to zero, and the output voltage V_{OH} is equal to the power supply voltage.

$$V_{out} = V_{OH} = V_{DD} \tag{5.70}$$

On the other hand, when the input voltage exceeds $(V_{DD} + V_{T0,p})$, the pMOS transistor is turned off. In this case, the nMOS transistor is operating in the linear region, but its drain-to-source voltage is equal to zero because condition (5.69) is satisfied. Consequently, the output voltage of the circuit is

$$V_{out} = V_{OL} = 0 \qquad\qquad (5.71)$$

Next, we examine the operating modes of the nMOS and the pMOS transistors as functions of the input and output voltages. The nMOS transistor operates in *saturation* if $V_{in} > V_{T0,n}$ and if the following condition is satisfied.

$$V_{DS,n} \geq V_{GS,n} - V_{T0,n} \qquad \Leftrightarrow \qquad V_{out} \geq V_{in} - V_{T0,n} \qquad\qquad (5.72)$$

The pMOS transistor operates in *saturation if $V_{in} < (V_{DD} + V_{T0,p})$, and if* :

$$V_{DS,p} \leq V_{GS,p} - V_{T0,p} \qquad \Leftrightarrow \qquad V_{out} \leq V_{in} - V_{T0,p} \qquad\qquad (5.73)$$

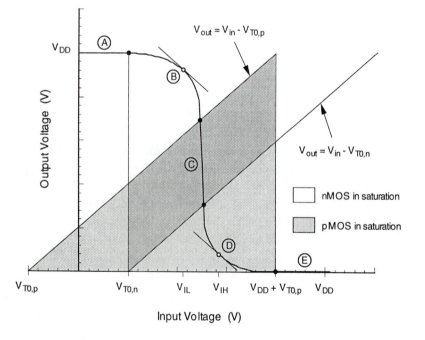

Figure 5.20.   Operating regions of the nMOS and the pMOS transistors.

Both of these conditions for device saturation are illustrated graphically as shaded areas on the V_{out} - V_{in} plane in Fig. 5.20. A typical CMOS inverter voltage transfer characteristic is also superimposed for easy reference. Here, we identify five distinct regions, labeled A through E, each corresponding to a different set of operating conditions. The table below lists these regions and the corresponding critical input and output voltage levels.

Region	V_{in}	V_{out}	nMOS	pMOS
A	$< V_{T0,n}$	V_{OH}	cut-off	linear
B	V_{IL}	$high \approx V_{OH}$	saturation	linear
C	V_{th}	V_{th}	saturation	saturation
D	V_{IH}	$low \approx V_{OL}$	linear	saturation
E	$> \left(V_{DD} + V_{T0,p} \right)$	V_{OL}	linear	cut-off

In Region A, where $V_{in} < V_{T0,n}$, the nMOS transistor is cut-off and the output voltage is equal to $V_{OH} = V_{DD}$. As the input voltage is increased beyond $V_{T0,n}$ (into Region B), the nMOS transistor starts conducting in saturation mode and the output voltage begins to decrease. Also note that the critical voltage V_{IL} which corresponds to $(dV_{out}/dV_{in}) = -1$ is located within Region B. As the output voltage further decreases, the pMOS transistor enters saturation at the boundary of Region C. It is seen from Fig. 5.20 that the inverter threshold voltage, where $V_{in} = V_{out}$, is located in Region C. When the output voltage V_{out} falls below $(V_{in} - V_{T0,n})$, the nMOS transistor starts to operate in linear mode. This corresponds to Region D in Fig. 5.20, where the critical voltage point V_{IH} with $(dV_{out}/dV_{in}) = -1$ is also located. Finally, in Region E, with the input voltage $V_{in} > (V_{DD} + V_{T0,p})$, the pMOS transistor is cut-off, and the output voltage is $V_{OL} = 0$.

In a simplistic analogy, the nMOS and the pMOS transistors can be seen as nearly ideal switches– controlled by the input voltage– that connect the output node to the power supply voltage or to the ground potential, depending on the input voltage level. The qualitative overview of circuit operation, illustrated in Fig. 5.20 and discussed above, also highlights the complementary nature of the CMOS inverter. The most significant feature of this circuit is that the current drawn from the power supply in both of these steady-state operating points, i.e., in Region A and in Region E, is nearly equal to zero. The only current that flows in either case is the very small leakage current of the reverse-biased source and drain junctions. The CMOS inverter can *drive* any load, such as interconnect capacitance or fanout logic gates which are connected to its output node, either by supplying current to, or by sinking current from, the load.

The steady-state input-output voltage characteristics of the CMOS inverter can be better visualized by considering the interaction of individual nMOS and pMOS transistor characteristics in the current-voltage space. We already know that the drain current $I_{D,n}$ of the nMOS transistor is a function of the voltages $V_{GS,n}$ and $V_{DS,n}$. Hence, the nMOS drain current is also a function of the inverter input and output voltages V_{in} and V_{out}, according to (5.67).

$$I_{D,n} = f\left(V_{in}, V_{out}\right)$$

This two-variable function, which is essentially described by the current equations (3.54)–(3.56), can be represented as a *surface* in the three-dimensional current-voltage space. Figure 5.21 shows this $I_{D,n}(V_{in}, V_{out})$ surface for the nMOS transistor.

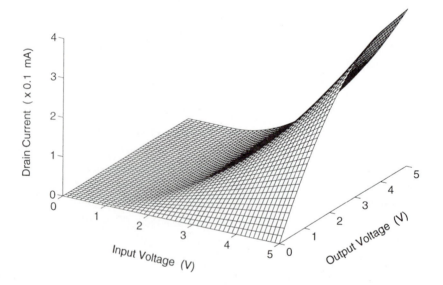

Figure 5.21.   Current-voltage surface representing the nMOS transistor characteristics.

Similarly, the drain current $I_{D,p}$ of the pMOS transistor is also a function of the inverter input and output voltages V_{in} and V_{out}, according to (5.68).

$$I_{D,p} = f\left(V_{in}, V_{out}\right)$$

This two-variable function, described by the current equations (3.57)–(3.59), can be represented as another surface in the three-dimensional current-voltage space. Figure 5.22 shows the corresponding $I_{D,p}(V_{in}, V_{out})$ surface for the pMOS transistor.

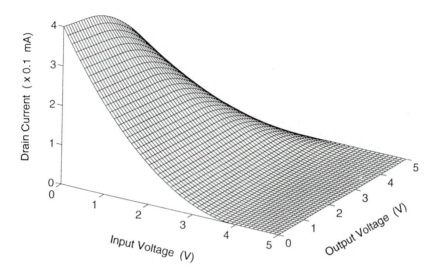

Figure 5.22.   Current-voltage surface representing the pMOS transistor characteristics.

Remember that in a CMOS inverter operating in steady-state, the drain current of the nMOS transistor is always equal to the drain current of the pMOS transistor, according to KCL:

$$I_{D,n} = I_{D,p}$$

Thus, the *intersection* of the two current-voltage surfaces shown in Figs. 5.21 and 5.22 will give the operating curve of the CMOS inverter circuit in the three-dimensional current-voltage space. The intersection of the two characteristic surfaces is shown in Fig. 5.23. The intersecting surfaces are shown from a different viewing angle in Fig. 5.24, with the intersection curve highlighted in bold.

It is clear that the vertical projection of the intersection curve on the V_{in} - V_{out} plane produces the typical CMOS inverter voltage transfer characteristic already shown in Fig. 5.20. Similarly, the horizontal projection of the intersection curve on the I_D - V_{in} plane gives the steady-state current drawn by the inverter from the power supply voltage as a function of the input voltage.

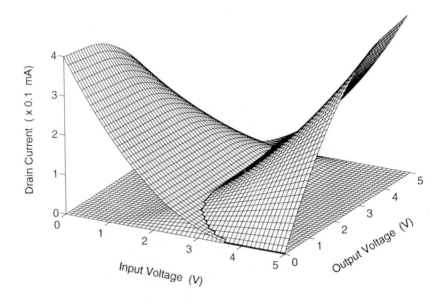

Figure 5.23.   Intersection of the current-voltage surfaces shown in Figures 5.21 and 5.22.

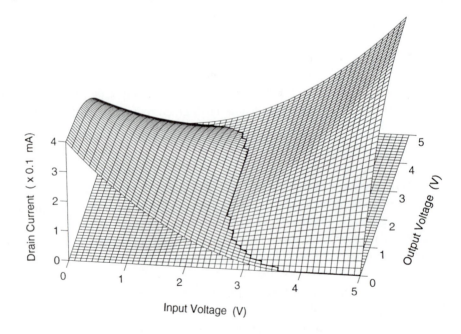

Figure 5.24.   The intersecting current-voltage surfaces shown from a different viewing angle. Notice that projection of the intersection curve on the voltage plane gives the VTC.

In the following, we will present an in-depth analysis of the CMOS inverter static characteristics, by calculating the critical voltage points on the VTC. It has already been established that $V_{OH} = V_{DD}$ and $V_{OL} = 0$ for this inverter; thus, we will devote our attention to V_{IL}, V_{IH} and the inverter switching threshold, V_{th}.

Calculation of V_{IL}

By definition, the slope of the VTC is equal to (-1), i.e., $dV_{out}/dV_{in} = -1$ when the input voltage is $V_{in} = V_{IL}$. Note that in this case, the nMOS transistor operates in saturation while the pMOS transistor operates in the linear region. From $I_{D,n} = I_{D,p}$, we obtain the following current equation.

$$\frac{k_n}{2} \cdot \left(V_{GS,n} - V_{T0,n}\right)^2 = \frac{k_p}{2} \cdot \left[2 \cdot \left(V_{GS,p} - V_{T0,p}\right) \cdot V_{DS,p} - V_{DS,p}^2\right] \qquad (5.74)$$

Using equations (5.67) and (5.68), this expression can be rewritten as

$$\frac{k_n}{2} \cdot \left(V_{in} - V_{T0,n}\right)^2 = \frac{k_p}{2} \cdot \left[2 \cdot \left(V_{in} - V_{DD} - V_{T0,p}\right) \right.$$
$$\left. \cdot \left(V_{out} - V_{DD}\right) - \left(V_{out} - V_{DD}\right)^2\right] \qquad (5.75)$$

To satisfy the derivative condition at V_{IL}, we differentiate both sides of (5.75) with respect to V_{in}.

$$k_n \cdot \left(V_{in} - V_{T0,n}\right) = k_p \cdot \left[\left(V_{in} - V_{DD} - V_{T0,p}\right) \cdot \left(\frac{dV_{out}}{dV_{in}}\right) + \left(V_{out} - V_{DD}\right) \right.$$
$$\left. - \left(V_{out} - V_{DD}\right) \cdot \left(\frac{dV_{out}}{dV_{in}}\right)\right] \qquad (5.76)$$

Substituting $V_{in} = V_{IL}$ and $(dV_{out}/dV_{in}) = -1$ in (5.76), we obtain

$$k_n \cdot \left(V_{IL} - V_{T0,n}\right) = k_p \cdot \left(2 V_{out} - V_{IL} + V_{T0,p} - V_{DD}\right) \qquad (5.77)$$

The critical voltage V_{IL} can now be found as a function of the output voltage V_{out}, as follows.

$$V_{IL} = \frac{2\,V_{out} + V_{T0,p} - V_{DD} + k_R\,V_{T0,n}}{1 + k_R} \tag{5.78}$$

where k_R is defined as

$$k_R = \frac{k_n}{k_p}$$

This equation must be solved together with the KCL equation (5.75) to obtain the numerical value of V_{IL} and the corresponding output voltage, V_{out}. Note that the solution is fairly straightforward and does not require numerical iterations as in the previous cases, since none of the transistors is subject to substrate-bias effects.

Calculation of V_{IH}

When the input voltage is equal to V_{IH}, the nMOS transistor operates in the linear region, and the pMOS transistor operates in saturation. Applying KCL to the output node, we obtain

$$\frac{k_n}{2} \cdot \left[2 \cdot \left(V_{GS,n} - V_{T0,n} \right) \cdot V_{DS,n} - V_{DS,n}^{\,2} \right] = \frac{k_p}{2} \cdot \left(V_{GS,p} - V_{T0,p} \right)^2 \tag{5.79}$$

Using equations (5.67) and (5.68), this expression can be rewritten as

$$\frac{k_n}{2} \cdot \left[2 \cdot \left(V_{in} - V_{T0,n} \right) \cdot V_{out} - V_{out}^{\,2} \right] = \frac{k_p}{2} \cdot \left(V_{in} - V_{DD} - V_{T0,p} \right)^2 \tag{5.80}$$

Now, differentiate both sides of (5.80) with respect to V_{in}.

$$k_n \cdot \left[2 \cdot \left(V_{in} - V_{T0,n} \right) \cdot \left(\frac{dV_{out}}{dV_{in}} \right) + V_{out} - V_{out} \cdot \left(\frac{dV_{out}}{dV_{in}} \right) \right]$$
$$= k_p \cdot \left(V_{in} - V_{DD} - V_{T0,p} \right) \tag{5.81}$$

Substituting $V_{in} = V_{IH}$ and $(dV_{out}/dV_{in}) = -1$ in (5.81), we obtain

$$k_n \cdot \left(-V_{IH} + V_{T0,n} + 2\,V_{out} \right) = k_p \cdot \left(V_{IH} - V_{DD} - V_{T0,p} \right) \tag{5.82}$$

The critical voltage V_{IH} can now be found as a function of V_{out}, as follows.

$$V_{IH} = \frac{V_{DD} + V_{T0,p} + k_R \cdot \left(2 V_{out} + V_{T0,n} \right)}{1 + k_R} \tag{5.83}$$

Again, this equation must be solved simultaneously with the KCL equation (5.80) to obtain the numerical values of V_{IH} and V_{out}.

Calculation of V_{th}

The inverter threshold voltage is defined as $V_{th} = V_{in} = V_{out}$. Since the CMOS inverter exhibits large noise margins and a very sharp VTC transition, the inverter threshold voltage emerges as an important parameter characterizing the DC performance of the inverter. For $V_{in} = V_{out}$, both transistors are expected to be in saturation; hence, we can write the following KCL equation.

$$\frac{k_n}{2} \cdot \left(V_{GS,n} - V_{T0,n} \right)^2 = \frac{k_p}{2} \cdot \left(V_{GS,p} - V_{T0,p} \right)^2 \tag{5.84}$$

Replacing $V_{GS,n}$ and $V_{GS,p}$ in (5.84) according to (5.67) and (5.68), we obtain

$$\frac{k_n}{2} \cdot \left(V_{in} - V_{T0,n} \right)^2 = \frac{k_p}{2} \cdot \left(V_{in} - V_{DD} - V_{T0,p} \right)^2 \tag{5.85}$$

This equation can be solved for the input voltage V_{in}:

$$V_{in} \cdot \left(1 + \sqrt{\frac{k_p}{k_n}} \right) = V_{T0,n} + \sqrt{\frac{k_p}{k_n}} \cdot \left(V_{DD} + V_{T0,p} \right) \tag{5.86}$$

Finally, the inverter threshold (switching threshold) voltage V_{th} is found as

$$V_{th} = \frac{V_{T0,n} + \sqrt{\dfrac{1}{k_R}} \cdot \left(V_{DD} + V_{T0,p} \right)}{\left(1 + \sqrt{\dfrac{1}{k_R}} \right)} \tag{5.87}$$

Note that the inverter threshold voltage is defined as $V_{th} = V_{in} = V_{out}$. When the input voltage is equal to V_{th}, however, we find that the output voltage can actually attain any value between $(V_{th} - V_{T0,n})$ and $(V_{th} - V_{T0,p})$, without violating the voltage conditions used in this analysis. This is due to the fact that the VTC segment corresponding to Region C in Fig. 5.20 becomes completely vertical if the channel-length modulation effect is neglected, i.e., if $\lambda = 0$. In more realistic cases with $\lambda > 0$, the VTC segment in Region C exhibits a finite, but very large, slope.

It has already been established that the CMOS inverter does not draw any significant current from the power source, except for small leakage and subthreshold currents, when the input voltage is either smaller than $V_{T0,n}$ or larger than $(V_{DD} + V_{T0,p})$. The nMOS and the pMOS transistors conduct a nonzero current, on the other hand, during low-to-high and high-to-low transitions, i.e., in Regions B, C, and D. It can be shown that the current being drawn from the power source during transition reaches its peak value when $V_{in} = V_{th}$. In other words, the maximum current is drawn when both transistors are operating in saturation. Figure 5.25 shows the voltage transfer characteristic of a typical CMOS inverter circuit and the power supply current, as a function of the input voltage.

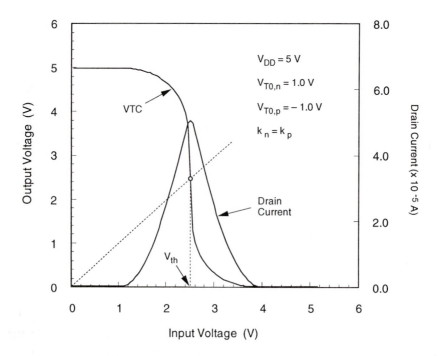

Figure 5.25   Typical VTC and the power supply current of a CMOS inverter circuit.

Design of CMOS Inverters

The inverter threshold voltage V_{th} was identified as one of the most important parameters that characterize the steady-state input-output behavior of the CMOS inverter circuit. The CMOS inverter can, by virtue of its complementary push-pull operating mode, provide a full output voltage swing between 0 and V_{DD}, and therefore, the noise margins are relatively wide. Thus, the problem of designing a CMOS inverter can be reduced to setting the inverter threshold to a desired voltage value.

Given the power supply voltage V_{DD}, the nMOS and the pMOS transistor threshold voltages, and the desired inverter threshold voltage V_{th}, the corresponding ratio k_R can be found as follows. Reorganizing (5.87) yields

$$\sqrt{\frac{1}{k_R}} = \frac{V_{th} - V_{T0,n}}{V_{DD} + V_{T0,p} - V_{th}} \tag{5.88}$$

Now solve for k_R that is required to achieve the given V_{th}:

$$k_R = \frac{k_n}{k_p} = \left(\frac{V_{DD} + V_{T0,p} - V_{th}}{V_{th} - V_{T0,n}} \right)^2 \tag{5.89}$$

Recall that the switching threshold voltage of an *ideal* inverter is defined as

$$V_{th,ideal} = \frac{1}{2} \cdot V_{DD} \tag{5.90}$$

Substituting (5.90) in (5.89) gives

$$\left(\frac{k_n}{k_p} \right)_{ideal} = \left(\frac{0.5 V_{DD} + V_{T0,p}}{0.5 V_{DD} - V_{T0,n}} \right)^2 \tag{5.91}$$

for a near-ideal CMOS VTC that satisfies the condition (5.90). Since the operations of the nMOS and the pMOS transistors of the CMOS inverter are fully complementary, we can achieve completely symmetric input-output characteristics by setting the threshold voltages as $V_{T0} = V_{T0,n} = |V_{T0,p}|$. This reduces (5.91) to:

$$\left(\frac{k_n}{k_p}\right)_{\substack{symmetric \\ inverter}} = 1 \tag{5.92}$$

Note that the ratio k_R is defined as

$$\frac{k_n}{k_p} = \frac{\mu_n\, C_{ox} \cdot \left(\dfrac{W}{L}\right)_n}{\mu_p\, C_{ox} \cdot \left(\dfrac{W}{L}\right)_p} = \frac{\mu_n \cdot \left(\dfrac{W}{L}\right)_n}{\mu_p \cdot \left(\dfrac{W}{L}\right)_p} \tag{5.93}$$

assuming that the gate oxide thickness t_{ox}, and hence, the gate oxide capacitance C_{ox} have the same value for both nMOS and pMOS transistors. The unity-ratio condition (5.92) for the ideal symmetric inverter requires that

$$\frac{\left(\dfrac{W}{L}\right)_n}{\left(\dfrac{W}{L}\right)_p} = \frac{\mu_p}{\mu_n} \approx \frac{230\ \text{cm}^2/\text{V·s}}{580\ \text{cm}^2/\text{V·s}} \tag{5.94}$$

Hence,

$$\left(\frac{W}{L}\right)_p \approx 2.5 \left(\frac{W}{L}\right)_n \tag{5.95}$$

It should be noted that the numerical values used in (5.94) for electron and hole mobilities are *typical* values, and that exact μ_n and μ_p values will vary with surface doping concentration of the substrate and the tub. The VTCs of three CMOS inverter circuits with different k_R ratios are shown in Fig. 5.26. It can be seen clearly that the inverter threshold voltage V_{th} shifts to lower values with increasing k_R ratio.

For a symmetric CMOS inverter with $V_{T0,n} = |V_{T0,p}|$ and $k_R = 1$, the critical voltage V_{IL} can be found, using (5.78), as follows.

$$V_{IL} = \frac{1}{8} \cdot \left(3\,V_{DD} - 2\,V_{T0,n}\right) \tag{5.96}$$

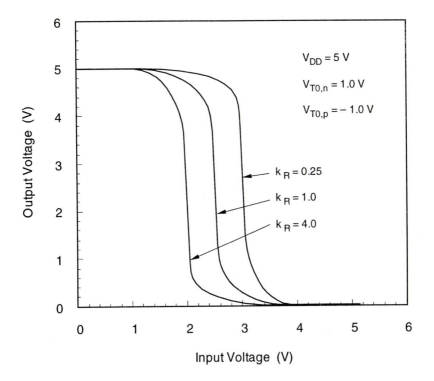

Figure 5.26.   Voltage transfer characteristics of three CMOS inverters, with different nMOS-to-pMOS ratios.

Also, the critical voltage V_{IH} is found as

$$V_{IH} = \frac{1}{8} \cdot \left(5\, V_{DD} - 2\, V_{T0,n}\right)$$

(5.97)

Note that the sum of V_{IL} and V_{IH} is always equal to V_{DD} in a symmetric inverter:

$$V_{IL} + V_{IH} = V_{DD}$$

(5.98)

The noise margins NM_L and NM_H for this symmetric CMOS inverter are now calculated using (5.3) and (5.4):

$$NM_L = V_{IL} - V_{OL} = V_{IL}$$
$$NM_H = V_{OH} - V_{IH} = V_{DD} - V_{IH}$$

$$(5.99)$$

which are equal to each other, and also to V_{IL}.

$$NM_L = NM_H = V_{IL}$$

$$(5.100)$$

Example 5.4

Consider a CMOS inverter circuit with the following parameters :

$$V_{DD} = 5 \text{ V}$$
$$V_{T0,n} = 1.0 \text{ V}$$
$$V_{T0,p} = -1.2 \text{ V}$$
$$k_n = 100 \text{ } \mu\text{A/V}^2$$
$$k_p = 40 \text{ } \mu\text{A/V}^2$$

Calculate the noise margins of the circuit. Notice that the CMOS inverter being considered here has $k_R = 2.5$ and $V_{T0,n} \neq |V_{T0,p}|$; hence, it is not a symmetric inverter.

First, the output low voltage V_{OL} and the output high voltage V_{OH} are found, using (5.70) and (5.71), as $V_{OL} = 0$ and $V_{OH} = 5$ V.

To calculate V_{IL} in terms of the output voltage, we use (5.78):

$$V_{IL} = \frac{2 V_{out} + V_{T0,p} - V_{DD} + k_R V_{T0,n}}{1 + k_R}$$

$$= \frac{2 V_{out} - 1.2 - 5 + 2.5}{1 + 2.5} = 0.57 V_{out} - 1.06$$

Now substitute this expression into the KCL equation (5.75).

$$2.5(0.57 V_{out} - 1.06 - 1)^2 = 2(0.57 V_{out} - 1.06 - 5 + 1.2)(V_{out} - 5) - (V_{out} - 5)^2$$

This expression yields a second-order polynomial in V_{out}, as follows.

$$0.66\,V_{out}^2 - 0.46\,V_{out} - 13 = 0$$

Only one root of this quadratic equation corresponds to a physically correct solution for V_{out} (i.e., $V_{out} > 0$).

$$V_{out} = 4.8 \text{ V}$$

From this value, we can calculate the critical voltage V_{IL} as:

$$V_{IL} = 0.57 \cdot 4.8 - 1.06 = \underline{\underline{1.68 \text{ V}}}$$

To calculate V_{IH} in terms of the output voltage, use (5.83):

$$V_{IH} = \frac{V_{DD} + V_{T0,p} + k_R \cdot (2\,V_{out} + V_{T0,n})}{1 + k_R}$$

$$= \frac{5 - 1.2 + 2.5(2\,V_{out} + 1)}{1 + 2.5} = 1.43\,V_{out} + 1.8$$

Next, substitute this expression into the KCL equation (5.80) to obtain a second-order polynomial in V_{out}.

$$2.5\left[2(1.43\,V_{out} + 1.8 - 1)\,V_{out} - V_{out}^2\right] = (1.43\,V_{out} - 2)^2$$

$$2.61\,V_{out}^2 + 9.72\,V_{out} - 4 = 0$$

Again, only one root of this quadratic equation corresponds to the physically correct solution for V_{out} at this operating point, i.e., when $V_{in} = V_{IH}$.

$$V_{out} = 0.37 \text{ V}$$

From this value, we can calculate the critical voltage V_{IH} as :

$$V_{IH} = 1.43 \cdot 0.37 + 1.8 = \underline{\underline{2.33 \text{ V}}}$$

Finally, we find the noise margins for low voltage levels and for high voltage levels using (5.3) and (5.4).

$$NM_L = V_{IL} - V_{OL} = 1.68 \text{ V}$$
$$NM_H = V_{OH} - V_{IH} = 2.66 \text{ V}$$

Power and Area Considerations

Since the CMOS inverter does not draw any significant current from the power source in both of its steady-state operating points ($V_{out} = V_{OH}$ and $V_{out} = V_{OL}$), the DC power dissipation of this circuit is almost negligible. The drain current that flows through the nMOS and the pMOS transistors in both cases is essentially limited to the reverse leakage current of the source and drain pn-junctions, and in short-channel MOSFETs, the relatively small subthreshold current. This unique property of the CMOS inverter was already identified as one of the most important advantages of this configuration. In many applications requiring a low overall power consumption, CMOS is preferred over other circuit alternatives for this reason. It must be noted, however, that the CMOS inverter does conduct a significant amount of current during a *switching event*, i.e., when the output voltage changes from a low to high state, or from a high to low state. The detailed calculation of this *dynamic power dissipation* will be examined in Chapter 6.

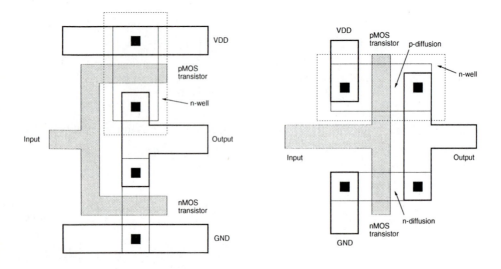

Figure 5.27.   Sample layouts of CMOS inverter circuits (for p-type substrate).

Figure 5.27 shows two layout examples for the simple CMOS inverter circuit. In both cases, it is assumed that the circuit is being built on a p-type wafer, which also provides the substrate for the nMOS transistor. The pMOS transistor, on the other hand, must be placed in an n-well (dotted lines), which becomes the substrate for this device. Also note that in Fig. 5.27 the channel width of the pMOS transistor is larger than that of the nMOS transistor. This is typical for symmetric inverter configurations, in which the k_R ratio is set approximately equal to unity.

Compared to other inverter layouts examined in previous sections, the CMOS inverters shown in Fig. 5.27 do not occupy significantly more area. The added complexity of the fabrication process (creating n-well diffusion, separate p-type and n-type source and drain diffusions, etc.) appears to be the only drawback for the inverter example. Because of the complementary nature of this circuit configuration, however, CMOS random logic circuits require significantly more transistors for the same function than their nMOS counterparts. Consequently, CMOS logic circuits tend to occupy more area than comparable nMOS logic circuits, which apparently affects the integration density of pure-CMOS logic. The actual integration density of nMOS logic, on the other hand, is limited by power dissipation and heat generation problems.

References

1.   N.H.E. Weste and K. Eshraghian, *Principles of CMOS VLSI Design– A Systems Perspective*, second edition, Reading, MA: Addison-Wesley, 1993.

2.   J.P. Uyemura, *Fundamentals of MOS Digital Integrated Circuits*, Reading, MA: Addison-Wesley, 1988.

3.   L.A. Glasser and D.W. Dobberpuhl, *The Design and Analysis of VLSI Circuits*, Reading, MA: Addison-Wesley, 1985.

4.   M. Annaratone, *Digital CMOS Circuit Design*, Norwell, MA: Kluwer, 1986.

5.   H. Haznedar, *Digital Microelectronics*, Redwood City, CA: Benjamin/ Cummings, 1991.

6.   M. Shoji, *CMOS Digital Circuit Technology*, Englewood Cliffs, NJ: Prentice Hall, 1988.

Exercise Problems

5.1   Consider an inverter with an enhancement-type nMOS driver transistor and an enhancement-type saturated nMOS load transistor with its gate and drain terminals tied. For the following parameters, calculate the noise margin, NM_L, i.e., the noise margin corresponding to the "0" level signal.

$$k_{driver}/k_{load} = 10.0$$
$$V_{T0} = 0.8 \text{ V}, \ V_{DD} = 5.0 \text{ V}$$
$$\gamma = \lambda = 0$$

5.2   For the saturated enhancement-load inverter in exercise problem (5.1), determine the highest output voltage V_{OH} when the input gate voltage is less than the threshold voltage by using the following device parameters:

$$\mu_n \cdot C_{ox} = 20 \ \mu A/V^2$$
$$V_{T0} = 1.0 \text{ V}, \quad \gamma = 0.4 \text{ V}^{1/2}$$
channel length modulation factor $(\lambda) = 0.0 \text{ V}^{-1}$
surface potential $|2\phi_F| = 0.64 \text{ V}$

5.3   Determine the steady-state voltage across the load capacitor shown in Fig. P5.3 below. You must include the body effect of both transistors. Use the following device parameters:

$$V_{T0} = 1.0 \text{ V}$$
$$\gamma \text{ (body effect coefficient)} = 0.2 \text{ V}^{1/2}$$
$$\phi_F = -0.3 \text{ V}$$

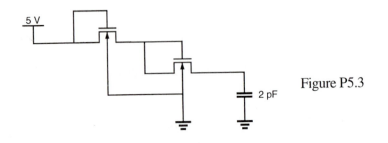

Figure P5.3

5.4   Consider a resistive-load inverter circuit with a 100 kΩ load resistor, where the input terminal (gate terminal of the nMOS transistor) is connected to its output (drain terminal of the nMOS transistor).

(a)   Sketch the load line for this circuit.

(b)   Is the transistor operating in the linear or saturation region?

(c)   For $V_T = 1$ V, $k' = 40$ μA/V^2 and $V_{OUT} = 2.5$ V, solve for W/L.

5.5   Refer to the CMOS fabrication process described in Chapter 2. Draw cross-sections of the following device along the lines A-A' and B-B'.

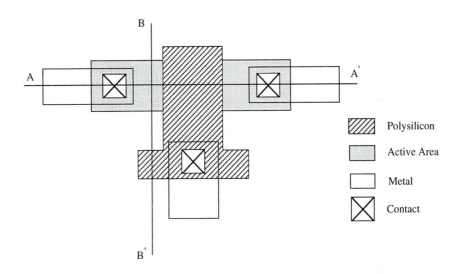

5.6   Design a resistive-load inverter with $R = 1$ kΩ such that $V_{OL} = 0.6$ V when the enhancement-type nMOS driver transistor has the following parameters:

$$V_{DD} = 5.0 \text{ V}$$
$$V_{T0} = 1.0 \text{ V}$$
$$\gamma = 0.2 \text{ V}^{1/2}$$
$$\lambda = 0.0 \text{ V}^{-1}$$
$$k' = 22.0 \text{ μA/V}^2$$

(a)   Determine the required aspect ratio, W/L.

(b)   Determine V_{IL} and V_{IH}.

(c)   Determine noise margins NM_L and NM_H.

5.7   Layout of the resistive-load inverter design.

(a)   Draw the layout of the R-load inverter designed in exercise problem (5.6) using a polysilicon resistor with sheet resistivity of 25 Ω/square and the minimum feature size of 2 µm. It should be noted that L stands for the effective channel length which is related to the mask channel length as $L = L_M + \delta - 2L_D$, where we assume δ (process error) $= 0$ and $L_D = 0.25$ µm. To save chip area, one should use the minimum sizes for L and W. Also, the resistor area can be reduced by using the folded layout (snake pattern) of the resistor.

(b)   Perform circuit extraction to obtain the SPICE input list from the layout.

(c)   Run the SPICE simulation of the circuit to obtain the DC voltage transfer characteristic (VTC) curve. Plot the VTC and check whether the calculated values in Problem 5.6 match with the SPICE results.

5.8   The nMOS and pMOS transistors in a CMOS inverter have the following parameters:

$$|V_{T0}| = 1.0 \text{ V for both nMOS and pMOS transistors}$$
$$\lambda = 0.0 \text{ V}^{-1}$$
$$k'_n = 50.0 \text{ µA/V}^2$$
$$k'_p = 20.0 \text{ µA/V}^2$$

The CMOS inverter is designed with $(W/L)_p = 20$ and $(W/L)_n = 10$, and its capacitive load is 2 pF. It is assumed that the load capacitance includes all parasitic capacitances connected to the drain node.

(a)   Calculate the average power dissipation in the inverter when its input signal is a rectangular pulse with 100 ns period which swings between 5 V and 0 V.

(b)   Repeat part (a) for the case when each transistor channel width is exactly twice as that used in (a), but the load capacitance and input signal remain the same.

(c)   What observations can you make from the results of part (b)? Can you explain the results?

(d)   Verify your answer using the power meter with SPICE simulation. For SPICE simulation, you can set the parasitic capacitances in drain and source regions to zero by setting AD, AS, PD, PS to zero.

CHAPTER 6

MOS INVERTERS: DYNAMIC CHARACTERISTICS

In this chapter, we will investigate the dynamic (time-domain) behavior of the inverter circuits. The switching characteristics of digital integrated circuits and, in particular, of inverter circuits, essentially determine the overall operating speed of digital systems. As already seen in the design example presented in Chapter 1, the dynamic performance requirements of a digital system are usually among the most important design specifications that must be met by the circuit designer. Therefore, the switching speed of the circuit must be estimated and optimized very early in the design phase.

The closed-form delay expressions will be derived under the assumption of pulse excitation and using lumped capacitances. While exact circuit simulation (SPICE) usually provides the most accurate estimation of the time-domain behavior of complex circuits, the delay expressions presented here can also be used in many cases to provide a fairly quick and accurate approximation of the switching characteristics. In the following, specific emphasis will be given to CMOS inverter circuit dynamic behavior.

6.1. Introduction

Consider the cascade connection of two CMOS inverter circuits shown in Fig. 6.1. The parasitic capacitances associated with each MOSFET are illustrated individually. Here, the capacitances C_{gd} and C_{gs} are primarily due to gate overlap, while C_{db} and C_{sb} are voltage-dependent junction capacitances, as discussed in Chapter 3. The capacitance component C_{gb} is due to the thin-oxide capacitance over the gate area. In addition, we also consider the lumped interconnect capacitance C_{int}, which represents the parasitic capacitance contribution of the metal or polysilicon connection between the two inverters. It is assumed that a pulse waveform is applied to the input of the first-stage inverter. We wish to derive the time-domain behavior of the first-stage output, V_{out}.

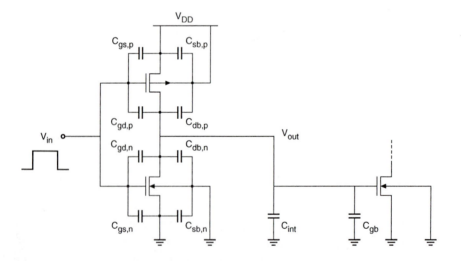

Figure 6.1.   Cascaded CMOS inverter stages.

Obviously, the problem of calculating the output voltage waveform is fairly complicated, even for this relatively simple circuit, because a number of nonlinear, voltage-dependent capacitances are involved. To simplify the problem, we first combine the capacitances seen in Fig. 6.1 into one lumped linear capacitance, connected between the output node of the inverter and the ground. This single capacitance at the output node will be called the load capacitance, C_{load}.

$$C_{load} = C_{gd,n} + C_{gd,p} + C_{db,n} + C_{db,p} + C_{int} + C_{gb} \qquad (6.1)$$

Note that some of the parasitic capacitance components shown in Fig. 6.1 do not appear in this lumped capacitance expression. In particular, $C_{sb,n}$ and $C_{sb,p}$ have no effect on the dynamic behavior of the circuit since the source-to-substrate voltages of both transistors are always equal to zero. The capacitances $C_{gs,n}$ and $C_{gs,p}$ are also not included in (6.1) because they are connected between the input node and the ground (or the power supply). The capacitance terms $C_{db,n}$ and $C_{db,p}$ in (6.1) are the equivalent junction capacitances calculated for a particular output voltage transition, according to (3.109) and (3.114). The reader is referred to Chapter 3 for details concerning the calculation of parasitic junction capacitances.

The first-stage CMOS inverter is shown with the single lumped output load capacitance C_{load} in Fig. 6.2. Now, the problem of calculating the dynamic response can be attacked more easily. In fact, the question of inverter dynamic response is reduced to finding the charge-up and charge-down times of a single capacitance which is charged and discharged through one transistor. The delay times calculated using C_{load} may slightly overestimate the actual inverter delay, but this is not considered a significant deficiency in a first-order approximation.

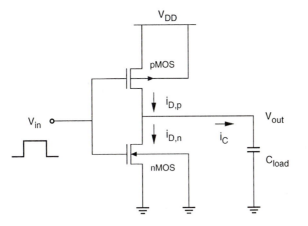

Figure 6.2.   First-stage CMOS inverter with lumped output load capacitance.

6.2. Delay-Time Definitions

Before we begin the derivation of delay expressions, we will briefly present some delay time definitions. The input and output voltage waveforms of a typical inverter circuit are shown in Fig. 6.3. The propagation delay times τ_{PHL} and τ_{PLH} determine the input-to-output signal delay during the high-to-low and low-to-high transitions of the output, respectively. By definition, τ_{PHL} is the time required for the output voltage to fall from V_{OH} to $V_{50\%}$, assuming a rising pulse input waveform with zero rise time. Similarly, the propagation delay τ_{PLH} is defined as the time required for the output voltage to rise from V_{OL} to $V_{50\%}$, assuming a falling pulse input. The voltage point $V_{50\%}$ is defined as follows.

$$V_{50\%} = V_{OL} + \frac{1}{2}\left(V_{OH} - V_{OL}\right) = \frac{1}{2}\left(V_{OL} + V_{OH}\right) \qquad (6.2)$$

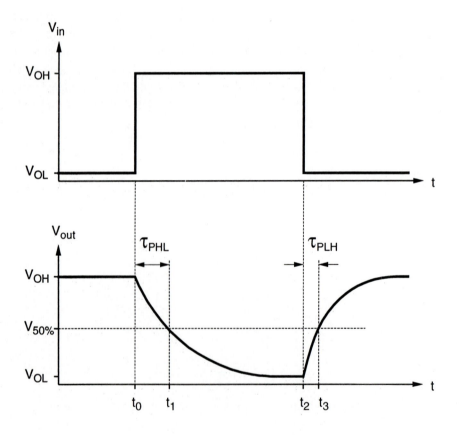

Figure 6.3.   Input and output voltage waveforms of a typical inverter.

Thus, the propagation delay times τ_{PHL} and τ_{PLH} are found from Fig. 6.3 as

$$\tau_{PHL} = t_1 - t_0$$
$$\tau_{PLH} = t_3 - t_2 \tag{6.3}$$

The average propagation delay τ_p of the inverter characterizes the average time required for the input signal to propagate through the inverter.

$$\tau_P = \frac{\tau_{PHL} + \tau_{PLH}}{2} \tag{6.4}$$

We will refer to Fig. 6.4 for the definition of output voltage rise and fall times. The rise time τ_{rise} is defined as the time required for the output voltage to rise from the $V_{10\%}$ level to $V_{90\%}$ level. Similarly, the fall time τ_{fall} is defined as the time required for the output voltage to drop from the $V_{90\%}$ level to $V_{10\%}$ level. The voltage levels $V_{10\%}$ and $V_{90\%}$ are defined as

$$V_{10\%} = V_{OL} + 0.1 \cdot (V_{OH} - V_{OL}) \tag{6.5}$$

$$V_{90\%} = V_{OL} + 0.9 \cdot (V_{OH} - V_{OL}) \tag{6.6}$$

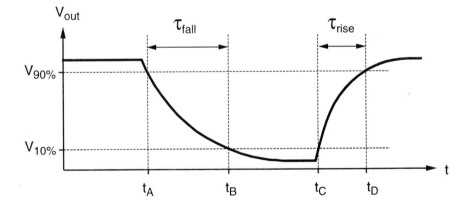

Figure 6.4.   Output voltage rise and fall times.

Thus, the output rise and fall times are found from Fig. 6.4 as follows.

$$\tau_{fall} = t_B - t_A$$
$$\tau_{rise} = t_D - t_C \tag{6.7}$$

6.3. Calculation of Delay Times

The simplest approach for calculating the propagation delay times τ_{PHL} and τ_{PLH} is based on estimating the average capacitance current during charge down and charge up, respectively. If the capacitance current during an output transition is approximated by a constant average current I_{avg}, the delay times are found as

$$\tau_{PHL} = \frac{C_{load} \cdot \Delta V_{HL}}{I_{avg,HL}} = \frac{C_{load} \cdot (V_{OH} - V_{50\%})}{I_{avg,HL}} \tag{6.8}$$

$$\tau_{PLH} = \frac{C_{load} \cdot \Delta V_{LH}}{I_{avg,LH}} = \frac{C_{load} \cdot (V_{50\%} - V_{OL})}{I_{avg,LH}} \tag{6.9}$$

Note that the average current during high-to-low transition can be calculated by using the current values at the beginning and at the end of the transition.

$$I_{avg,HL} = \frac{1}{2}\left[i_C\left(V_{in} = V_{OH}, V_{out} = V_{OH}\right) + i_C\left(V_{in} = V_{OH}, V_{out} = V_{50\%}\right)\right] \tag{6.10}$$

Similarly, the average capacitance current during low-to-high transition is

$$I_{avg,LH} = \frac{1}{2}\left[i_C\left(V_{in} = V_{OL}, V_{out} = V_{50\%}\right) + i_C\left(V_{in} = V_{OL}, V_{out} = V_{OL}\right)\right] \tag{6.11}$$

While the average-current method is relatively simple and requires minimal calculation, it neglects the variations of the capacitance current between the beginning and end points of the transition. Therefore, we do not expect the average-current method to provide a very accurate estimate of the delay times. Still, this approach can provide rough, first-order estimates of the charge-up and charge-down times.

The propagation delay times can be found more accurately by solving the state equation of the output node in the time domain. The differential equation associated with the output node is given below. Note that the capacitance current is also a function of the output voltage.

$$C_{load} \frac{dV_{out}}{dt} = i_C = i_{D,p} - i_{D,n} \tag{6.12}$$

First, we consider the rising-input case for a CMOS inverter. Initially, the output voltage is assumed to be equal to V_{OH}. When the input voltage switches from low (V_{OL}) to high (V_{OH}), the nMOS transistor is turned on and it starts to discharge the load capacitance. At the same time, the pMOS transistor is switched off; thus,

$$i_{D,p} \approx 0 \tag{6.13}$$

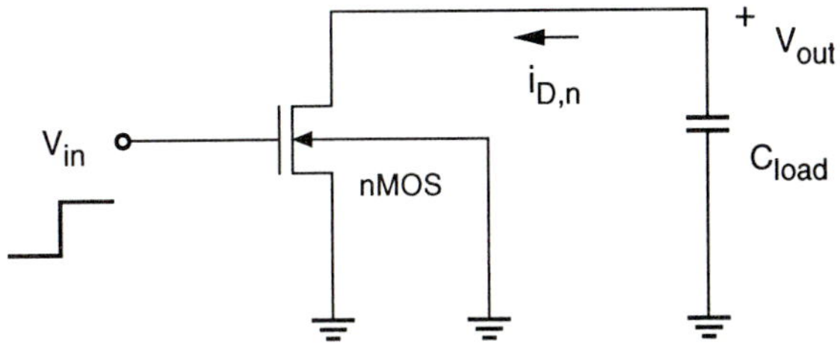

Figure 6.5. Equivalent circuit of the CMOS inverter during high-to-low transition.

The circuit given in Fig. 6.2 can now be reduced to a single nMOS transistor and a capacitor, as shown in Fig. 6.5. The differential equation describing the discharge event is then

$$C_{load} \frac{dV_{out}}{dt} = -i_{D,n} \tag{6.14}$$

Note that in other types of inverter circuits, such as the resistive-load inverter or the depletion-load inverter, the load device continues to conduct a nonzero current when the input is switched from low to high. However, the load current is usually negligible in comparison to the driver current. Therefore, (6.14) can be used to calculate the charge-down time not only in CMOS inverters, but also in almost all common types of inverter circuits.

The input and output voltage waveforms during this high-to-low transition are illustrated in Fig. 6.6. When the nMOS transistor starts conducting, it initially operates in the saturation region. Eventually, the output voltage falls below ($V_{DD} - V_{T,n}$), so that the nMOS transistor continues to conduct in the linear region. These two operating regions are also shown in Fig. 6.6. First, consider the nMOS transistor operating in saturation.

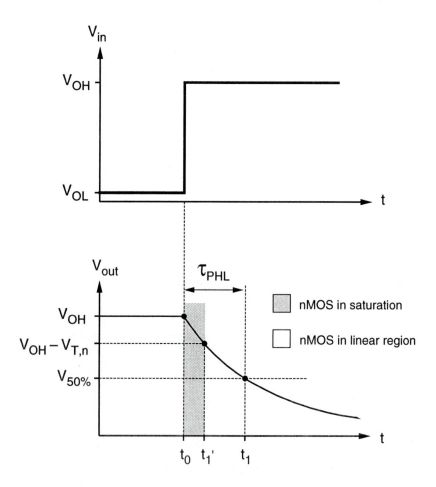

Figure 6.6.   Input and output voltage waveforms during high-to-low transition.

$$i_{D,n} = \frac{k_n}{2}\left(V_{in} - V_{T,n}\right)^2$$

$$= \frac{k_n}{2}\left(V_{OH} - V_{T,n}\right)^2, \quad for \quad V_{OH} - V_{T,n} < V_{out} \le V_{OH} \tag{6.15}$$

Since the saturation current is practically independent of the output voltage (neglecting channel-length modulation), the solution of (6.14) in the time interval between t_0 and

t_1' can be found as

$$\int_{t=0}^{t=t_1'} dt = -C_{load} \int_{V_{out}=V_{OH}}^{V_{out}=V_{OH}-V_{T,n}} \left(\frac{1}{i_{D,n}}\right) dV_{out}$$

$$= -\frac{2 C_{load}}{k_n \left(V_{OH}-V_{T,n}\right)^2} \int_{V_{out}=V_{OH}}^{V_{out}=V_{OH}-V_{T,n}} dV_{out} \qquad (6.16)$$

Evaluating this integral yields

$$t_1' = \frac{2 C_{load} V_{T,n}}{k_n \left(V_{OH}-V_{T,n}\right)^2} \qquad (6.17)$$

At $t = t_1'$, the output voltage will be equal to $(V_{DD} - V_{T,n})$ and the transistor will be at the saturation-linear region boundary. Next, consider the nMOS transistor operating in the linear region.

$$i_{D,n} = \frac{k_n}{2}\left[2\left(V_{in}-V_{T,n}\right)V_{out}-V_{out}^2\right]$$

$$= \frac{k_n}{2}\left[2\left(V_{OH}-V_{T,n}\right)V_{out}-V_{out}^2\right], \qquad for \qquad V_{out} \le V_{OH}-V_{T,n} \qquad (6.18)$$

The solution of (6.14) in the time interval between t_1' and t_1 can be found as

$$\int_{t=t_1'}^{t=t_1} dt = -C_{load} \int_{V_{out}=V_{OH}-V_{T,n}}^{V_{out}=V_{50\%}} \left(\frac{1}{i_{D,n}}\right) dV_{out}$$

$$= -2 C_{load} \int_{V_{out}=V_{OH}-V_{T,n}}^{V_{out}=V_{50\%}} \left(\frac{1}{k_n\left[2\left(V_{OH}-V_{T,n}\right)V_{out}-V_{out}^2\right]}\right) dV_{out} \qquad (6.19)$$

Evaluating this integral yields

$$t_1 - t_1' = -\frac{2C_{load}}{k_n} \frac{1}{2(V_{OH} - V_{T,n})} \ln\left(\frac{V_{out}}{2(V_{OH} - V_{T,n}) - V_{out}}\right)\Bigg|^{V_{out}=V_{50\%}}_{V_{out}=V_{OH}-V_{T,n}} \qquad (6.20)$$

$$t_1 - t_1' = \frac{C_{load}}{k_n(V_{OH} - V_{T,n})} \ln\left(\frac{2(V_{OH} - V_{T,n}) - V_{50\%}}{V_{50\%}}\right) \qquad (6.21)$$

Finally, the propagation delay time for high-to-low output transition (τ_{PHL}) can be found by combining (6.17) and (6.21):

$$\tau_{PHL} = \frac{C_{load}}{k_n(V_{OH} - V_{T,n})}\left[\frac{2V_{T,n}}{V_{OH} - V_{T,n}} + \ln\left(\frac{4(V_{OH} - V_{T,n})}{V_{OH} + V_{OL}} - 1\right)\right] \qquad (6.22a)$$

For $V_{OH} = V_{DD}$ and $V_{OL} = 0$, as is the case for the CMOS inverter, (6.22a) becomes:

$$\tau_{PHL} = \frac{C_{load}}{k_n(V_{DD} - V_{T,n})}\left[\frac{2V_{T,n}}{V_{DD} - V_{T,n}} + \ln\left(\frac{4(V_{DD} - V_{T,n})}{V_{DD}} - 1\right)\right] \qquad (6.22b)$$

Example 6.1.

Consider the CMOS inverter circuit shown in Fig. 6.2. The I-V characteristics of the nMOS transistor are specified as follows: when $V_{GS} = 5$ V, the drain current reaches its saturation level $I_{sat} = 5$ mA for $V_{DS} \geq 4$ V. Assume that the input signal applied to the gate is a step pulse that switches instantaneously from 0 V to 5 V. Using the data above, calculate the delay time necessary for the output to fall from its initial value of 5 V to 2.5 V, assuming an output load capacitance of 1.0 pF.

For the solution, consider the simplified pull-down circuit shown in Fig. 6.5. We will assume that the nMOS transistor operates in saturation from $t = 0$ to $t = t_1' = t_{sat}$, and that it will operate in the linear region from $t = t_1' = t_{sat}$ to $t = t_2 = t_{delay}$. We can also deduce from the I-V characteristics that $V_{T,n} = 1.0$ V, since the nMOS transistor enters saturation when $V_{DS} \geq V_{GS} - V_{T,n}$. The voltage V_{GS} is equal to 5 V for $t \geq 0$.

The current equation for the saturation region can be written as

$$C\frac{dV_{out}}{dt} = -I_D = -I_{sat} = -\frac{1}{2}k_n(V_{OH} - V_{T,n})^2$$

We can calculate the amount of time in which the nMOS transistor operates in saturation (t_{sat}), by integrating this equation.

$$\int_{t=0}^{t=t_{sat}} dt = -\int_{V_{out}=5}^{V_{out}=4} \frac{C}{I_D} dV_{out}$$

$$t_{sat} = \frac{V_{T,n}C}{I_{sat}} = \frac{1 \text{ V} \cdot 1 \text{ pF}}{5 \text{ mA}} = 0.2 \text{ [ns]}$$

The transconductance k_n of the nMOS transistor can be found as follows:

$$k_n = \frac{2I_{sat}}{(V_{OH} - V_{T,n})^2} = \frac{2 \cdot 5 \times 10^{-3}}{4^2} = 0.625 \times 10^{-3} \text{ [A/V}^2]$$

Now, the current equation for the linear operating region is written as

$$C\frac{dV_{out}}{dt} = -I_D = -\frac{1}{2}k_n\left[2(V_{OH} - V_{T,n})V_{out} - V_{out}^2\right]$$

Integrating this differential equation between the two voltage boundary conditions yields the amount of time in which the nMOS transistor operates in the linear region.

$$\int_{t=t_{sat}}^{t=t_{delay}} dt = -2C \int_{V_{out}=4}^{V_{out}=2.5} \frac{dV_{out}}{k_n\left[2(V_{OH} - V_{T,n})V_{out} - V_{out}^2\right]}$$

$$t_{delay} - t_{sat} = -\frac{C}{k_n}\frac{1}{(V_{OH} - V_{T,n})}\ln\left(\frac{V_{out}}{2(V_{OH} - V_{T,n}) - V_{out}}\right)\Bigg|_{V_{out}=4}^{V_{out}=2.5}$$

$$t_{delay} - t_{sat} = \frac{C}{k_n \left(V_{OH} - V_{T,n} \right)} \frac{1}{} \ln\left(\frac{2\left(V_{OH} - V_{T,n} \right) - V_{2.5}}{V_{2.5}} \right)$$

$$= \frac{1 \times 10^{-12}}{0.625 \times 10^{-3}} \cdot \frac{1}{4} \ln\left(\frac{8 - 2.5}{2.5} \right) = 1.26 \ [ns]$$

Thus, the total delay time is found to be

$$t_{delay} = 1.26 + 0.2 = 1.46 \ [ns]$$

Note that t_{delay} corresponds to the propagation delay time τ_{PHL} for falling output.

Example 6.2.

For the CMOS inverter shown in Fig. 6.2, determine the fall time τ_{fall}, which is defined as the time elapsed between the time point at which $V_{out} = 4.5$ V and the time point at which $V_{out} = 0.5$ V. Use both the average-current method and the differential-equation method for calculating τ_{fall}. The output load capacitance is 1 pF. The nMOS transistor parameters are given as

$$\mu_n C_{ox} = 20 \ \mu A/V^2$$
$$(W/L)_n = 10$$
$$V_{T,n} = 1.0 \ V$$

Using a simple expression similar to (6.10), we can determine the average capacitor current during the charge-down event described above.

$$I_{avg} = \frac{1}{2}\left[I(V_{in} = 5 \ V, V_{out} = 4.5 \ V) + I(V_{in} = 5 \ V, V_{out} = 0.5 \ V) \right]$$

$$= \frac{1}{2}\left[\frac{1}{2}k_n\left(V_{in} - V_{T,n} \right)^2 + \frac{1}{2}k_n\left(2(V_{in} - V_{T,n})V_{out} - V_{out}^2 \right) \right]$$

$$= \frac{1}{2} \cdot \frac{1}{2} \cdot 20 \times 10^{-6} \cdot 10\left[(5-1)^2 - \left(2(5-1)0.5 - 0.5^2 \right) \right] = 0.9875 \ [mA]$$

The fall time is then found as

$$\tau_{fall} = \frac{C \cdot \Delta V}{I_{avg}} = \frac{1 \times 10^{-12}(4.5-0.5)}{0.9875 \times 10^{-3}} = 4.05 \times 10^{-9} \; [\text{s}] = 4.05 \; [\text{ns}]$$

Now, we will recalculate the fall time using the differential equation approach. It is seen that the nMOS transistor operates in the saturation region for $4.0 \; \text{V} \le V_{out} \le 4.5 \; \text{V}$. Writing the current equation for the saturation region, we obtain

$$C\frac{dV_{out}}{dt} = -\frac{1}{2}k_n(V_{in}-V_{T,n})^2, \; \text{where} \;\; k_n = \mu_n C_{ox}\left(\frac{W}{L}\right)_n$$

$$\frac{dV_{out}}{dt} = \frac{-20 \times 10^{-6} \cdot 10 \cdot (5-1)^2}{2 \cdot 1 \times 10^{-12}} = -1.6 \times 10^9$$

Integrating this simple expression yields the time during which the nMOS transistor operates in saturation.

$$\int_{t=0}^{t=t_{sat}} dt = -\frac{1}{1.6 \times 10^9} \int_{V_{out}=4.5}^{V_{out}=4} dV_{out}$$

$$t_{sat} = \frac{0.5}{1.6 \times 10^9} = 0.3125 \times 10^{-9} \; [\text{s}] = 0.3125 \; [\text{ns}]$$

The nMOS transistor operates in the linear region for $0.5 \; \text{V} \le V_{out} \le 4.0 \; \text{V}$. The current equation for this operating region is written as follows:

$$C\frac{dV_{out}}{dt} = -\frac{1}{2}k_n\left[2(V_{in}-V_{T,n})V_{out}-V_{out}^2\right]$$

Integrating this equation, we obtain the time during which the nMOS transistor operates in the linear region.

$$\int_{t=t_{sat}}^{t=t_{delay}} dt = -2C \int_{V_{out}=4}^{V_{out}=0.5} \frac{dV_{out}}{k_n\left[2(V_{in}-V_{T,n})V_{out}-V_{out}^2\right]}$$

$$\tau_{fall} - t_{sat} = \frac{C}{k_n} \frac{1}{\left(V_{in} - V_{T,n}\right)} \ln\left(\frac{2\left(V_{in} - V_{T,n}\right) - V_{0.5}}{V_{0.5}}\right)$$

$$= \frac{1 \times 10^{-12}}{20 \times 10^{-6} \cdot 10 \cdot 4} \ln\frac{8 - 0.5}{0.5} = 3.385 \times 10^{-9} \, [\text{s}] = 3.385 \, [\text{ns}]$$

Thus, the fall time of the CMOS inverter is found as follows:

$$\tau_{fall} = 3.6975 \, [\text{ns}]$$

In a CMOS inverter, the charge-up event of the output load capacitance for falling input transition is completely analogous to the charge-down event for rising input. When the input voltage switches from high (V_{OH}) to low (V_{OL}), the nMOS transistor is cut off, and the load capacitance is being charged up through the pMOS transistor. Following a very similar derivation procedure, the propagation delay time τ_{PLH} can be found as

$$\tau_{PLH} = \frac{C_{load}}{k_p\left(V_{OH} - V_{OL} - |V_{T,p}|\right)}\left[\frac{2|V_{T,p}|}{V_{OH} - V_{OL} - |V_{T,p}|} +\right.$$
$$\left. \ln\left(\frac{2\left(V_{OH} - V_{OL} - |V_{T,p}|\right)}{V_{OH} - V_{50\%}} - 1\right)\right] \qquad (6.23a)$$

For $V_{OH} = V_{DD}$ and $V_{OL} = 0$, (6.23) becomes

$$\tau_{PLH} = \frac{C_{load}}{k_p\left(V_{DD} - |V_{T,p}|\right)}\left[\frac{2|V_{T,p}|}{V_{DD} - |V_{T,p}|} + \ln\left(\frac{4\left(V_{DD} - |V_{T,p}|\right)}{V_{DD}} - 1\right)\right] \qquad (6.23b)$$

It can be noted that the necessary conditions for $\tau_{PHL} = \tau_{PLH}$ in a CMOS inverter are $V_{T,n} = |V_{T,p}|$ and $k_n = k_p$ (or $W_p/W_n = \mu_n/\mu_p$).

The calculation of τ_{PLH} in different types of inverters obviously depends on the load device and its operation. We will consider the nMOS depletion-load inverter case as an example. When the input voltage switches from high to low, the enhancement-type nMOS driver transistor is turned off. The output load capacitance is then being charged up through the depletion-type load transistor. The differential equation describing this event is

$$C_{load} \frac{dV_{out}}{dt} = i_{D,load}(V_{out}) \qquad (6.24)$$

Note that the load device is initially in saturation, and that it enters the linear region when the output voltage rises above $(V_{DD} + V_{T,load})$.

$$i_{D,load} = \frac{k_{n,load}}{2} \left(|V_{T,load}| \right)^2, \quad for \quad V_{out} \le V_{DD} - |V_{T,load}| \qquad (6.25)$$

$$i_{D,load} = \frac{k_{n,load}}{2} \left[2|V_{T,load}|(V_{DD} - V_{out}) - (V_{DD} - V_{out})^2 \right]$$
$$for \quad V_{out} > V_{DD} - |V_{T,load}| \qquad (6.26)$$

The delay time τ_{PLH} can be found as follows:

$$\tau_{PLH} = C_{load} \left[\int_{V_{out}=V_{OL}}^{V_{out}=V_{DD}-|V_{T,load}|} \left(\frac{dV_{out}}{i_{D,load}(sat)} \right) \right.$$
$$\left. + \int_{V_{out}=V_{DD}-|V_{T,load}|}^{V_{out}=V_{50\%}} \left(\frac{dV_{out}}{i_{D,load}(linear)} \right) \right] \qquad (6.27)$$

$$\tau_{PLH} = \frac{C_{load}}{k_{n,load}|V_{T,load}|} \left[\frac{2(V_{DD}-|V_{T,load}|-V_{OL})}{|V_{T,load}|} \right.$$
$$\left. + \ln \left(\frac{2|V_{T,load}|-(V_{DD}-V_{50\%})}{V_{DD}-V_{50\%}} \right) \right] \qquad (6.28)$$

Finally, we consider the case where the input voltage waveform is not an ideal (step) pulse waveform, but has finite rise and fall times, τ_r and τ_f. The exact calculation of the output voltage delay times will be obviously more complicated under this more realistic assumption. However, we can utilize the propagation delay times calculated using step input to estimate the actual propagation delays, using the following empirical expressions:

$$\tau_{PHL}(actual) = \sqrt{\tau_{PHL}^2(step\,input) + \left(\frac{\tau_r}{2}\right)^2} \qquad (6.29)$$

$$\tau_{PLH}(actual) = \sqrt{\tau_{PLH}^2(step\,input) + \left(\frac{\tau_f}{2}\right)^2} \qquad (6.30)$$

Example 6.3.

A company in Urbana called Prairie Technology has a CMOS fabrication process with the device parameters listed below.

$$\mu_n\,C_{ox} = 30\ \mu A/V^2$$
$$\mu_p\,C_{ox} = 10\ \mu A/V^2$$
$L = 1\ \mu m$ for both nMOS and pMOS devices
$$V_{T0,n} = 1.0\ V$$
$$V_{T0,p} = -1.5\ V$$
$$W_{min} = 2\ \mu m$$

Design a CMOS inverter circuit by determining the channel widths W_n and W_p of the nMOS and pMOS transistor, to meet the following performance specifications.

- $V_{th} = 2\ V$ for $V_{DD} = 5\ V$
- Delay time of 2 ns for an output transition from 4 V to 1 V, with a load capacitance of 1 pF and ideal step input

We start our design by satisfying the time delay constraint. During the output transition described above, the nMOS transistor of the CMOS inverter will operate exclusively in the linear region. The current equation for this region is

$$C \frac{dV_{out}}{dt} = -\frac{1}{2} \mu_n C_{ox} \frac{W_n}{L_n} \left[2(V_{OH} - V_{T0,n}) V_{out} - V_{out}^2 \right]$$

By integrating this expression, we obtain the following relationship.

$$t_{delay} = 2 \times 10^{-9} = -2C \int_{V_{out}=4}^{V_{out}=1} \frac{dV_{out}}{\mu_n C_{ox} \frac{W_n}{L_n} \left[2(V_{OH} - V_{T0,n}) V_{out} - V_{out}^2 \right]}$$

$$t_{delay} = \frac{C}{\mu_n C_{ox} \left(\frac{W_n}{L_n} \right)} \cdot \frac{1}{(V_{OH} - V_{T0,n})} \ln \left(\frac{2(V_{OH} - V_{T0,n}) - V_{1.0}}{V_{1.0}} \right)$$

$$2 \times 10^{-9} = \frac{1 \times 10^{-12}}{30 \times 10^{-6} \left(\frac{W_n}{L_n} \right) 4} \ln \left(\frac{8-1}{1} \right)$$

Now we solve this equation for the nMOS transistor (W/L) ratio:

$$\frac{W_n}{L_n} = 8.1$$

Thus, we can determine the gate dimensions of the nMOS transistor as $L_n = 1$ μm and $W_n = 8.1$ μm. Next, the logic threshold constraint of $V_{th} = 2$ V will help determine the pMOS transistor dimensions. We have seen in Chapter 5 that the logic threshold voltage of a CMOS inverter is found as

$$V_{th} = \frac{V_{T0,n} + \sqrt{\frac{1}{k_R}} (V_{DD} + V_{T0,p})}{1 + \sqrt{\frac{1}{k_R}}} = 2$$

where the ratio k_R is defined as

$$k_R = \frac{\mu_n C_{ox} \dfrac{W_n}{L_n}}{\mu_p C_{ox} \dfrac{W_p}{L_p}} = \frac{\mu_n W_n}{\mu_p W_p} = \frac{9}{4}$$

The channel width W_p of the pMOS transistor can be determined from this equation. Note that both the nMOS and the pMOS transistors are assumed to have the same channel length of $L_n = L_p = 1\ \mu m$.

$$W_p = \frac{1}{k_R} \frac{\mu_n W_n}{\mu_p} = \frac{4}{9} \cdot \frac{3}{3} \cdot 8.1\ \mu m$$

$$W_p = 10.8\ \mu m$$

Thus, $L = 1\ \mu m$, $W_n = 8.1\ \mu m$, and $W_p = 10.8\ \mu m$ will satisfy the constraints given here.

6.4. CMOS Ring Oscillator Circuit

The following circuit example illustrates some of the basic notions associated with the dynamic operation characteristics of inverters, which were introduced in the preceding sections. At the same time, this example will provide us with a simple demonstration of astable behavior in digital circuits.

Consider the cascade connection of three identical CMOS inverters, as shown in Fig. 6.7, where the output node of the third inverter is connected to the input node of the first inverter. As such, the three inverters form a voltage feedback loop. It can be found by simple inspection that this circuit does not have a stable operating point, and the only possible operating point, at which the input and output voltages of all inverters are equal to the logic threshold V_{th}, is inherently unstable. In fact, a closed-loop cascade connection of any *odd* number of inverters will display astable behavior, i.e., such a circuit will oscillate once any of the inverter input or output voltages diverge from the unstable operating point, V_{th}. Therefore, the circuit is called a *ring oscillator*. A more detailed analysis of closed-loop cascade circuits consisting of identical inverters will be presented in Chapter 8. For this analysis, a qualitative understanding of the circuit behavior will be sufficient.

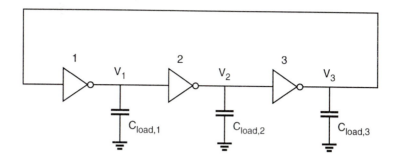

Figure 6.7.   Three-stage ring oscillator circuit consisting of identical inverters.

Figure 6.8 shows the typical output voltage waveforms of the three inverters during oscillation. As the output voltage V_1 of the first inverter stage rises from V_{OL} to V_{OH}, it triggers the second inverter output V_2 to fall, from V_{OH} to V_{OL}. Note that the difference between the $V_{50\%}$-crossing times of V_1 and V_2 is defined as the signal propagation delay τ_{PHL2}, for the second inverter. As the output voltage V_2 of the second

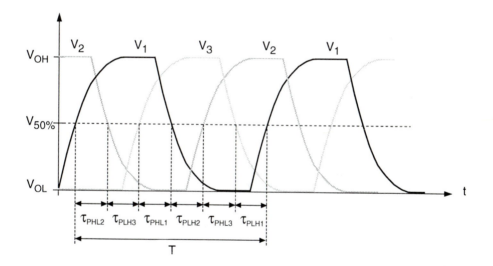

Figure 6.8.   Typical voltage waveforms of the three inverters shown in Fig. 6.7.

inverter falls, it triggers the output voltage V_3 of the third inverter to rise from V_{OL} to V_{OH}. Again, the difference between the $V_{50\%}$-crossing times of V_2 and V_3 is defined as the signal propagation delay τ_{PLH3}, for the third inverter. It can be seen from Fig. 6.8 that each inverter triggers the next inverter in the cascade connection, and the last inverter again triggers the first, continuing the oscillation.

In this three-stage circuit, the oscillation period T of any of the inverter output voltages can be expressed as the sum of six propagation delay times (Fig. 6.8). Since the three inverters in the closed-loop cascade connection are assumed to be identical, and since the output load capacitances are equal to each other ($C_{load1}=C_{load2}=C_{load3}$), we can also express the oscillation period T in terms of the average propagation delay τ_p, as follows:

$$T = \tau_{PHL1} + \tau_{PLH1} + \tau_{PHL2} + \tau_{PLH2} + \tau_{PHL3} + \tau_{PLH3}$$
$$= 2\tau_p + 2\tau_p + 2\tau_p \qquad\qquad (6.31)$$
$$= 3 \cdot 2\tau_p = 6\tau_p$$

Generalizing this relationship for an arbitrary odd number of cascade-connected inverters, we obtain

$$f = \frac{1}{T} = \frac{1}{2 \cdot n \cdot \tau_p} \qquad\qquad (6.32)$$

Thus, the oscillation frequency is found to be a very simple function of the average propagation delay of an inverter stage. This relationship can also be utilized to *measure* the average propagation delay of a typical inverter with minimum capacitive loading, simply by fabricating a ring oscillator circuit consisting of n identical inverters, and by accurately determining its oscillation frequency.

$$\tau_p = \frac{1}{2 \cdot n \cdot f} \qquad\qquad (6.33)$$

Typically, the number n is made much larger than just 3 or 5, in order to keep the oscillation frequency of the circuit within an easily measurable range. The ring oscillator frequency measurements are routinely utilized to characterize a particular design and/or a new fabrication process.

6.5. Estimation of Interconnect Parasitics

It was mentioned in Section 6.1 that the parasitic interconnect capacitance of metal and polysilicon lines (C_{int}) is one of the important capacitance components which constitute the total lumped output load capacitance C_{load} of an inverter. In fact, in many cases the interconnect capacitance may become the dominant term in the total capacitance expression (6.1), compared to other transistor parasitics. Until this point, however, we have not examined the calculation of interconnect parasitics in detail. The following discussion will present some of the basic concepts on the estimation of interconnect parasitic resistance and capacitance values.

Consider the section of an interconnect which is illustrated in Fig. 6.9. It is assumed that this wire segment has a length of (l) in the current direction, a width of (w), and a thickness of (t). Moreover, we assume that the interconnect segment runs parallel to the chip surface and is separated from the ground plane by a dielectric layer which has a height of (h).

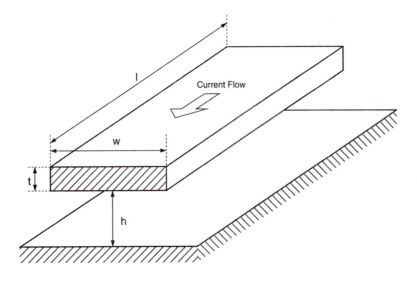

Figure 6.9.   Interconnect segment running parallel to the surface, which is used for parasitic resistance and capacitance estimations.

First, consider the parasitic resistance of this wire segment. The total resistance in the indicated current direction can be found as follows:

$$R_{metal} = \rho \cdot \frac{l}{w \cdot t} = R_{sheet}\left(\frac{l}{w}\right) \tag{6.34}$$

where ρ represents the characteristic resistivity of the interconnect material, and R_{sheet} represents the sheet resistivity of the line, in (Ω/square).

$$R_{sheet} = \frac{\rho}{t} \tag{6.35}$$

For a typical polysilicon layer, the sheet resistivity is between 20–40 Ω/square, whereas the sheet resistivity of silicide is about 2–4 Ω/square. Using (6.34), we can estimate the total parasitic resistance of a wire segment based on its geometry. In most short-distance aluminum and silicide interconnects, the amount of parasitic wire resistance is usually negligible. Especially for longer polysilicon wire segments, on the other hand, the effect of the parasitic resistance must be taken into account. As a first-order approximation in simulations, the total lumped resistance may be assumed to be connected in series with the total lumped capacitance of the wire. A much better approximation of the influence of distributed parasitic resistance can be obtained by using an RC-ladder network model to represent the interconnect segment.

The correct estimation of the parasitic interconnect capacitance with respect to ground is also an important issue. Using the basic geometry given in Fig. 6.9, one can simply calculate the parallel-plate capacitance C_{pp} of the interconnect segment. It can be shown, however, that in interconnect structures where the wire thickness (t) is comparable in magnitude to the ground-plane distance (h), *fringing electric fields* significantly increase the total parasitic capacitance (Fig. 6.10).

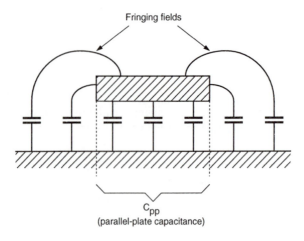

Figure 6.10.   Influence of fringing electric fields upon the parasitic wire capacitance.

Figure 6.11 shows the variation of the fringing-field factor $FF = C_{total}/C_{pp}$ with (t/h), (w/h), and (w/l). It can be seen that the influence of fringing fields increases with the (w/h) ratio, and that the fringing-field capacitance can be as much as 10 times larger than the parallel plate capacitance for $(t/h) = 0.4$, $(w/h) = 0.25$, and $(w/l) = 0$, i.e., for a long interconnection wire.

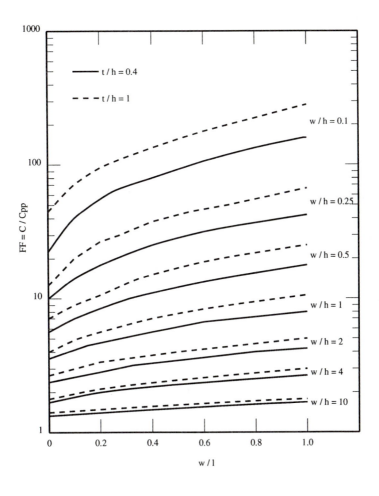

Figure 6.11.   Variation of the fringing-field factor with the interconnect geometry.

A set of simple formulas developed by Yuan and Trick can also be used to estimate the capacitance of interconnect structures in which fringing electric fields complicate the parasitic capacitance calculation. The following two cases are considered for two different ranges of width w.

$$C = \varepsilon \left[\frac{\left(w - \dfrac{t}{2}\right)}{h} + \frac{2\pi}{\ln\left(1 + \dfrac{2h}{t} + \sqrt{\dfrac{2h}{t}\left(\dfrac{2h}{t} + 2\right)}\right)} \right] \qquad for \qquad w \geq \frac{t}{2} \qquad (6.36)$$

$$C = \varepsilon \left[\frac{w}{h} + \frac{\pi\left(1 - 0.0543 \cdot \dfrac{t}{2h}\right)}{\ln\left(1 + \dfrac{2h}{t} + \sqrt{\dfrac{2h}{t}\left(\dfrac{2h}{t} + 2\right)}\right)} + 1.47 \right] \qquad for \qquad w < \frac{t}{2} \qquad (6.37)$$

The formulas given above permit the accurate approximation of the parasitic capacitance values to within 10% error for very small values of (t/h).

So far, we considered the modeling of interconnection lines by RC circuits. In general, if the time of flight across the interconnection line (as determined by the speed of light) is shorter than the signal rise/fall times, then the wire can be modeled as a capacitive load, or as a lumped or distributed RC network. If the interconnection lines are sufficiently long and the rise times of the signal waveforms are comparable to the time of flight across the line, then the *inductance* also becomes important, and the interconnection lines must be modeled as *transmission lines*.

For example, the longest wire on a VLSI chip may be about 2 cm. The time of flight of a signal across this wire, assuming $\varepsilon_r = 4$, is approximately 133 ps, which is shorter than typical on-chip signal rise/fall times. Thus, either capacitive or RC model is adequate for the wire. On the other hand, the time of flight across a 10 cm multichip module (MCM) interconnect in an alumina substrate is approximately 1 ns, which is of the same order of magnitude as the rise times of signals generated by some drivers. In this case, the interconnection lines can be modeled with RLCG (resistance, inductance, capacitance, and conductance) parasitics as shown in Fig. 6.12.

Note that the signal integrity can be significantly degraded due to the transmission effect, especially when the output impedance of the driver is significantly lower than the characteristic impedance of the transmission line. Most CMOS chips at present are not sufficiently big or fast enough to warrant transmission line analysis of on-chip wires, but this situation is being changed as deep-submicron chips become larger and operate with increasingly shorter signal rise/fall times.

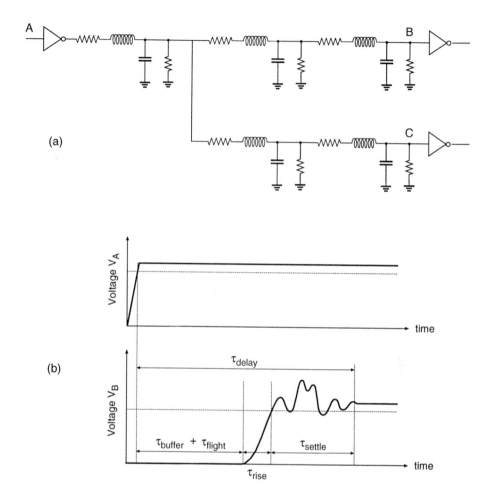

Figure 6.12.   (a) An RLCG interconnection tree. (b) Typical signal waveforms at the nodes A and B, showing the signal delay and the various delay components.

6.6. Dynamic Power Dissipation of CMOS Inverters

It was shown in Chapter 5 that the static power dissipation of the CMOS inverter is quite negligible; therefore, the CMOS inverter has a significant advantage in comparison to resistive-load and depletion-load nMOS inverter circuits. During switching events where the output load capacitance is alternatingly charged up and charged down, on the other hand, the CMOS inverter inevitably dissipates power. In the following section, we will derive the expressions for the dynamic power consumption of the CMOS inverter.

Consider the simple CMOS inverter circuit shown in Fig. 6.13. We will assume that the input voltage is an ideal step waveform with negligible rise and fall times. Typical input and output voltage waveforms and the expected load capacitor current waveform are shown in Fig. 6.14. When the input voltage switches from low to high, the pMOS transistor in the circuit is turned off, and the nMOS transistor starts conducting. During this phase, the output load capacitance C_{load} is being discharged through the nMOS transistor. Thus, the capacitor current equals the instantaneous drain current of the nMOS transistor. When the input voltage switches from high to low, the nMOS transistor in the circuit is turned off, and the pMOS transistor starts conducting. During this phase, the output load capacitance C_{load} is being charged up through the pMOS transistor; therefore, the capacitor current equals the instantaneous drain current of the pMOS transistor.

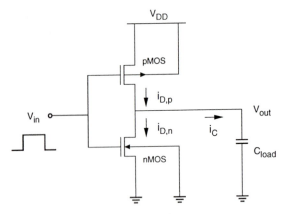

Figure 6.13.   CMOS inverter used in the dynamic power-dissipation analysis.

Assuming periodic input and output waveforms, the average power dissipated by any device over one period can be found as follows:

$$P_{avg} = \frac{1}{T} \int_0^T v(t) \cdot i(t) \, dt \qquad (6.38)$$

Since during switching, the nMOS transistor and the pMOS transistor in a CMOS inverter conduct current for one-half period each, the average power dissipation of the

CMOS inverter can be calculated as the power required to charge up and charge down the output load capacitance.

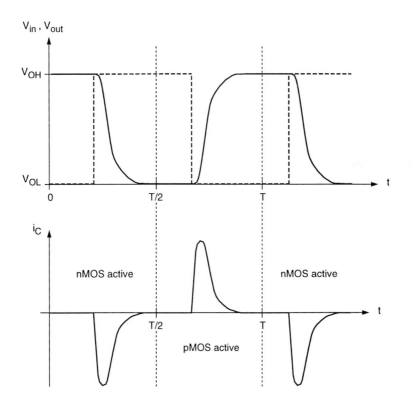

Figure 6.14. Typical input and output voltage waveforms and the capacitor current waveform during switching of the CMOS inverter.

$$P_{avg} = \frac{1}{T}\left[\int_0^{T/2} V_{out}\left(-C_{load}\frac{dV_{out}}{dt}\right)dt + \int_{T/2}^{T}\left(V_{DD}-V_{out}\right)\left(C_{load}\frac{dV_{out}}{dt}\right)dt\right] \quad (6.39)$$

Evaluating the integrals in (6.39), we obtain

$$P_{avg} = \frac{1}{T}\left[\left(-C_{load}\frac{V_{out}^2}{2}\right)\Big|_0^{T/2} + \left(V_{DD}\cdot V_{out}\cdot C_{load} - \frac{1}{2}C_{load}V_{out}^2\right)\Big|_{T/2}^{T}\right] \quad (6.40)$$

$$P_{avg} = \frac{1}{T} C_{load} V_{DD}^2 \qquad\qquad (6.41)$$

Noting that $f = 1/T$, this expression can also be written as:

$$P_{avg} = C_{load} \cdot V_{DD}^2 \cdot f \qquad\qquad (6.42)$$

It is clear that the average dynamic power dissipation of the CMOS inverter is proportional to the switching frequency f. Therefore, the low-power advantage of CMOS circuits becomes less prominent in high-speed operation, where the switching frequency is higher. Also note that the average power dissipation is independent of all transistor characteristics and transistor sizes. Consequently, the switching delay times have no relevance to the amount of power consumption during the switching events. The reason for this is that the dynamic power is solely dissipated for charging and discharging the output capacitance from V_{OL} to V_{OH}, and vice versa.

For this reason, the dynamic power expression derived for the CMOS inverter also applies to all general CMOS circuits, as shown in Fig. 6.15. A general CMOS logic circuit consists of an nMOS logic block between the output node and the ground, and a pMOS logic block between the output and V_{DD}. As in the simple CMOS inverter case, either the pMOS block or the nMOS block can conduct depending on the input voltage combination, but not both at the same time. Therefore, dynamic power is again dissipated solely for charging and discharging the output capacitance.

To summarize, if the total parasitic capacitance in the circuit can be lumped at the output node with reasonable accuracy, if the output voltage swing is between 0 and V_{DD}, and if the input voltage waveforms are assumed to be ideal step inputs, the dynamic power expression (6.42) will hold for any CMOS logic circuit.

Note that under realistic conditions, when the input voltage waveform deviates from ideal step input and has nonzero rise and fall times, for example, both the nMOS and the pMOS transistor will simultaneously conduct a certain amount of current during the switching event. This is called the short-circuit current, since in this case, the two transistors temporarily form a conducting path between the V_{DD} and the ground. The additional power dissipation, which is due to the short-circuit current, cannot be predicted by the dynamic power-dissipation formula (6.42) derived above, since the short-circuit current is not being utilized to charge or discharge the output load capacitor. We must be aware that this additional dynamic power-dissipation term can be quite significant under some nonideal conditions. If the load capacitance is increased, on the other hand, the short-circuit dissipation term usually becomes negligible in comparison to the purely capacitive power dissipation.

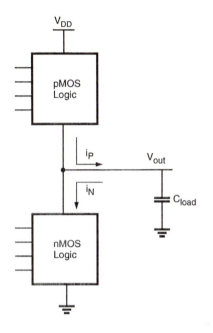

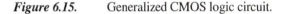

Figure 6.15.   Generalized CMOS logic circuit.

Power Meter Simulation

In the following, we present a simple circuit simulation approach which can be used to estimate the average dynamic power dissipation of arbitrary circuits, under realistic operating conditions. According to (6.38), the average power dissipation of any device or circuit which is driven by a periodic input waveform can be found by integrating the product of its instantaneous terminal voltage and its instantaneous terminal current over one period. If we have to determine the amount of P_{avg} drawn from the power supply over one period, the problem is reduced to finding only the time-average of the power supply current, since the power supply voltage is a constant.

Using a simple simulation model called the *power meter*, we can estimate the average power dissipation of an arbitrary device or circuit driven by a periodic input, through transient circuit simulation. Consider the circuit structure shown in Fig. 6.16, in which a zero-volt independent voltage source is connected in series with the power supply voltage source V_{DD} of the device or circuit in question. Consequently, the instantaneous power supply current $i_{DD}(t)$ which is being drawn by the circuit will also pass through the zero-volt voltage source, $i_s(t) = i_{DD}(t)$.

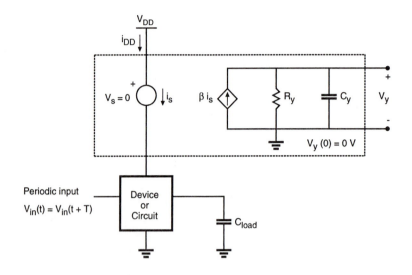

Figure 6.16.   The power meter circuit used for the simulation of average dynamic power dissipation of an arbitrary device or circuit.

The power meter circuit consists of three elements: a linear current-controlled current source, a capacitor, and a resistor, all connected in parallel. The current equation for the common node of the power meter circuit can be written as follows:

$$C_y \frac{dV_y}{dt} = \beta i_S - \frac{V_y}{R_y}$$

(6.43)

The initial condition of the node voltage V_y is set as $V_y(0) = 0$ V. Then, the time-domain solution of $V_y(t)$ can be found by integrating (6.43).

$$V_y(t) = \frac{\beta}{C_y} \int_0^t e^{-\frac{1}{R_y C_y}(t-\tau)} i_{DD}(\tau) d\tau$$

(6.44)

Assuming $R_y C_y \gg T$, the voltage value $V_y(T)$ at the end of one period can be approximated as follows.

$$V_y(T) \approx \frac{\beta}{C_y} \int_0^T i_{DD}(\tau) d\tau$$

(6.45)

If the constant coefficient value of the current-controlled current source is set to be

$$\beta = V_{DD} \frac{C_y}{T} \tag{6.46}$$

the voltage value $V_y(T)$ at the end of one period will be found by transient simulation as:

$$V_y(T) = V_{DD} \cdot \frac{1}{T} \int_0^T i_{DD}(\tau) d\tau \tag{6.47}$$

Note that the right-hand side of (6.47) corresponds to the average dynamic power drawn from the power supply source over one period. Thus, the value of the node voltage V_y at $t = T$ gives the average dynamic power.

The power meter circuit shown in Fig. 6.16 can be easily simulated using a conventional circuit simulation program such as SPICE, and it enables us to accurately estimate the average power dissipation of any circuit with arbitrary complexity. Also note that the power meter circuit inherently takes into account the additional power dissipation due to the short-circuit currents, which may arise because of nonideal input conditions. In the following example, we present a sample SPICE simulation of the power meter for estimating the dynamic power dissipation of a CMOS inverter circuit.

===

Example 6.4.

Consider the simple CMOS inverter circuit shown in Fig. 6.13. We will assume that the circuit is being driven by a square-wave input signal with period $T = 20$ ns, and that the total output load capacitance is equal to 1 pF. The power supply voltage is 5 V. Using the average dynamic power-dissipation formula (6.42) derived earlier, we can calculate the expected power dissipation to be $P_{avg} = 1.25$ mW.

Now, the circuit with an attached power meter will be simulated using SPICE. The corresponding circuit input file is listed here for reference. The controlled current source coefficient is calculated as 0.025, according to (6.46). The resistance and capacitance values R_y and C_y are chosen as 100 kΩ and 100 pF, respectively.

```
Power meter simulation:
mn 3 2 0 0 nmod w=10u l=1u
mp 3 2 4 1 pmod w=20u l=1u
```

```
vdd 1 0 5
vtstp 1 4 0
.model nmod nmos(vto=1 kp=20u)
.model pmod pmos(vto=-1 kp=10u)
vin 2 0 pulse(0 5 8n 2n 2n 8n 20n)
cl 3 0 1p
fp 0 9 vtstp 0.025
rp 9 0 100k
cp 9 0 100p
.tran 1n 60n uic
.print tran v(3) v(2)
.print tran i(vtstp)
.print tran v(9)
.end
```

The simulation results are plotted on the following page. It can be seen that the power supply current is being drawn from the voltage source V_{DD} only during the charge-up phase of the output capacitor. The power meter output voltage by the end of the first period corresponds to exactly 1.25 mW, as expected.

References

1.   N.H.E. Weste and K. Eshraghian, *Principles of CMOS VLSI Design– A Systems Perspective*, second edition, Reading, MA: Addison-Wesley, 1993.

2.   J.P. Uyemura, *Fundamentals of MOS Digital Integrated Circuits*, Reading, MA: Addison-Wesley, 1988.

3.   L.A. Glasser and D.W. Dobberpuhl, *The Design and Analysis of VLSI Circuits*, Reading, MA: Addison-Wesley, 1985.

4.   C.P. Yuan and T.N. Trick, "A simple formula for the estimation of capacitance of two-dimensional interconnects in VLSI circuits," *IEEE Electron Device Letters*, vol. EDL-3, no. 12, pp. 391-393, December 1982.

5.   D. Zhou, F. P. Preparata, and S. M. Kang, "Interconnection delay in very high-speed VLSI," *IEEE Transactions on Circuits and Systems*, vol. 38, no. 7, pp. 779-790, July 1991.

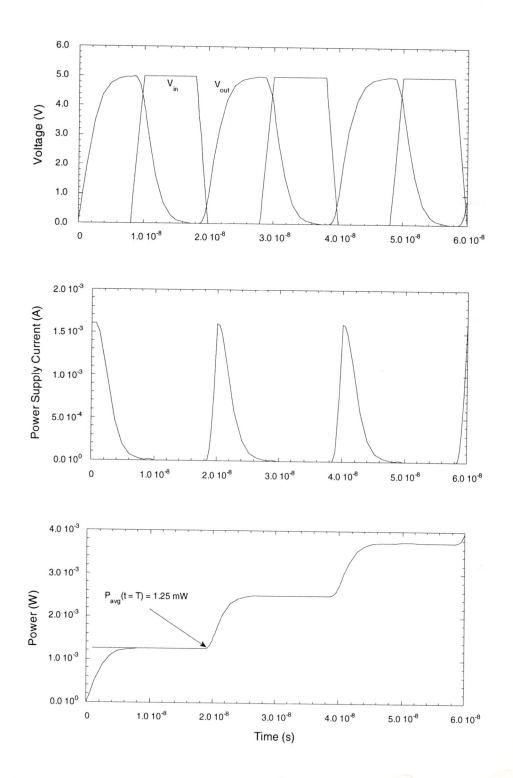

6.   G. Bilardi, M. Pracchi, and F. P. Preparata, "A critique of network speed in VLSI models of computation," *IEEE Journal of Solid-State Circuits*, vol. DC-17, no. 4, pp. 696-702, August 1982

7.   S. M. Kang, "Accurate simulation of power dissipation in VLSI circuits," *IEEE Journal of Solid-State Circuits*, vol. SC-21, no. 10, pp. 889-891, October 1986.

8.   M. Horowitz and R. W. Dutton, "Resistance extraction from mask layout data," *IEEE Transactions on Computer-Aided Design*, vol. CAD-2, no. 3, pp. 145-150, July 1983.

9.   A. E. Ruehli and P. A. Brennan, "Efficient capacitance calculations for three-dimensional multiconductor systems," *IEEE Transactions on Microwave Theory and Applications*, vol. MTT-21, no. 2, pp. 76-82, February 1973.

10.   P. DeWilde, "New algebraic methods for modelling large-scale integrated circuits," *International Journal of Circuit Theory and Applications*, vol. 16, no. 4, pp. 473-503, October 1988.

11.   T. Sakurai and A. R. Newton, "Alpha-power law MOSFET model and its application to CMOS inverter delay and other formulas," *IEEE Journal of Solid-State Circuits*, vol. 25, no. 2, pp. 584-594, April 1990.

12.   H.B. Bakoglu, *Circuits, Interconnections and Packaging for VLSI*, Reading, MA: Addison-Wesley, 1990.

Exercise Problems

6.1   Does the inverter with a lower V_{OL} always have the shorter high-to-low switching time? Justify your answer.

6.2   Consider switching delays in a 10-kΩ resistive-load inverter circuit, where

$$\mu_n \cdot C_{ox} = 25 \ \mu A/V^2$$
$$W/L = 10$$
$$V_{T0} = 1.0 \ V$$

(a) Find t_{PHL} (50% high-to-low transition delay) by using the average-current method. Assume that the input signal is an ideal rectangular pulse switching between 0 and 5 V with zero rise/fall times. You will have to calculate V_{OL} to solve this problem.

(b) By using an appropriate differential equation and the proper initial voltage across the capacitor (when the input voltage is at V_{OH}) which is V_{OL} and not 0 V, calculate t_{PLH}. Use the same input voltage as in part (a).

6.3 Consider a CMOS ring oscillator consisting of an odd number (n) of identical inverters connected in a ring configuration as shown in Fig. 6.7. The layout of the ring oscillator is such that the interconnection (wiring) parasitics can be assumed to be zero. Therefore, the delay of each stage is the same and the average gate delay is called the intrinsic delay (τ_p) as long as identical gates are used. The ring oscillator circuit is often used to quote the circuit speed of a particular technology using the ring oscillator frequency (f).

(a) Derive an expression for the intrinsic delay (τ_p) in terms of the number of stages n.

(b) Show that τ_p is independent of the transistor sizes, i.e., it remains the same when all the gates are scaled uniformly up or down.

6.4 Suppose that the resistive-load inverter examined in Exercise 6.2 is connected to a load capacitance of 1.0 pF which is initially discharged. The gate of the nMOS transistor is driven by a rectangular pulse which changes from high to low at $t=0$. As a result, the nMOS transistor begins to charge up the capacitor. Solve the following two parts by using the differential equations and not by using the average-current methods.

(a) Determine the 50% low-to-high delay time, which is defined as the time difference between 50% points of input and output waveforms when the output waveform switches from low to high.

(b) Determine the 50% high-to-low delay time, which is defined as the time difference between 50% points of input and output waveforms when the output waveform switches from high to low when the capacitor is initially charged to 5.0 V.

6.5   A resistive-load inverter with $R_L = 50 \text{ k}\Omega$ has the following device parameters:

$V_{TN}(V_{SB}=0) = 1.0 \text{ V}$
$\gamma = 0.5 \text{ V}^{1/2}$
$t_{ox} = 0.05 \text{ } \mu\text{m}$
$\mu_n = 500 \text{ cm}^2/\text{V·s}$
$W = 10 \text{ } \mu\text{m}$ and $L = 1.0 \text{ } \mu\text{m}$

A ring oscillator is formed by connecting nine identical inverters in a closed loop. We are interested in finding the resulting oscillation frequency.

(a)   Find the delay times of the inverter for an ideal step pulse whose voltage swing is between V_{OL} and V_{OH}, i.e., t_{PHL} and t_{PLH}. It should be noted that the loading capacitance of each inverter is strictly due to the drain parasitic capacitance and the gate capacitance of the following stage. For simplicity, neglect the drain parasitics and assume C_{load} is equal to gate capacitance.

(b)   The rise and fall times are defined between 10% and 90% of the full voltage swing. But for simplicity, we will assume that $t_{fall} = 2 \cdot t_{PHL}$ and $t_{rise} = 2 \cdot t_{PLH}$.

Estimate the actual propagation delays t_{PHL} and t_{PLH} by using the rise and fall times of the inverter and the ideal delays found in (a).

(c)   Find the oscillation frequency from the information in part (b).

6.6   The layout of a CMOS inverter is shown in Fig. P6.6. This inverter is driving another inverter, which is identical to the one shown below except that the transistor widths are three times larger. Calculate t_{PLH} and t_{PHL}. Assume that the interconnect capacitance is negligible. The parameters are given as:

$V_{TP} = -1.0 \text{ V}$ $V_{TN} = 1.0 \text{ V}$
$k'_n = 40 \text{ } \mu\text{A/V}^2$ $k'_p = 20 \text{ } \mu\text{A/V}^2$
$C_{ox} = 69 \text{ nF/cm}^2$ $C_{jsw} = 2.2 \text{ pF/cm}$
$C_{j0} = 7 \text{ nF/cm}^2$ $\phi_0 = 0.86 \text{ V}$
$L_D = 1 \text{ } \mu\text{m}$ $L_{mask} = 5 \text{ } \mu\text{m}$

The source and drain region length is $Y = 12 \text{ } \mu\text{m}$ and the channel width is $W = 10 \text{ } \mu\text{m}$ for both transistors.

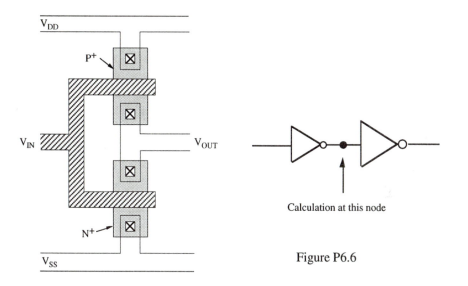

Figure P6.6

Calculation at this node

6.7   For an nMOS depletion-load inverter circuit, calculate the propagation delay
times t_{PLH} and t_{PHL} assuming that:

- the inverter is driving an identical gate (fanout = 1)
- the interconnect capacitance is negligible
- the lateral diffusion for both transistors is $L_D = 0.25 \ \mu m$
- $L_{mask} = 2 \ \mu m$, $W_{mask} = 10 \ \mu m$, $Y = 10 \ \mu m$ for the depletion-type nMOS
- $L_{mask} = 2 \ \mu m$, $W_{mask} = 15 \ \mu m$, $Y = 10 \ \mu m$ for the enhancement-type nMOS

Use the device parameters given in Example 5.3 and the following information for calculating junction capacitances.

$t_{ox} = 0.1 \ \mu m$
$x_j = 1.0 \ \mu m$
$N_A = 10^{16} \ cm^{-3}$
$N_D = 10^{19} \ cm^{-3}$
$N_A \ (sw) = 10^{17} \ cm^{-3}$

CHAPTER 7

COMBINATIONAL MOS LOGIC CIRCUITS

7.1. Introduction

Combinational logic circuits, or gates, which perform Boolean operations on multiple input variables and determine the outputs as Boolean functions of the inputs, are the basic building blocks of all digital systems. In this chapter, we will examine the static and dynamic characteristics of various combinational MOS logic circuits. It will be seen that many of the basic principles used in the design and analysis of MOS inverters in Chapters 5 and 6 can be directly applied to combinational logic circuits as well.

The first major class of combinational logic circuits to be presented in this chapter is the nMOS depletion-load gates. Our purpose for including nMOS depletion-load circuits here is mainly pedagogical, to emphasize the *load* concept, which is still being widely used in many areas in digital circuit design. We will examine simple circuit configurations such as two-input NAND and NOR gates and then expand our analysis to more general cases of multiple-input circuit structures. Next, the CMOS logic circuits will be presented in a similar fashion. We will stress the similarities and differences between the nMOS depletion-load logic and CMOS logic circuits and point out the advantages of CMOS gates with examples. The design of complex logic gates, which allows the realization of complex Boolean functions of multiple variables, will be examined in detail. Finally, we will devote the last section to CMOS transmission gates and to transmission gate (TG) logic circuits.

In its most general form, a combinational logic circuit, or gate, performing a Boolean function can be represented as a multiple-input single-output system, as depicted in Fig. 7.1. All input variables are represented by node voltages, referenced to the ground potential. Using *positive logic convention*, the Boolean (or logic) value of "1" can be represented by a high voltage of V_{DD}, and the Boolean (or logic) value of "0" can be represented by a low voltage of 0. The output node is loaded with a capacitance C_L, which represents the combined parasitic device capacitances in the circuit and the interconnect capacitance components *seen* by the output node. This output load capacitance certainly plays a very significant role in the dynamic operation of the logic gate.

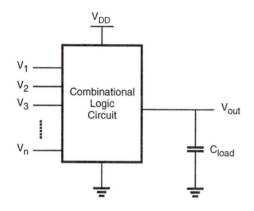

Figure 7.1.   Generic combinational logic circuit (gate).

As in the simple inverter case, the voltage transfer characteristic (VTC) of a combinational logic gate provides valuable information on the DC operating performance of the circuit. Critical voltage points such as V_{OL} or V_{th} are considered to be important design parameters for combinational logic circuits. Other design parameters and concerns include the dynamic (transient) response characteristics of the circuit, the silicon area occupied by the circuit, and the amount of static and dynamic power dissipation.

7.2. MOS Logic Circuits with Depletion nMOS Loads

The first circuit to be examined in this section is the two-input NOR gate. The circuit diagram, the logic symbol, and the corresponding truth table of the gate are given in Fig. 7.2. The Boolean OR operation is performed by the parallel connection of the two enhancement-type nMOS driver transistors. If the input voltage V_A *or* the input voltage V_B is equal to the logic-high level, the corresponding driver transistor turns on and provides a conducting path between the output node and the ground. Hence, the output voltage becomes low. In this case, the circuit operates like a depletion-load inverter with respect to its static behavior. A similar result is achieved when both V_A and V_B are high, in which case two parallel conducting paths are created between the output node and the ground. If, on the other hand, both V_A and V_B are low, both driver transistors remain cut-off. The output node voltage is pulled to a logic-high level by the depletion-type nMOS load transistor.

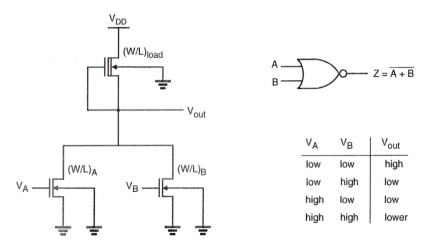

V_A	V_B	V_{out}
low	low	high
low	high	low
high	low	low
high	high	lower

Figure 7.2.   A two-input depletion-load NOR gate, its logic symbol, and the corresponding truth table. Note that the substrates of all transistors are connected to ground.

The DC analysis of the circuit can be simplified significantly by considering the structural similarities between this circuit and the simple nMOS depletion-load inverter. In the following, the calculation of output low and output high voltages will be examined.

Calculation of V_{OH}

When both input voltages V_A and V_B are lower than the corresponding driver threshold voltage, the driver transistors are turned off and conduct no drain current. Consequently, the load device, which operates in the linear region, also has zero drain current. In particular, its linear region current equation becomes

$$I_{D,load} = \frac{k_{n,load}}{2} \cdot \left[2\left| V_{T,load}(V_{OH}) \right| \cdot (V_{DD} - V_{OH}) - (V_{DD} - V_{OH})^2 \right] = 0 \qquad (7.1)$$

The solution of this equation gives $V_{OH} = V_{DD}$.

Calculation of V_{OL}

To calculate the output low voltage V_{OL}, we must consider three different cases, i.e., three different input voltage combinations, which produce a conducting path from the output node to the ground. These cases are

(i) $V_A = V_{OH}$ $V_B = V_{OL}$
(ii) $V_A = V_{OL}$ $V_B = V_{OH}$
(iii) $V_A = V_{OH}$ $V_B = V_{OH}$

For the first two cases, (i) and (ii), the NOR circuit reduces to a simple nMOS depletion-load inverter. Assuming that the threshold voltages of the two enhancement-type driver transistors are identical ($V_{T0,A} = V_{T0,B} = V_{T0}$), the driver-to-load ratio of the corresponding inverter can be found as follows. In case (i), where the driver transistor A is on, the ratio is

$$k_R = \frac{k_{driver,A}}{k_{load}} = \frac{k'_{n,driver}\left(\dfrac{W}{L}\right)_A}{k'_{n,load}\left(\dfrac{W}{L}\right)_{load}} \tag{7.2}$$

In case (ii), where the driver transistor B is on, the ratio is

$$k_R = \frac{k_{driver,B}}{k_{load}} = \frac{k'_{n,driver}\left(\dfrac{W}{L}\right)_B}{k'_{n,load}\left(\dfrac{W}{L}\right)_{load}} \tag{7.3}$$

The output low voltage level V_{OL} in both cases is found by using (5.54), as follows:

$$V_{OL} = V_{OH} - V_{T0} - \sqrt{\left(V_{OH} - V_{T0}\right)^2 - \left(\frac{k_{load}}{k_{driver}}\right) \cdot \left|V_{T,load}\left(V_{OL}\right)\right|^2} \tag{7.4}$$

Note that if the (W/L) ratios of both drivers are identical, i.e., $(W/L)_A = (W/L)_B$, the output low voltage (V_{OL}) values calculated for case (i) and case (ii) will be identical.

In case (iii), where both driver transistors are turned on, the saturated load current is the sum of the two linear-mode driver currents.

$$I_{D,load} = I_{D,driverA} + I_{D,driverB} \tag{7.5}$$

$$
\begin{aligned}
\frac{k_{load}}{2}\left|V_{T,load}(V_{OL})\right|^2 = {} & \frac{k_{driver,A}}{2}\left[2(V_A - V_{T0})V_{OL} - V_{OL}^2\right] \\
+ {} & \frac{k_{driver,B}}{2}\left[2(V_B - V_{T0})V_{OL} - V_{OL}^2\right]
\end{aligned}
\tag{7.6}
$$

Since the gate voltages of both driver transistors are equal ($V_A = V_B = V_{OH}$), we can devise an *equivalent* driver-to-load ratio for the NOR structure:

$$k_R = \frac{k_{driver,A} + k_{driver,B}}{k_{load}} = \frac{k'_{n,driver}\left[\left(\dfrac{W}{L}\right)_A + \left(\dfrac{W}{L}\right)_B\right]}{k'_{n,load}\left(\dfrac{W}{L}\right)_{load}} \tag{7.7}$$

Thus, the NOR gate with both of its inputs tied to a logic-high voltage is replaced with an nMOS depletion-load inverter circuit with the driver-to-load ratio given by (7.7). The output voltage level in this case is

$$V_{OL} = V_{OH} - V_{T0} - \sqrt{(V_{OH} - V_{T0})^2 - \left(\frac{k_{load}}{k_{driver,A} + k_{driver,B}}\right)\cdot\left|V_{T,load}(V_{OL})\right|^2} \tag{7.8}$$

Note that the V_{OL} given by (7.8) is *lower* than the V_{OL} values calculated for case (i) and for case (ii), when only one input is logic-high. We conclude that the worst-case condition from the static operation viewpoint, i.e., the highest possible V_{OL} value, is observed in case (i) or in (ii).

This result also suggests a simple design strategy for NOR gates. Usually, we have to achieve a certain maximum V_{OL} for the worst case, i.e., when only one input is high. Thus, we assume that one input (either V_A or V_B) is logic-high and determine the driver-to-load ratio of the resulting inverter using (7.4). Then set

$$k_{driver,A} = k_{driver,B} = k_R\, k_{load} \tag{7.9}$$

This design choice yields two identical driver transistors, which guarantee the required value of V_{OL} in the worst case. When both inputs are logic-high, the output voltage is even lower than the required maximum V_{OL}, thus the design constraint is satisfied.

Exercise 7.1.

Consider the depletion-load nMOS NOR2 gate shown in Fig. 7.2, with the following parameters: $\mu_n C_{ox} = 25\ \mu A/V^2$, $V_{T0,driver} = 1.0\ V$, $V_{T0,load} = -3.0\ V$, $\gamma = 0.4\ V^{1/2}$, and $|2\phi_F| = 0.6\ V$. The transistor dimensions are given as $(W/L)_A = 2$, $(W/L)_B = 4$, and $(W/L)_{load} = 1/3$. The power supply voltage is $V_{DD} = 5\ V$.

Calculate the output voltage levels for all four valid input voltage combinations.

Generalized NOR Structure

At this point, we can expand our analysis to generalized n-input NOR gates, which consist of n parallel driver transistors, as shown in Fig. 7.3. Note that the combined current I_D in this circuit is supplied by the driver transistors which are turned on, i.e., transistors which have gate voltages higher than the threshold voltage V_{T0}.

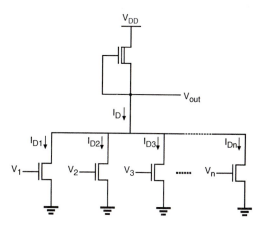

Figure 7.3.   Generalized n-input NOR gate.

The combined pull-down current can then be expressed as follows:

$$I_D = \sum_{k(on)} I_{D,k} = \begin{cases} \sum_{k(on)} \dfrac{\mu_n C_{ox}}{2}\left(\dfrac{W}{L}\right)_k \left[2\left(V_{GS,k} - V_{T0}\right)V_{out} - V_{out}^2\right], & linear \\[4ex] \sum_{k(on)} \dfrac{\mu_n C_{ox}}{2}\left(\dfrac{W}{L}\right)_k \left(V_{GS,k} - V_{T0}\right)^2, & saturation \end{cases} \quad (7.10)$$

Assuming that the input voltages of all driver transistors are identical,

$$V_{GS,k} = V_{GS} \qquad for \qquad k = 1, 2, \ldots, n \tag{7.11}$$

the pull-down current expression can be rewritten as

$$
I_D =
\begin{cases}
\dfrac{\mu_n C_{ox}}{2} \left(\displaystyle\sum_{k(on)} \left(\dfrac{W}{L}\right)_k \right) \left[2(V_{GS} - V_{T0})V_{out} - V_{out}^2 \right], & linear \\[4ex]
\dfrac{\mu_n C_{ox}}{2} \left(\displaystyle\sum_{k(on)} \left(\dfrac{W}{L}\right)_k \right) (V_{GS} - V_{T0})^2, & saturation
\end{cases}
\tag{7.12}
$$

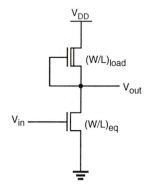

Figure 7.4.   Equivalent inverter circuit corresponding to the *n*-input NOR gate.

Thus, the multiple-input NOR gate can also be reduced to an equivalent inverter, shown in Fig. 7.4, for static analysis. The (*W/L*) ratio of the driver transistor here is

$$\left(\frac{W}{L}\right)_{equivalent} = \sum_{k(on)} \left(\frac{W}{L}\right)_k \tag{7.13}$$

Note that the source terminals of all enhancement-type nMOS driver transistors in the NOR gate are connected to ground. Thus, the drivers do not experience any substrate-bias effect. The depletion-type nMOS load transistor, however, is subject to substrate-bias effect, since its source is connected to the output node, and its source-to-substrate voltage is $V_{SB} = V_{out}$.

Transient Analysis of NOR Gate

Figure 7.5 shows the two-input NOR (NOR2) gate with all of its relevant parasitic device capacitances. As in the inverter case, we can combine the capacitances seen in Fig. 7.5 into one lumped capacitance, connected between the output node and the ground.

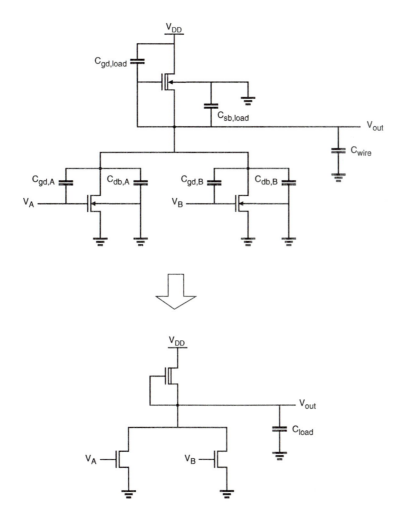

Figure 7.5. Parasitic device capacitances in the NOR2 gate and the lumped equivalent load capacitance. The gate-to-source capacitances of the driver transistors are included in the load of the previous stages driving the inputs A and B.

The value of this combined load capacitance, C_{load}, can be found as

$$C_{load} = C_{gd,A} + C_{gd,B} + C_{gd,load} + C_{db,A} + C_{db,B} + C_{sb,load} + C_{wire} \qquad (7.14)$$

Note that the output load capacitance given in (7.14) is valid for simultaneous as well as for single-input switching, i.e., the load capacitance C_{load} will be present at the output node even if only one input is active and all other inputs are low. This fact must be taken into account in calculations using the inverter equivalent of the NOR gate. The load capacitance at the output node of the *equivalent* inverter corresponding to a NOR gate is always *larger* than the total lumped load capacitance of an actual inverter with the same dimensions. Hence, while the static (DC) behaviors of the NOR gate and the inverter are essentially equivalent in this case, the actual transient response of the NOR gate will be slower than that of the inverter.

Next, we will examine the two-input NAND (NAND2) gate. The circuit diagram, the logic symbol, and the corresponding truth table of the gate are given in Fig. 7.6. The Boolean AND operation is performed by the series connection of the two enhancement-type nMOS driver transistors. There is a conducting path between the output node and the ground only if the input voltage V_A and the input voltage V_B are

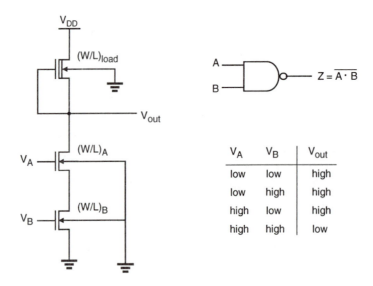

Figure 7.6.   A two-input depletion-load NAND gate, its logic symbol, and the corresponding truth table. Notice the substrate-bias effect for all nMOS transistors except one.

equal to logic-high, i.e., only if both of the series-connected drivers are turned on. In this case, the output voltage will be low, which is the complemented result of the AND operation. Otherwise, either one or both of the driver transistors will be off, and the output voltage will be pulled to a logic-high level by the depletion-type nMOS load transistor.

Figure 7.6 shows that all transistors except the one closest to the ground are subject to substrate-bias effect, since their source voltages are larger than zero. We have to consider this fact in detailed calculations. For all of the three input combinations which produce a logic-high output voltage, the corresponding V_{OH} value can easily be found as $V_{OH} = V_{DD}$. The calculation of the logic-low voltage V_{OL}, on the other hand, requires a closer investigation.

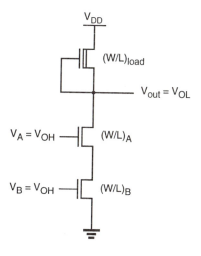

Figure 7.7.   The NAND2 gate with both of its inputs at logic-high level.

Consider the NAND2 gate with both of its inputs equal to V_{OH}, as shown in Fig. 7.7. It can easily be seen that the drain currents of all transistors in the circuit are equal to each other.

$$I_{D,load} = I_{D,driverA} = I_{D,driverB} \tag{7.15}$$

$$\frac{k_{load}}{2} \left| V_{T,load}(V_{OL}) \right|^2 = \frac{k_{driver,A}}{2} \left[2(V_{GS,A} - V_{T,A})V_{DS,A} - V_{DS,A}^2 \right]$$

$$= \frac{k_{driver,B}}{2} \left[2(V_{GS,B} - V_{T,B})V_{DS,B} - V_{DS,B}^2 \right] \tag{7.16}$$

The gate-to-source voltages of both driver transistors can be assumed to be approximately equal to V_{OH}. Also, we may neglect, for simplicity, the substrate-bias effect for driver transistor A, and assume $V_{T,A} = V_{T,B} = V_{T0}$, since the source-to-substrate voltage of driver A is relatively low. The drain-to-source voltages of both driver transistors can then be solved from (7.16) as

$$V_{DS,A} = V_{OH} - V_{T0} - \sqrt{\left(V_{OH} - V_{T0}\right)^2 - \left(\frac{k_{load}}{k_{driver,A}}\right) \cdot \left|V_{T,load}(V_{OL})\right|^2} \quad (7.17)$$

$$V_{DS,B} = V_{OH} - V_{T0} - \sqrt{\left(V_{OH} - V_{T0}\right)^2 - \left(\frac{k_{load}}{k_{driver,B}}\right) \cdot \left|V_{T,load}(V_{OL})\right|^2} \quad (7.18)$$

Let the two driver transistors be identical, i.e., $k_{driver,A} = k_{driver,B} = k_{driver}$. Noting that the output voltage V_{OL} is equal to the sum of the drain-to-source voltages of both drivers, we obtain

$$V_{OL} \approx 2\left(V_{OH} - V_{T0} - \sqrt{\left(V_{OH} - V_{T0}\right)^2 - \left(\frac{k_{load}}{k_{driver}}\right) \cdot \left|V_{T,load}(V_{OL})\right|^2}\right) \quad (7.19)$$

The following analysis gives a better and more accurate view of the operation of two series-connected driver transistors. Consider the two identical enhancement-type nMOS transistors with their gate terminals connected. At this point, the only simplifying assumption will be $V_{T,A} = V_{T,B} = V_{T0}$. When both driver transistors are in the linear region, the drain currents can be written as

$$I_{D,A} = \frac{k_{driver}}{2}\left[2\left(V_{GS,A} - V_{T0}\right)V_{DS,A} - V_{DS,A}^2\right] \quad (7.20)$$

$$I_{D,B} = \frac{k_{driver}}{2}\left[2\left(V_{GS,B} - V_{T0}\right)V_{DS,B} - V_{DS,B}^2\right] \quad (7.21)$$

Since $I_{D,A} = I_{D,B}$, this current can also be expressed as

$$I_D = I_{D,A} = I_{D,B} = \frac{I_{D,A} + I_{D,B}}{2} \quad (7.22)$$

Using $V_{GS,A} = V_{GS,B} - V_{DS,B}$, (7.22) yields

$$I_D = \frac{k_{driver}}{4}\left[2\left(V_{GS,B} - V_{T0}\right)\left(V_{DS,A} + V_{DS,B}\right) - \left(V_{DS,A} + V_{DS,B}\right)^2\right] \quad (7.23)$$

Now let $V_{GS} = V_{GS,B}$ and $V_{DS} = V_{DS,A} + V_{DS,B}$. The drain-current expression can then be written as follows.

$$I_D = \frac{k_{driver}}{4}\left[2\left(V_{GS} - V_{T0}\right)V_{DS} - V_{DS}^2\right] \quad (7.24)$$

Thus, two nMOS transistors connected in series and with the same gate voltage behave like *one* nMOS transistor with $k_{eq} = 0.5\, k_{driver}$.

Generalized NAND Structure

At this point, we expand our analysis to generalized n-input NAND gates, which consist of n series-connected driver transistors, as shown in Fig. 7.8. Neglecting the substrate-bias effect, and assuming that the threshold voltages of all transistors are equal to V_{T0}, the driver current I_D in this circuit can be expressed as

$$I_D = \frac{\mu_n C_{ox}}{2}\left(\frac{1}{\sum\limits_{k(on)}\dfrac{1}{\left(\dfrac{W}{L}\right)_k}}\right)\begin{cases}\left[2\left(V_{in} - V_{T0}\right)V_{out} - V_{out}^2\right], & linear \\ \left(V_{in} - V_{T0}\right)^2, & saturation\end{cases} \quad (7.25)$$

Hence, the (W/L) ratio of the equivalent driver transistor is

$$\left(\frac{W}{L}\right)_{equivalent} = \frac{1}{\sum\limits_{k(on)}\dfrac{1}{\left(\dfrac{W}{L}\right)_k}} \quad (7.26)$$

If the series-connected transistors are identical, i.e., $(W/L)_1 = (W/L)_2 = \ldots = (W/L)$, the width-to-length ratio of the equivalent transistor becomes

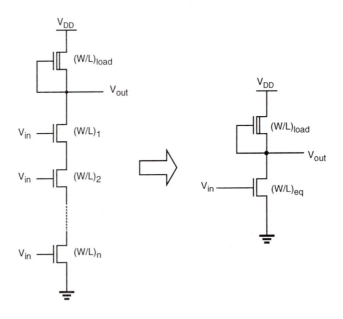

Figure 7.8.   The generalized NAND structure and its inverter equivalent.

$$\left(\frac{W}{L}\right)_{equivalent} = \frac{1}{n}\left(\frac{W}{L}\right) \tag{7.27}$$

The NAND design strategy which emerges from this analysis is summarized as follows, for an n-input NAND. First, we determine the (W/L) ratios for an *equivalent inverter* that satisfies the required V_{OL} value. This gives us the driver transistor ratio $(W/L)_{driver}$ and the load transistor ratio $(W/L)_{load}$. Then, we set the (W/L) ratios of all NAND driver transistors as $(W/L)_1 = (W/L)_2 = \ldots = n\,(W/L)_{driver}$. This guarantees that the series structure consisting of n driver transistors has an equivalent (W/L) ratio of $(W/L)_{driver}$ when all inputs are logic-high.

For a two-input NAND gate, this means that each driver transistor must have a (W/L) ratio twice that of the equivalent inverter driver. If the area occupied by the depletion-type load transistor is negligible, the resulting NAND2 structure will occupy approximately four times the area occupied by the equivalent inverter which has the same static characteristics.

Transient Analysis of NAND Gate

Figure 7.9 shows a NAND2 gate with all parasitic device capacitances. As in the inverter case, we can combine the capacitances seen in Fig. 7.9 into one lumped capacitance, connected between the output node and the ground. The value of the lumped capacitance C_{load}, however, depends on the input voltage conditions.

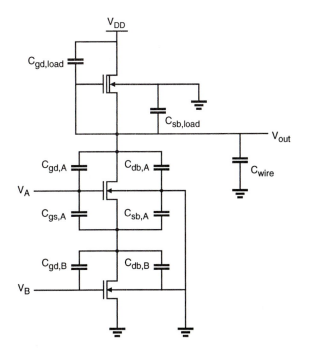

Figure 7.9.   Parasitic device capacitances in the NAND2 gate.

Assume, for example, that the input V_A is equal to V_{OH} and the other input V_B is switching from V_{OH} to V_{OL}. In this case, both the output voltage V_{out} and the internal node voltage V_x will rise, resulting in

$$C_{load} = C_{gd,load} + C_{gd,A} + C_{gd,B} + C_{gs,A} \qquad (7.28)$$
$$+ C_{db,A} + C_{db,B} + C_{sb,A} + C_{sb,load} + C_{wire}$$

Note that this value is quite conservative and fully reflects the internal node capacitances into the lumped output capacitance C_{load}. In reality, only a fraction of the internal node capacitance is reflected into C_{load}.

Now consider another case where V_B is equal to V_{OH} and V_A switches from V_{OH} to V_{OL}. In this case, the output voltage V_{out} will rise, but the internal node voltage V_x will remain low because the bottom driver transistor is on. Thus, the lumped output capacitance is

$$C_{load} = C_{gd,load} + C_{gd,A} + C_{db,A} + C_{sb,load} + C_{wire} \qquad (7.29)$$

It should be noted that the load capacitance in this case is smaller than the load capacitance found in the previous case. Thus, it is expected that the high-to-low switching delay from signal B connected to the bottom transistor is larger than the high-to-low switching delay from signal A connected to the top transistor.

Example 7.1.

A depletion-load nMOS NAND2 gate is simulated with SPICE for the two different input switching events described above. The SPICE input file of the circuit is listed in the following. Note that the total capacitance between the intermediate node X and the ground is assumed to be half of the total capacitance appearing between the output node and the ground.

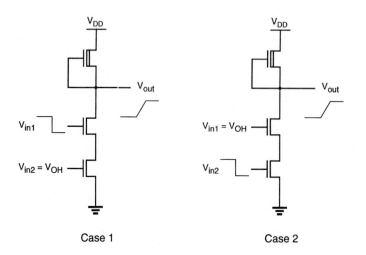

Case 1   Case 2

```
NAND2 circuit delay analysis
m1 3 1 0 0 mn w=5u l=1u
m2 4 2 3 0 mn w=5u l=1u
m3 5 4 4 0 mnd w=1u l=3u
cl 4 0 0.1p
cp 3 0 0.05p
vdd 5 0 dc 5.0
* case 1 (upper input switching from high to low)
vin1 2 0 dc pulse (5.0 0.0 1ns 1ns 2ns 40ns 50ns)
vin2 1 0 dc 5.0
* case 2 (lower input switching from high to low)
* vin1 2 0 dc 5.0
* vin2 1 0 dc pulse (5.0 0.0 1ns 1ns 2ns 40ns 50ns)
.model mn nmos (vto=1.0 kp=25u gamma=0.4)
.model mnd nmos (vto=-3.0 kp=25u gamma=0.4)
.tran 0.1ns 40ns
.print tran v(1) v(2) v(4)
.end
```

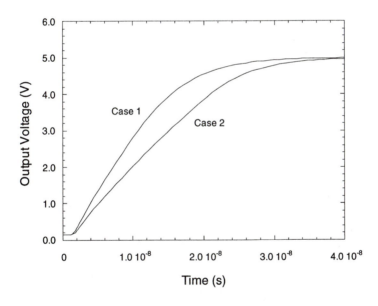

The simulated transient response of the NAND2 gate for both cases is plotted against time above. The time delay difference between the two cases is clearly visible. In fact, the propagation delay time in Case 2 is about 30% larger than that in Case 1, which proves that the input switching order has a significant influence on speed.

7.3. CMOS Logic Circuits

The design and analysis of CMOS co.nbinational logic circuits can be based on the basic principles developed for the nMOS depletion-load logic circuits in the previous section. Figure 7.10 shows the circuit diagram of a two-input CMOS NOR gate. Note that the circuit consists of a parallel-connected n-net and a series-connected complementary p-net. The input voltages V_A and V_B are applied to the gates of one nMOS and one pMOS transistor. The complementary nature of the operation can be summarized as follows.

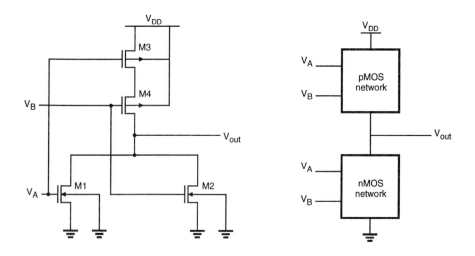

Figure 7.10.　　A CMOS NOR2 gate and its complementary operation: Either the nMOS network is on and the pMOS network is off, or the pMOS network is on and the nMOS network is off.

When either one or both inputs are high, i.e., when the n-net creates a conducting path between the output node and the ground, the p-net is cut-off. On the other hand, if both input voltages are low, i.e., the n-net is cut-off, then the p-net creates a conducting path between the output node and the supply voltage V_{DD}. Thus, the dual or complementary circuit structure allows that, for any given input combination, the output is connected either to V_{DD} or to ground via a low-resistance path. A DC current path between the V_{DD} and ground is not established for any of the input combinations. This yields the low-power operation mode already examined for the simple CMOS inverter circuit in Chapter 5.

The output voltage of the CMOS NOR2 gate will attain a logic-low voltage of $V_{OL} = 0$ and a logic-high voltage of $V_{OH} = V_{DD}$. For circuit design purposes, the switching threshold voltage V_{th} of the CMOS gate emerges as an important design criterion. We start our analysis of the switching threshold by assuming that both input voltages switch simultaneously, i.e., $V_A = V_B$. Furthermore, it is assumed that the device sizes in each block are identical, $(W/L)_{n,A} = (W/L)_{n,B}$ and $(W/L)_{p,A} = (W/L)_{p,B}$, and the substrate-bias effect for the pMOS transistors is neglected for simplicity.

By definition, the output voltage is equal to the input voltage at the switching threshold.

$$V_A = V_B = V_{out} = V_{th} \tag{7.30}$$

It is obvious that the two parallel nMOS transistors are saturated at this point, because $V_{GS} = V_{DS}$. The combined drain current of the two nMOS transistors is

$$I_D = k_n \left(V_{th} - V_{T,n} \right)^2 \tag{7.31}$$

Thus, we obtain the first equation for the switching threshold V_{th}.

$$V_{th} = V_{T,n} + \sqrt{\frac{I_D}{k_n}} \tag{7.32}$$

A quick examination of the p-net shows that the pMOS transistor M3 operates in the linear region, while the other pMOS transistor, M4, is in saturation for $V_{in} = V_{out}$. Thus,

$$I_{D3} = \frac{k_p}{2} \left[2 \left(V_{DD} - V_{th} - \left| V_{T,p} \right| \right) V_{SD3} - V_{SD3}^2 \right] \tag{7.33}$$

$$I_{D4} = \frac{k_p}{2} \left(V_{DD} - V_{th} - \left| V_{T,p} \right| - V_{SD3} \right)^2 \tag{7.34}$$

The drain currents of both pMOS transistors are identical, i.e., $I_{D3} = I_{D4} = I_D$. Thus,

$$V_{DD} - V_{th} - \left| V_{T,p} \right| = 2 \sqrt{\frac{I_D}{k_p}} \tag{7.35}$$

This yields the second equation of the switching threshold voltage V_{th}. Combining (7.32) and (7.35), we obtain

$$V_{th}(\text{NOR2}) = \cfrac{V_{T,n} + \cfrac{1}{2}\sqrt{\cfrac{k_p}{k_n}}\left(V_{DD} - |V_{T,p}|\right)}{1 + \cfrac{1}{2}\sqrt{\cfrac{k_p}{k_n}}} \tag{7.36}$$

Now compare this expression with the switching threshold voltage of the CMOS inverter, which was derived in Chapter 5.

$$V_{th}(\text{INR}) = \cfrac{V_{T,n} + \sqrt{\cfrac{k_p}{k_n}}\left(V_{DD} - |V_{T,p}|\right)}{1 + \sqrt{\cfrac{k_p}{k_n}}} \tag{7.37}$$

If $k_n = k_p$ and $V_{T,n} = |V_{T,p}|$, the switching threshold of the CMOS inverter is equal to $V_{DD}/2$. Using the same parameters, the switching threshold of the NOR2 gate is

$$V_{th}(\text{NOR2}) = \frac{V_{DD} + V_{T,n}}{3} \tag{7.38}$$

which is not equal to $V_{DD}/2$. For example, when $V_{DD} = 5$ V and $V_{T,n} = |V_{T,p}| = 1$ V, the switching threshold voltages of the NOR2 gate and the inverter are

$V_{th}(\text{NOR2}) = 2$ V,   and

$V_{th}(\text{INR}) = 2.5$ V.

The switching threshold voltage of the NOR2 gate can also be obtained by using the equivalent-inverter approach. When both inputs are identical, the parallel-connected nMOS transistors can be represented by a single nMOS transistor with $2k_n$. Similarly, the series-connected pMOS transistors are represented by a single pMOS transistor with $k_p/2$. The resulting equivalent CMOS inverter is shown in Fig. 7.11.

Using the inverter switching threshold expression (7.37) for the equivalent inverter circuit, we obtain

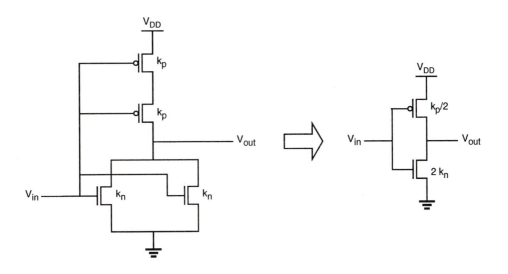

Figure 7.11.   A CMOS NOR2 gate and its inverter equivalent.

$$V_{th}(\text{NOR2}) = \frac{V_{T,n} + \sqrt{\dfrac{k_p}{4k_n}}\left(V_{DD} - |V_{T,p}|\right)}{1 + \sqrt{\dfrac{k_p}{4k_n}}} \tag{7.39}$$

which is identical to (7.36).

From (7.36), we can easily derive simple design guidelines for the NOR2 gate. For example, in order to achieve a switching threshold voltage of $V_{DD}/2$ for simultaneous switching, we have to set $V_{T,n} = |V_{T,p}|$ and $k_p = 4\,k_n$.

Transient Analysis of CMOS NOR Gate

Figure 7.12 shows the CMOS NOR2 gate with the parasitic device capacitances, the inverter equivalent, and the corresponding lumped output load capacitance. In the worst case, the total lumped load capacitance is assumed to be equal to the sum of all internal parasitic device capacitances seen in Fig. 7.12.

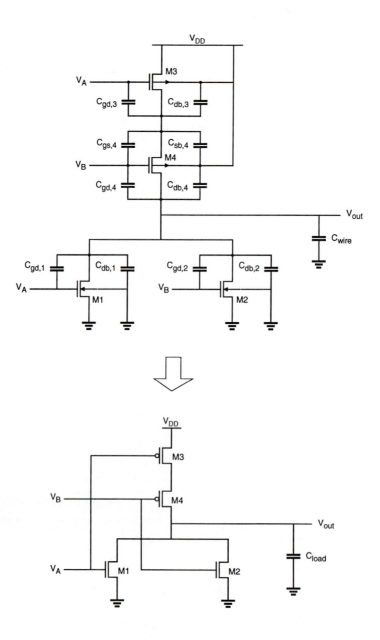

Figure 7.12.   Parasitic device capacitances of the CMOS NOR2 circuit and the simplified equivalent with the lumped output load capacitance.

Figure 7.13 shows a two-input CMOS NAND (NAND2) gate. The operation principle of this circuit is the exact dual of the CMOS NOR2 operation examined earlier. The n-net consisting of two series-connected nMOS transistors creates a conducting path between the output node and the ground only if both input voltages are logic-high, i.e., are equal to V_{OH}. In this case, both of the parallel-connected pMOS transistors in the p-net will be off. For all other input combinations, either one or both of the pMOS transistors will be turned on, while the n-net is cut-off, thus creating a current path between the output node and the power supply voltage.

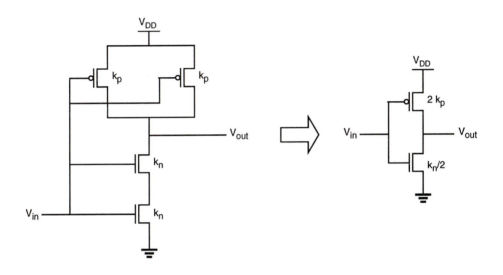

Figure 7.13.   A CMOS NAND2 gate and its inverter equivalent.

Using an analysis similar to the one developed for the NOR2 gate, we can easily calculate the switching threshold for the CMOS NAND2 gate. Again, we will assume that the device sizes in each block are identical, with $(W/L)_{n,A} = (W/L)_{n,B}$ and $(W/L)_{p,A} = (W/L)_{p,B}$. The switching threshold for this gate is then found as

$$V_{th}(\text{NAND2}) = \frac{V_{T,n} + 2\sqrt{\dfrac{k_p}{k_n}}\left(V_{DD} - |V_{T,p}|\right)}{1 + 2\sqrt{\dfrac{k_p}{k_n}}} \qquad (7.40)$$

It is seen from (7.40) that to achieve a switching threshold voltage of $V_{DD}/2$ for simultaneous switching, we have to set $V_{T,n} = |V_{T,p}|$ and $k_n = 4\,k_p$ in the NAND2.

At this point, we can state the following observation about the area requirements of CMOS combinational logic gates. In comparison with equivalent nMOS
depletion-load logic, the total number of transistors in CMOS gates is about twice the
number of transistors in nMOS gates ($2n$ vs. ($n+1$) for n inputs). The silicon area
occupied by the CMOS gate, however, is not necessarily twice the area occupied by
the nMOS depletion-load gate, since a significant portion of the silicon area must be
reserved for signal routing and contacts in both cases. Thus, the area disadvantage of
CMOS logic may actually be smaller than the simple transistor count suggests.

Layout of Simple CMOS Logic Gates

In the following, we will examine simplified layout examples for CMOS
NOR2 and NAND2 gates. Figure 7.14 shows a sample layout of a CMOS NOR2 gate,
using single-layer metal and single-layer polysilicon.

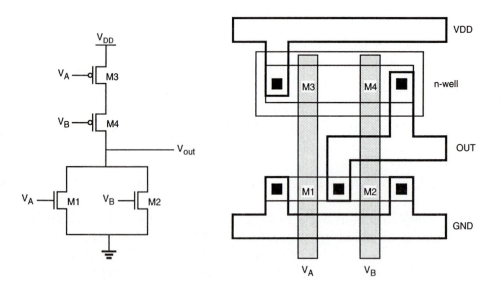

Figure 7.14.   Sample layout of the CMOS NOR2 gate.

In this example, the p-type diffusion area for pMOS transistors and the n-type
diffusion area for nMOS transistors are aligned in parallel to allow simple routing of
the gate signals via two parallel polysilicon lines running vertically. Figure 7.15 shows
the layout of a CMOS NAND2 gate, using the same basic layout principles as in the
NOR2 layout example.

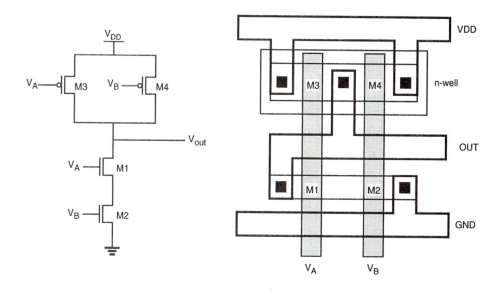

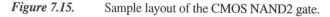

Figure 7.15.   Sample layout of the CMOS NAND2 gate.

Finally, Fig. 7.16 shows a simplified (stick diagram) view of the CMOS NOR2 gate layout given in Fig. 7.14. Here, the diffusion areas are depicted by rectangles, the metal connections and contacts are represented by solid lines and circles, respectively, and the polysilicon columns are represented by cross-hatched strips. The stick-diagram layout does not carry any information on the actual geometry relations of the individual features, but it conveys valuable information on the relative placement of the transistors and their interconnections.

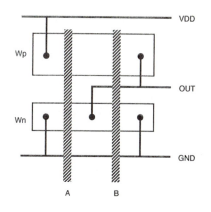

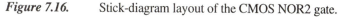

Figure 7.16.   Stick-diagram layout of the CMOS NOR2 gate.

7.4. Complex Logic Circuits

To realize arbitrary Boolean functions of multiple input variables, the basic circuit structures and design principles developed for simple NOR and NAND gates in the previous sections can easily be extended to complex logic gates. The ability to realize complex logic functions using a small number of transistors is one of the most attractive features of nMOS and CMOS logic circuits.

Consider the following Boolean function as an example.

$$Z = \overline{A(D+E) + BC} \tag{7.41}$$

The nMOS depletion-load complex logic gate that is used to realize this function is shown in Fig. 7.17. Inspection of the circuit topology reveals the simple design principle of the pull-down network:

- "OR" operations are performed by parallel-connected drivers.
- "AND" operations are performed by series-connected drivers.
- Inversion is provided by the nature of MOS circuit operation.

The design principles stated here for individual inputs and corresponding driver transistors can also be extended to circuit sub-blocks, so that Boolean OR and AND operations can be performed in a nested circuit structure. Thus, we obtain a circuit topology which consists of series- and parallel-connected branches, as shown below.

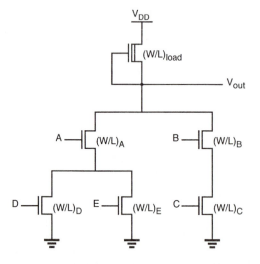

Figure 7.17.   nMOS complex logic gate realizing the Boolean function given in (7.41).

Here, the left-hand side nMOS driver branch consisting of three driver transistors is used to perform the logic function $A(D+E)$, while the right-hand side branch performs the function BC. By connecting the two branches in parallel, and by placing the load transistor between the output node and the power supply voltage V_{DD}, we obtain the complex function given in (7.41). Each input variable is assigned to only one driver.

For the analysis and design of complex logic gates, we can employ the equivalent-inverter approach already used for the simpler NOR and NAND gates. It can be shown for the circuit in Fig. 7.17 that, if all input variables are logic-high, the equivalent-driver (W/L) ratio of the pull-down network consisting of five nMOS transistors is

$$\left(\frac{W}{L}\right)_{equivalent} = \frac{1}{\dfrac{1}{\left(\dfrac{W}{L}\right)_B}+\dfrac{1}{\left(\dfrac{W}{L}\right)_C}} + \frac{1}{\dfrac{1}{\left(\dfrac{W}{L}\right)_A}+\dfrac{1}{\left(\dfrac{W}{L}\right)_D}+\left(\dfrac{W}{L}\right)_E} \qquad (7.42)$$

For calculating the logic-low voltage level V_{OL}, we have to consider various cases, since the value of V_{OL} actually depends on the number *and* the configuration of the conducting nMOS transistors in each case. All possible output-to-ground paths are tabulated below. Also, each path is assigned a class number which reflects the total resistance of the current path.

A - D	Class 1
A - E	Class 1
B - C	Class 1
A - D - E	Class 2
A - D - B - C	Class 3
A - E - B - C	Class 3
A - D - E - B - C	Class 4

Assuming that all driver transistors have the same (W/L) ratio, a Class 1 path such as (B-C) has the highest series resistance, followed by Class 2, Class 3, etc. Consequently, the logic-low voltage levels corresponding to each class have the following order, where each subscript numeral represents the class number.

$$V_{OL1} > V_{OL2} > V_{OL3} > V_{OL4} \qquad (7.43)$$

The design of complex logic gates is based on the same ideas as the design of NOR and NAND gates. We usually start by specifying a maximum V_{OL} value. The design objective is to determine the driver and load transistor sizes so that the complex logic gate achieves the specified V_{OL} value even in the worst case. The given V_{OL} value first allows us to find the $(W/L)_{load}$ and $(W/L)_{driver}$ ratios for an equivalent inverter. Next, we have to identify all worst-case (Class 1) paths in the circuit, and determine the transistor sizes in these worst-case paths such that each Class 1 path has the equivalent driver ratio of $(W/L)_{driver}$.

In this example, this design strategy yields the following ratios for the three worst-case paths.

$$\left(\frac{W}{L}\right)_A = \left(\frac{W}{L}\right)_D = 2\left(\frac{W}{L}\right)_{driver}$$

$$\left(\frac{W}{L}\right)_A = \left(\frac{W}{L}\right)_E = 2\left(\frac{W}{L}\right)_{driver} \tag{7.44}$$

$$\left(\frac{W}{L}\right)_B = \left(\frac{W}{L}\right)_C = 2\left(\frac{W}{L}\right)_{driver}$$

The transistor sizes found above guarantee that, for all other input combinations, the logic-low output voltage level will be less than the specified V_{OL}.

Complex CMOS Logic Gates

The realization of the n-net, or pull-down network, is based on the same basic design principles examined earlier. The pMOS pull-up network, on the other hand, must be the dual network of the n-net. This means that all parallel connections in the nMOS pull-down network will correspond to a series connection in the pMOS pull-up network, and all series connections in the pull-down network correspond to a parallel connection in the pull-up network.

Figure 7.18 shows the simple construction of the dual p-net (pull-up) graph from the n-net (pull-down) graph. Each driver transistor in the pull-down network is represented by an edge, and each node is represented by a vertex in the pull-down graph. Next, a new vertex is created within each confined area in the pull-down graph, and neighboring vertices are connected by edges which cross each edge in the pull-down graph only once. This new graph represents the pull-up network. The resulting CMOS complex logic gate is shown in Fig. 7.19.

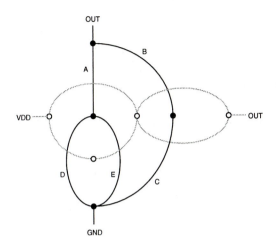

Figure 7.18.   Construction of the dual pull-up graph from the pull-down graph, using the dual-graph concept.

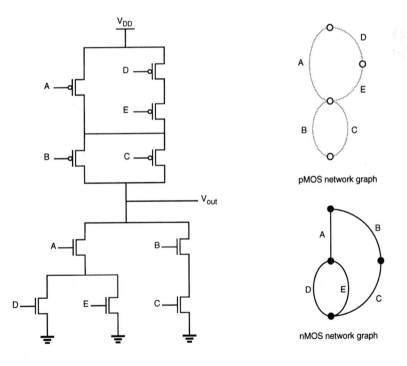

pMOS network graph

nMOS network graph

Figure 7.19.   A complex CMOS logic gate realizing the Boolean function (7.41).

Layout of Complex CMOS Logic Gates

Now, we will investigate the problem of constructing a minimum-area layout for the complex CMOS logic gate. Figure 7.20 shows the stick-diagram layout of a "first attempt," using an arbitrary ordering of the polysilicon gate columns. Note that in this case, the separation Δd between the polysilicon columns must allow for one diffusion-to-diffusion separation and two metal-to-diffusion contacts on both sides. This certainly consumes a considerable amount of extra silicon area.

If we can minimize the number of diffusion-area breaks both for nMOS and for pMOS transistors, the separation between the polysilicon gate columns can be made smaller, which will reduce the overall horizontal dimension and, hence, the circuit layout area. The number of diffusion breaks can be minimized by changing the *ordering* of the polysilicon columns.

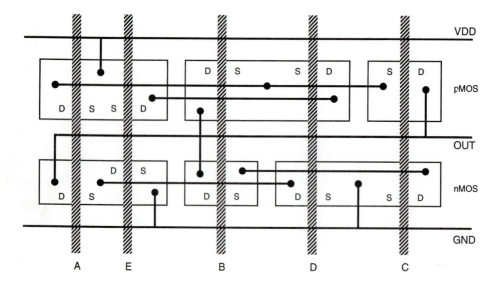

Figure 7.20.   Stick-diagram layout of the complex CMOS logic gate, with an arbitrary ordering of the polysilicon gate columns.

A simple method for finding the optimum gate ordering is the Euler-path approach: find a Euler path in the pull-down graph and a Euler path in the pull-up graph with the identical ordering of input labels, i.e., find a common Euler path for both graphs. The Euler path is defined as an uninterrupted path that traverses each edge (branch) of the graph exactly once. Figure 7.21 shows the construction of a common Euler path for both graphs in our example.

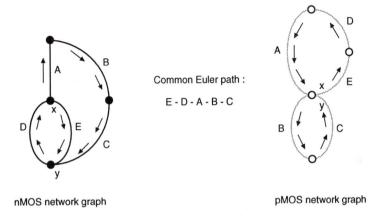

Common Euler path :

E - D - A - B - C

nMOS network graph   pMOS network graph

Figure 7.21. Finding a common Euler path in both graphs for n-net and p-net provides a gate ordering that minimizes the number of diffusion breaks and, thus, minimizes the logic-gate layout area. In both cases, the Euler path starts at (x) and ends at (y).

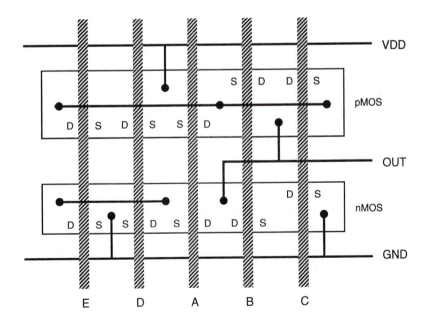

Figure 7.22. Optimized stick-diagram layout of the complex CMOS logic gate.

It is seen that there is a common sequence (E - D - A - B - C) in both graphs, i.e., a Euler path. The polysilicon gate columns can be arranged according to this sequence, which results in uninterrupted p-type and n-type diffusion areas. The stick diagram of the new layout is shown in Fig. 7.22. In this case, the polysilicon column separation Δd has to allow for only one metal-to-diffusion contact. The advantages of this new layout are more compact (smaller) layout area, simple routing of signals, and consequently, less parasitic capacitance.

As a further example of complex CMOS gates, the full-CMOS implementation of the exclusive-OR (XOR) function is shown in Fig. 7.23. Note that two additional inverters are also needed to obtain the inverse of both input variables (A and B). With these inverters, the CMOS XOR circuit in Fig. 7.23 requires a total of 12 transistors. Other CMOS realizations of the XOR gate that can be implemented with fewer transistors will be examined later.

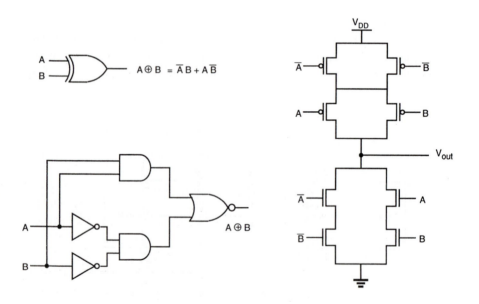

Figure 7.23.   Full-CMOS implementation of the XOR function.

AOI and OAI Gates

While theoretically there are no strict limitations on the topology of the pull-down and the corresponding pull-up networks in a complex CMOS logic gate, we may recognize two important circuit categories as subsets of the general complex CMOS gate topology. These are the AND-OR-INVERT (AOI) gates and the OR-AND-

INVERT (OAI) gates. The AOI gate, as its name suggests, enables the sum-of-products realization of a Boolean function in one logic stage (Fig. 7.24). The pull-down net of the AOI gate consists of parallel branches of series-connected nMOS driver transistors. The corresponding p-type pull-up network can simply be found using the dual-graph concept.

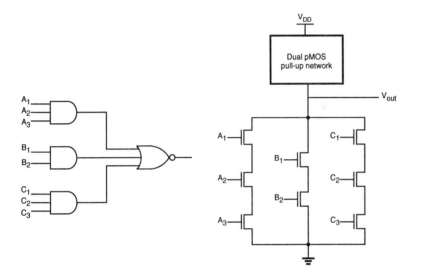

Figure 7.24.   An AND-OR-INVERT (AOI) gate and the corresponding pull-down net.

The OAI gate, on the other hand, enables the product-of-sums realization of a Boolean function in one logic stage (Fig. 7.25). The pull-down net of the OAI gate consists of series branches of parallel-connected nMOS driver transistors, while the corresponding p-type pull-up network can be found using the dual-graph concept.

Pseudo-nMOS Gates

The large area requirements of complex CMOS gates present a problem in high-density designs, since two complementary transistors, one nMOS and one pMOS, are needed for every input. One possible approach to reduce the number of transistors is to use a single pMOS transistor, with its gate terminal connected to ground, as the *load* device (Fig. 7.26). With this simple pull-up arrangement, the complex gate can be implemented with much fewer transistors. The similarities of pseudo-nMOS gates to depletion-load nMOS logic gates are obvious.

The most significant disadvantage of using a pseudo-nMOS gate instead of a full-CMOS gate is the nonzero static power dissipation, since the always-on pMOS load device conducts a steady-state current when the output voltage is lower than V_{DD}. Also, the value of V_{OL} and the noise margins are now determined by the *ratio* of the pMOS load transconductance to the pull-down or driver transconductance.

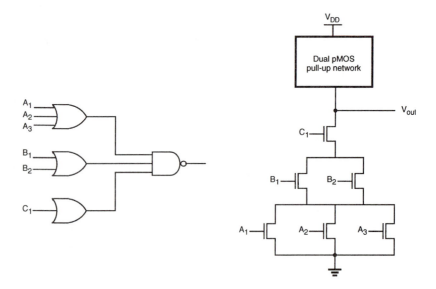

Figure 7.25. An OR-AND-INVERT (OAI) gate, and the corresponding pull-down net.

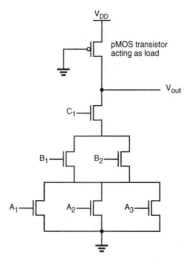

Figure 7.26. The pseudo-nMOS implementation of the OAI gate in Fig. 7.25.

Example 7.2.

The simplified layout of a CMOS complex logic circuit is given below. Draw the corresponding circuit diagram, and find an equivalent CMOS inverter circuit for simultaneous switching of all inputs, assuming that $(W/L)_p = 15$ for all pMOS transistors and $(W/L)_n = 10$ for all nMOS transistors.

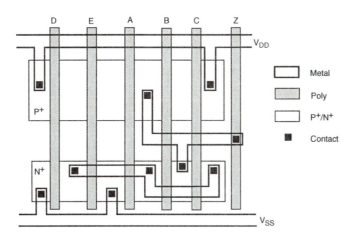

The circuit diagram can be found from the layout by inspection:

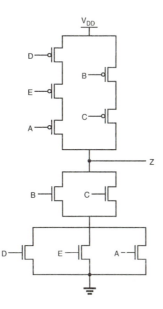

The Boolean function realized by this circuit is

$$Z = \overline{(D + E + A)(B + C)}$$

The equivalent (W/L) ratios of the nMOS network and the pMOS network are determined by using the series-parallel equivalency rules discussed earlier in this chapter, as follows.

$$\left(\frac{W}{L}\right)_{n,eq} = \frac{1}{\dfrac{1}{\left(\dfrac{W}{L}\right)_D + \left(\dfrac{W}{L}\right)_E + \left(\dfrac{W}{L}\right)_A} + \dfrac{1}{\left(\dfrac{W}{L}\right)_B + \left(\dfrac{W}{L}\right)_C}}$$

$$= \frac{1}{\dfrac{1}{30} + \dfrac{1}{20}} = 12$$

$$\left(\frac{W}{L}\right)_{p,eq} = \frac{1}{\dfrac{1}{\left(\dfrac{W}{L}\right)_D} + \dfrac{1}{\left(\dfrac{W}{L}\right)_E} + \dfrac{1}{\left(\dfrac{W}{L}\right)_A}} + \frac{1}{\dfrac{1}{\left(\dfrac{W}{L}\right)_B} + \dfrac{1}{\left(\dfrac{W}{L}\right)_C}}$$

$$= \frac{1}{\dfrac{1}{15} + \dfrac{1}{15} + \dfrac{1}{15}} + \frac{1}{\dfrac{1}{15} + \dfrac{1}{15}} = 12.5$$

7.5. CMOS Transmission Gates (TGs) and TG Logic

In this section, we will examine a simple switch circuit called the CMOS transmission gate (TG) or pass gate, and present a new class of logic circuits which use the TGs as their basic building blocks. As shown in Fig. 7.27, the CMOS transmission gate consists of one nMOS and one pMOS transistor, connected in parallel. The gate voltages applied to these two transistors are also set to be complementary signals. As such, the CMOS TG operates as a bidirectional switch between the nodes A and B which is controlled by the signal C.

If the control signal C is logic-high, i.e., equal to V_{DD}, then both transistors are turned on and provide a low-resistance current path between the nodes A and B. If, on the other hand, the control signal C is low, then both transistors will be off, and the path between the nodes A and B will be an open circuit. This condition is also called the high-impedance state.

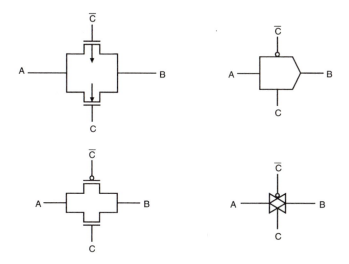

Figure 7.27.   Four different representations of the CMOS transmission gate (TG).

Note that the substrate terminal of the nMOS transistor is connected to ground and the substrate terminal of the pMOS transistor is connected to V_{DD}. Thus, we must take into account the substrate-bias effect for both transistors, depending on the bias conditions. Figure 7.27 also shows three other commonly used symbolic representations of the CMOS transmission gate.

For a detailed DC analysis of the CMOS transmission gate, we will consider the following bias condition, shown in Fig. 7.28. The input node (A) is connected to a constant logic-high voltage, $V_{in} = V_{DD}$. The control signal is also logic-high, thus ensuring that both transistors are turned on. The output node (B) may be connected to a capacitor, which represents capacitive loading of the subsequent logic stages driven by the transmission gate. We will now investigate the input-output current-voltage relationship of the CMOS TG as a function of the output voltage V_{out}.

It can be seen from Fig. 7.28 that the drain-to-source and the gate-to-source voltages of the nMOS transistor are

$$V_{DS,n} = V_{DD} - V_{out}$$
$$V_{GS,n} = V_{DD} - V_{out}$$

(7.45)

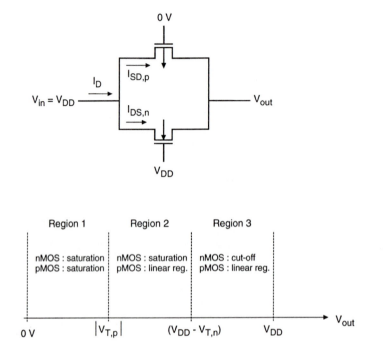

Figure 7.28.   Bias conditions and operating regions of the CMOS transmission gate, shown as functions of the output voltage.

Thus, the nMOS transistor will be turned off for $V_{out} > V_{DD} - V_{T,n}$ and will operate in the saturation mode for $V_{out} < V_{DD} - V_{T,n}$. The V_{DS} and V_{GS} voltages of the pMOS transistor are

$$V_{DS,p} = V_{out} - V_{DD}$$
$$V_{GS,n} = -V_{DD}$$

(7.46)

Consequently, the pMOS transistor is in saturation for $V_{out} < |V_{T,p}|$, and it operates in the linear region for $V_{out} > |V_{T,p}|$. Note that, unlike the nMOS transistor, the pMOS transistor remains turned on, regardless of the output voltage level V_{out}.

This analysis has shown that we can identify three operating regions for the CMOS transmission gate, depending on the output voltage level. These operating regions are depicted in Fig. 7.28 as functions of V_{out}.

The total current flowing through the transmission gate is the sum of the nMOS drain current and the pMOS drain current.

$$I_D = I_{DS,n} + I_{SD,p}$$ (7.47)

At this point, we may devise an *equivalent resistance* for each transistor in this structure, as follows.

$$R_{eq,n} = \frac{V_{DD} - V_{out}}{I_{DS,n}}$$

$$R_{eq,p} = \frac{V_{DD} - V_{out}}{I_{SD,p}}$$ (7.48)

The total equivalent resistance of the CMOS TG will then be the parallel equivalent of these two resistances, $R_{eq,n}$ and $R_{eq,p}$. Now, we will calculate the equivalent resistance values for the three operating regions of the transmission gate.

Region 1

Here, the output voltage is smaller than the absolute value of the pMOS transistor threshold voltage, i.e., $V_{out} < |V_{T,p}|$. According to Fig. 7.28, both transistors are in saturation. We obtain the series resistance of both devices as

$$R_{eq,n} = \frac{2(V_{DD} - V_{out})}{k_n (V_{DD} - V_{out} - V_{T,n})^2}$$ (7.49)

$$R_{eq,p} = \frac{2(V_{DD} - V_{out})}{k_p (V_{DD} - |V_{T,p}|)^2}$$ (7.50)

Note that the source-to-substrate voltage of the nMOS transistor is equal to the output voltage V_{out}, while the source-to-substrate voltage of the pMOS transistor is equal to zero. Thus, we have to take into account the substrate-bias effect for the nMOS transistor in our calculations.

Region 2

In this region, $|V_{T,p}| < V_{out} < (V_{DD} - V_{T,n})$. Thus, the pMOS transistor now operates in the linear region, while the nMOS transistor continues to operate in saturation.

$$R_{eq,n} = \frac{2(V_{DD} - V_{out})}{k_n (V_{DD} - V_{out} - V_{T,n})^2} \tag{7.51}$$

$$
\begin{aligned}
R_{eq,p} &= \frac{2(V_{DD} - V_{out})}{k_p \left[2(V_{DD} - |V_{T,p}|)(V_{DD} - V_{out}) - (V_{DD} - V_{out})^2 \right]} \\
&= \frac{2}{k_p \left[2(V_{DD} - |V_{T,p}|) - (V_{DD} - V_{out}) \right]}
\end{aligned} \tag{7.52}
$$

Region 3

Here, the output voltage is $V_{out} > (V_{DD} - V_{T,n})$. Consequently, the nMOS transistor will be turned off, which results in an open-circuit equivalent. The pMOS transistor will continue to operate in the linear region.

$$R_{eq,p} = \frac{2}{k_p \left[2(V_{DD} - |V_{T,p}|) - (V_{DD} - V_{out}) \right]} \tag{7.53}$$

Combining the equivalent resistance values found for the three operating regions, we can now plot the total resistance of the CMOS transmission gate as a function of the output voltage V_{out}, as shown in Fig. 7.29.

It can be seen that the total equivalent resistance of the TG remains relatively constant, i.e., its value is almost independent of the output voltage, whereas the individual equivalent resistances of both the nMOS and the pMOS transistors are strongly dependent on V_{out}. This property of the CMOS TG is naturally quite desirable. A CMOS pass gate which is turned on by a logic-high control signal can be replaced by its simple equivalent resistance for dynamic analysis, as shown in Fig. 7.30.

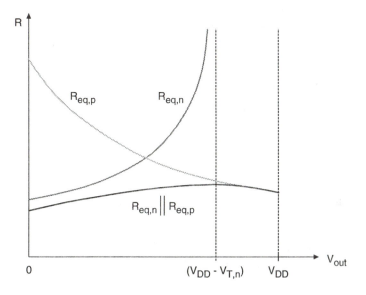

Figure 7.29. Equivalent resistance of the CMOS transmission gate, plotted as a function of the output voltage.

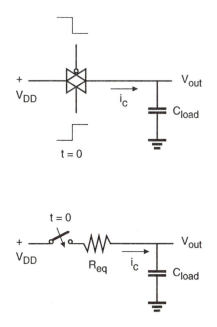

Figure 7.30. Replacing the CMOS TG with its resistor equivalent for transient analysis.

The implementation of CMOS transmission gates in logic circuit design usually results in compact circuit structures which may even require a smaller number of transistors than their standard CMOS counterparts. Note that the control signal *and* its complement must be available simultaneously for TG applications. Figure 7.31 shows a two-input multiplexor circuit consisting of two CMOS transmission gates. The operation of the multiplexor can be understood quite easily: If the control input S is logic-high, then the bottom TG will conduct, and the output will be equal to the input B. If the control signal is low, the bottom TG will turn off and the top TG will connect the input A to the output node.

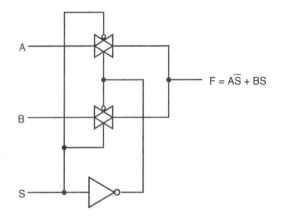

Figure 7.31.   Two-input multiplexor circuit implemented using two CMOS TGs.

Figure 7.32 shows an eight-transistor implementation of the logic XOR function, using two CMOS TGs and two CMOS inverters. The same function can also be implemented using only six transistors, as shown in Fig. 7.33.

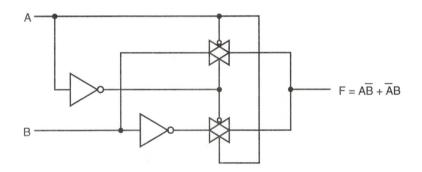

Figure 7.32.   Eight-transistor CMOS TG implementation of the XOR function.

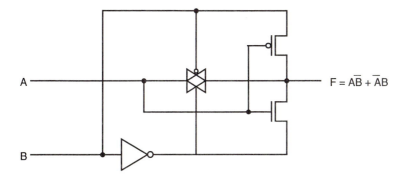

$$F = A\overline{B} + \overline{A}B$$

Figure 7.33.   Six-transistor CMOS TG implementation of the XOR function.

Using the generalized multiplexor approach, each Boolean function can be realized with a TG logic circuit. As an example, Figure 7.34(a) shows the TG logic implementation of a three-variable Boolean function. Note that the three input variables *and* their inverses must be used to control the CMOS transmission gates. Including the three inverters not shown here, the TG implementation requires a total of 14 transistors. An important point in TG logic design is that a conducting TG network (*low-impedance* path) should always be provided between the output node and one of the inputs, for all possible input combinations. This is to make sure that the output node with its capacitive load is never left in a *high-impedance* state.

If each CMOS transmission gate in TG logic circuits is realized with a full nMOS/pMOS pair, the disjoint n-well structures of the pMOS transistors and the diffusion contacts may cause a significant overall area increase. In an attempt to reduce the silicon area occupied by TG circuits, the transmission gates can be laid out as separated nMOS/pMOS pairs with all pMOS transistors placed in one single n-well, as shown in Fig. 7.34(b). However, the *routing* area required for connecting the p-type diffusion regions to input signals must be carefully considered.

References

1.   N.H.E. Weste and K. Eshraghian, *Principles of CMOS VLSI Design– A Systems Perspective*, second edition, Reading, MA: Addison-Wesley, 1993.

2.   J.P. Uyemura, *Fundamentals of MOS Digital Integrated Circuits*, Reading, MA: Addison-Wesley, 1988.

3.   L.A. Glasser and D.W. Dobberpuhl, *The Design and Analysis of VLSI Circuits*, Reading, MA: Addison-Wesley, 1985.

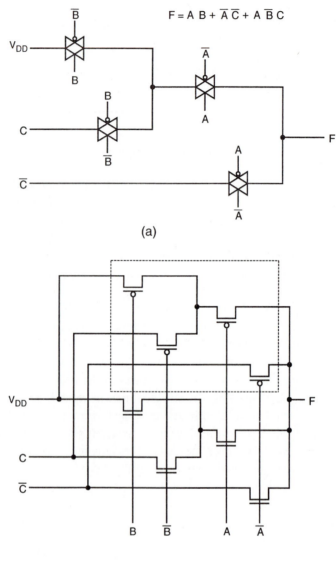

$$F = A\,B + \overline{A}\,\overline{C} + A\,\overline{B}\,C$$

(a)

(b)

Figure 7.34. (a) CMOS TG realization of a three-variable Boolean function. (b) All pMOS transistors can be placed into one n-well to save area.

4.   F. J. Hill and G. R. Peterson, *Computer-Aided Logical Design with Emphasis on VLSI*, fourth edition, New York, NY: John Wiley & Sons, Inc., 1993.

5.   T. Sakurai and A. R. Newton, "Delay analysis of series-connected MOSFET circuits," *IEEE Journal of Solid-State Circuits*, vol. 26, no. 2, pp. 122-131, February 1991.

6.   T. Uehara and W. M. Van Cleemput, "Optimal layout of CMOS functional arrays," *IEEE Transactions on Computers*, vol. C-30, no. 5, pp. 305-313, May 1981.

Exercise Problems

7.1   A CMOS circuit was laid out based on company X's 3 μm design rules as shown in Fig. P7.1 with $W_N = 7$ μm and $W_P = 15$ μm.

   (a)   From Fig. P7.1, determine the circuit configuration and draw the circuit diagram.

   (b)   For simple hand analysis, make the following assumptions:

   i) Wiring parasitic capacitances and resistances are negligible.

   ii) Device parameters are

	nMOST	pMOST
V_{T0}	1.0 V	−1.0 V
t_{ox}	500 Å	500 Å
k'	20 μA/V²	10 μA/V²
X_j	0.5 μm	0.5 μm
L_D	0.5 μm	0.5 μm

   iii) The total capacitance at node I is 0.6 pF.

   iv) An ideal step-pulse signal is applied to the CK terminal such that

$$V_{CK} = 5 \text{ V}, \qquad t < 0$$
$$V_{CK} = 0 \text{ V}, \qquad 0 \le t < T$$
$$V_{CK} = 5 \text{ V}, \qquad t \ge T_w$$
$$V_{DD} = 5 \text{ V}$$

v) At $t = 0$, the node voltage at I is zero.

vi) The input voltages at A_1, B_1, and B_2 are zero for $0 \le t \le T_W$. Find the minimum T_W that allows V_I to reach 2.5 V.

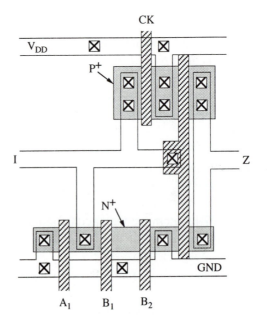

Figure P7.1

7.2   Calculate the equivalent W/L of the two nMOSTs with W_1/L and W_2/L connected in series. For simplicity, neglect the body effect, i.e., the threshold voltages of individual transistors are constant and do not depend on the source voltages. Although this is not true in reality, such an assumption is necessary for simple analysis with a reasonably good approximation.

7.3   Analytical expressions for V_{th} (logic) have been derived in Chapter 7 for the CMOS NOR2 gate. Now consider the CMOS NAND2 gate for the following cases and using $k_p = k_n = 100 \ \mu A/V^2$:

*   two inputs switching simultaneously
*   top nMOST switching while the bottom nMOST's gate is tied to V_{DD}
*   top nMOST gate is tied to V_{DD} and the gate input of the bottom nMOST is changing

(a) Derive an analytical expression for V_{th} corresponding to the first case. Also find the V_{th} value for the first case for $V_{DD} = 5$ V when the magnitudes of the threshold voltages are 1 V with $\gamma = 0$.

(b) Determine V_{th} for all three cases by using SPICE.

(c) For $C_{load} = 0.2$ pF, calculate 50% delays (low-to-high and high-to-low propagation delays) for an ideal pulse input signal for each of the three cases by assuming that C_{load} includes all of the internal parasitic capacitances. Verify the results using SPICE.

7.4 Write down the SPICE input description for transistor connections, source and drain parasitics in terms of areas, and perimeters for the layout shown in Example 7.2. Neglect the wiring capacitances in the polysilicon and metal runners. Default model names to be used for pMOS and nMOS are MODP and MODN. Assume $L = 1$ μm and $Y = 10$ μm for all transistors.

7.5 For the gate shown in Fig. P7.5,

- Pull-up transistor ratio is 5/5
- Pull-down transistor ratios are 100/5
- $V_{T0} = 1.0$ V
- $\gamma = 0.4$ V$^{1/2}$
- $|2\phi_F| = 0.6$ V

(a) Identify the worst-case input combination(s) for V_{OL}.
(b) Calculate the worst-case value of V_{OL}. (Assume that all pull-down transistors have the same body bias and initially, that $V_{OL} \approx 5\% V_{DD}$.)

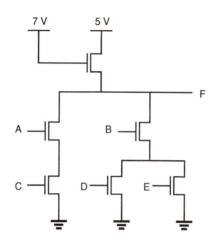

Figure P7.5

7.6　A store has one express register and three regular ones. It is the store policy that the express register be open only when two or more of the other registers are busy. Assume that the Boolean variables A, B, and C reflect the status of each of the regular registers (1 busy, 0 idle). Design the logic circuit, with A, B, and C as inputs and F as output, to automatically notify the manager (by setting $F = 1$) to open the express register. Present two solutions, the first using only NAND gates, the second using only NOR gates.

7.7　Calculate V_{OL}, V_{OH}, V_{IL}, V_{IH}, NM_L, and NM_H for a two-input NOR gate fabricated in a CMOS technology.

- $(W/L)_p = 4$
- $(W/L)_n = 1$
- $V_{Tn} = 0.7$ V
- $V_{Tp} = -0.7$ V
- $k_n' = 40\ \mu A/V^2$
- $k_p' = 20\ \mu A/V^2$
- $V_{DD} = 5$ V

Compare your answers with SPICE simulation results.

7.8　Use a layout editor (e.g., Magic) to lay out a two-input CMOS NAND gate. All devices have $W = 10\ \mu m$. N-channel transistors have $L_{eff} = 1\ \mu m$ and p-channel transistors have $L_{eff} = 2\ \mu m$. You can calculate the drawn channel lengths by assuming that $L_D = 0.25\ \mu m$. After you lay out the gate use the design rule checker. Finally, you should have the layout editor perform parasitic capacitance extraction.

7.9　Assume that the 2-input NAND gate in Problem 7.8 is driving a 0.01 pF load. Use hand calculations to estimate t_{PLH} and t_{PHL}. Do not forget to add in the parasitic capacitances extracted from your layout! Check your answer with SPICE. Use:
- $k_n' = 20\ \mu A/V^2$
- $k_p' = 10\ \mu A/V^2$
- $V_{Tn} = |V_{Tp}| = 1.0$ V

7.10　Consider the circuit shown in Fig. P7.10.

(a)　Determine the logic function F.
(b)　Calculate W_L/L_L such that V_{OL} does not exceed 0.4 V.
(c)　Qualitatively, would W_L/L_L increase or decrease if the same conditions of (b) are to be achieved but $\gamma = 0.4$ $V^{1/2}$?

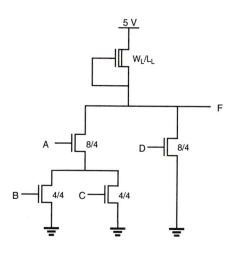

Figure P7.10

7.11   Consider the circuit shown in Fig. P7.11.

(a)   Determine the logic function F.

(b)   Design a circuit to implement the same logic function, but using NOR gates. Draw a transistor-level schematic and use nMOS E-D technology.

(c)   Design a circuit to implement the same logic function, but use an AOI (AND-OR-INVERT) gate. Draw a transistor-level schematic and use CMOS technology.

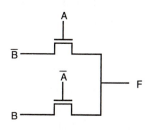

Figure P7.11

7.12   The enhancement-type MOS transistors have the following parameters:

- $V_{DD} = 5$ V
- $|V_{T0}| = 1.0$ V for both nMOS and pMOS transistor
- $\lambda = 0.0$ V^{-1}
- $k_p' = 20$ μA/V^2
- $k_n' = 50$ μA/V^2

For a CMOS complex gate OAI432 with $(W/L)_p = 30$ and $(W/L)_n = 40$,

(a)   Calculate the W/L sizes of an equivalent inverter with the weakest pull-down and pull-up. Such an inverter can be used to calculate worst-case pull-up and pull-down delays, with proper incorporation of parasitic capacitances at internal nodes into the total load capacitance. In this problem, you are asked to calculate only $(W/L)_{worse-case}$ for both p-channel and n-channel MOSFETs by neglecting the parasitic capacitances.

(b)   Do the layout of OAI432 with minimal diffusion breaks to reduce the number of polysilicon column pitches. With proper ordering of polysilicon gate columns, the number of diffusion breaks can be minimized. One way of achieving such a goal is to find a Euler path common to both p-channel and n-channel nets using graph models. Symbolic layout that shows source and drain connections is sufficient to answer this problem.

7.13   Consider a fully complementary CMOS transmission gate with its input terminal tied to ground (0 V) while the other non-gate terminal is tied to a 1 pF load capacitor initially charged to 5 V. Use the V_{T0}, k_p', and k_n' values in Problem 7.12. At $t = 0$, both transistors are fully turned on by clock signals to start the discharge of the capacitor.

(a)   Plot the effective resistance of this transmission gate as a function of capacitor voltage when $(W/L)_p = 50$ and $(W/L)_n = 40$. From the plot find the average value of the resistance. Then calculate the RC delay for the capacitor voltage to change from 5 V to 2.5 V. This can be found by solving the RC-circuit differential equation.

(b)   Verify your answer to part (a) by using SPICE simulation. The source/drain parasitic capacitances can be neglected.

CHAPTER 8

SEQUENTIAL
MOS LOGIC CIRCUITS

8.1. Introduction

In all of the combinational logic circuits examined in Chapter 7, if we neglect the propagation delay time, the output levels at any given time point are directly determined as Boolean functions of the input variables applied at that time. Thus, the combinational circuits lack the capability of storing any previous events, or displaying an output behavior which is dependent upon the previously applied inputs. Circuits of this type are also classified as non-regenerative circuits, since there is no feedback relationship between the output and the input.

The other major class of logic circuits is called sequential circuits, in which the output is determined by the current inputs as well as the previously applied input variables. Figure 8.1(a) shows a sequential circuit consisting of a combinational circuit and a memory block in the feedback loop. In most cases, regenerative behavior of sequential circuits is due to either a direct or indirect feedback connection between the output and the input. Regenerative operation can, under certain conditions, also be interpreted as a simple memory function. The critical components of sequential systems are the basic regenerative circuits, which can be classified into three main groups: bistable circuits, monostable circuits, and astable circuits. The general classification of non-regenerative and regenerative logic circuits is shown in Fig. 8.1.

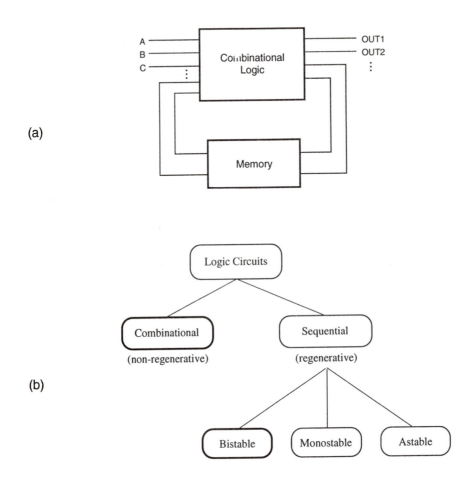

Figure 8.1. (a) Sequential circuit consisting of a combinational logic block and a memory block in the feedback loop. (b) Classification of logic circuits based on their temporal behavior.

Bistable circuits have, as their name implies, two stable states or operation modes, each of which can be attained under certain input and output conditions. Monostable circuits, on the other hand, have only one stable operating point (state). Even if the circuit experiences an external perturbation, the output eventually returns to the single stable state after a certain time period. Finally, in astable circuits, there is no stable operating point or state which the circuit can preserve for a certain time period. Consequently, the output of an astable circuit must oscillate without settling into a stable operating mode. The ring oscillator circuit examined in Chapter 6 would be a typical example for astable regenerative circuits.

Among these three main groups of regenerative circuit types, the bistable circuits are by far the most widely used and the most important class. All basic latch and flip-flop circuits, registers, and memory elements used in digital systems fall into this category. In the following, we will first examine the electrical behavior of the simple bistable element, and then present some of its useful applications.

8.2. Behavior of Bistable Elements

The basic bistable element to be examined in this section consists of two identical cross-coupled inverter circuits, as shown in Fig. 8.2(a). Here, the output voltage of inverter (1) is equal to the input voltage of inverter (2), i.e., $v_{o1} = v_{i2}$, and the output voltage of inverter (2) is equal to the input voltage of inverter (1), i.e., $v_{o2} = v_{i1}$. In order to investigate the static input-output behavior of both inverters, we start by plotting the voltage transfer characteristic of inverter (1) with respect to the v_{o1} - v_{i1} axis pair. Notice that the input and output voltages of inverter (2) correspond to the output and input voltages of inverter (1), respectively. Consequently, we can also plot the voltage transfer characteristic of inverter (2) using the same axis pair, as shown in Fig. 8.2(b).

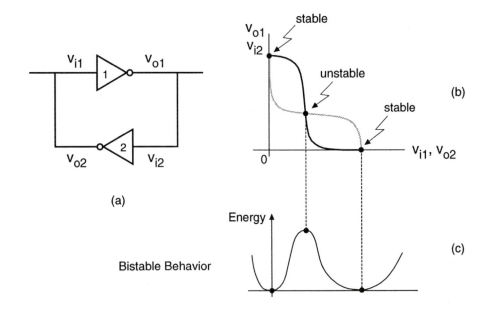

Figure 8.2.   Static behavior of the two-inverter basic bistable element: (a) Circuit schematic. (b) Intersecting voltage transfer curves of the two inverters, showing the three possible operating points. (c) Qualitative view of the potential energy levels corresponding to the three operating points.

It can be seen that the two voltage transfer characteristics intersect at three points. Simple reasoning can help us to conclude that two of these operating points are stable, as indicated in Fig. 8.2(b). If the circuit is initially operating at one of these two stable points, it will preserve this state unless it is forced externally to change its operating point. Note that the gain of each inverter circuit, i.e., the slope of the respective voltage transfer curves, is smaller than unity at the two stable operating points. Thus, in order to change the state by moving the operating point from one stable point to the other, a sufficiently large external voltage perturbation must be applied so that the voltage gain of the inverter loop becomes larger than unity.

On the other hand, the voltage gains of both inverters are larger than unity at the third operating point. Consequently, even if the circuit is biased at this point initially, a small voltage perturbation at the input of any of the inverters will be amplified, causing the operating point to move to one of the stable operating points. This leads to the conclusion that the third operating point is unstable. The circuit has two stable operating points, hence, it is called bistable.

The bistable behavior of the cross-coupled inverter circuit can also be visualized qualitatively by examining the total potential energy level at each of the three possible operating points (Fig. 8.2(c)). It is seen that the potential energy is at its minimum at two of the three operating points, since the voltage gains of both inverters are equal to zero. By contrast, the energy attains a maximum at the operating point at which the voltage gains of both inverters are maximum. Thus, the circuit has two stable operating points corresponding to the two energy minima, and one unstable operating point corresponding to the potential energy maximum.

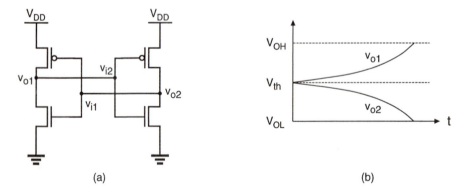

(a)   (b)

Figure 8.3.   (a) Circuit diagram of a CMOS bistable element. (b) One possibility for the expected time-domain behavior of the output voltages, if the circuit is initially set at its unstable operating point.

Figure 8.3(a) shows the circuit diagram of a CMOS two-inverter bistable element. Note that at the unstable operating point of this circuit, all four transistors are in saturation, resulting in maximum loop gain for the circuit. If the initial operating condition is set at this point, any small voltage perturbation will cause significant changes in the operating modes of the transistors. Thus, we expect the output voltages of the two inverters to diverge and eventually settle at V_{OH} and V_{OL}, respectively, as illustrated in Fig. 8.3(b). The direction in which each output voltage diverges is determined by the initial perturbation polarity. In the following, we will examine this event more closely using a small-signal analysis approach.

Consider the bistable circuit shown in Fig. 8.4, which is initially operating at $v_{o1} = v_{o2} = V_{th}$, i.e., at the unstable operating point. For our analysis, we will assume that the input (gate) capacitance C_g of each inverter is much larger than its output (drain) capacitance C_d, i.e., $C_g \gg C_d$.

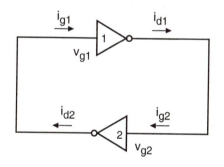

Figure 8.4.   Small-signal input and output currents of the inverters.

The small-signal drain current supplied by each inverter (1 and 2) can be expressed, in terms of the small-signal gate voltage of that inverter, as follows. Note that the drain current of each inverter is also equal to the gate current of the other inverter.

$$i_{g1} = i_{d2} = g_m v_{g2}$$
$$i_{g2} = i_{d1} = g_m v_{g1} \tag{8.1}$$

Here, g_m represents the small-signal transconductance of the inverter. The gate voltages of both inverters can be expressed in terms of the gate charges, q_1 and q_2.

$$v_{g1} = \frac{q_1}{C_g} \qquad\qquad v_{g2} = \frac{q_2}{C_g} \tag{8.2}$$

Note that the small-signal gate current of each inverter can be written as a function of the time derivative of its small-signal gate voltage, as follows.

$$i_{g1} = C_g \frac{dv_{g1}}{dt}$$
$$i_{g2} = C_g \frac{dv_{g2}}{dt}$$

(8.3)

Combining (8.1) with (8.3), we obtain:

$$g_m v_{g2} = C_g \frac{dv_{g1}}{dt}$$
$$g_m v_{g1} = C_g \frac{dv_{g2}}{dt}$$

(8.4)

Expressing the gate voltages in terms of the gate charges, these two differential equations can also be written as

$$\frac{g_m}{C_g} q_2 = \frac{dq_1}{dt}$$
$$\frac{g_m}{C_g} q_1 = \frac{dq_2}{dt}$$

(8.5)

Both differential equations given in (8.5) can now be combined to yield a second-order differential equation describing the time behavior of gate charge q_1.

$$\frac{g_m}{C_g} q_1 = \frac{C_g}{g_m} \frac{d^2 q_1}{dt^2} \qquad \Rightarrow \qquad \frac{d^2 q_1}{dt^2} = \left(\frac{g_m}{C_g}\right)^2 q_1$$

(8.6)

This equation can also be expressed in a more simplified form by using τ_0, the transit time constant.

$$\frac{d^2 q_1}{dt^2} = \frac{1}{\tau_0^2} q_1 \qquad \text{with } \tau_0 = \frac{C_g}{g_m}$$

(8.7)

In order to find the time-domain solution of (8.7), we first take its Laplace transform in the s-domain, as follows.

$$s^2 Q_1(s) - s q_1(0) - q_1'(0) = \frac{1}{\tau_0^2} Q_1(s) \tag{8.8}$$

$$Q_1(s) = \frac{s q_1(0) + q_1'(0)}{\left(s + \dfrac{1}{\tau_0}\right)\left(s - \dfrac{1}{\tau_0}\right)} \tag{8.9}$$

The reverse Laplace transform of (8.9) yields the time-domain solution for q_1.

$$q_1(t) = \frac{q_1(0) - \tau_0 q_1'(0)}{2} e^{-\frac{t}{\tau_0}} + \frac{q_1(0) + \tau_0 q_1'(0)}{2} e^{+\frac{t}{\tau_0}} \tag{8.10}$$

where the initial condition is given as:

$$q_1(0) = C_g \cdot v_{g1}(0) \tag{8.11}$$

Note that $v_{g1} = v_{o2}$ and $v_{g2} = v_{o1}$. Replacing the gate charge of both inverters with the corresponding output-voltage variables, we obtain,

$$v_{o2}(t) = \frac{1}{2}\left(v_{o2}(0) - \tau_0 v_{o2}'(0)\right) e^{-\frac{t}{\tau_0}} + \frac{1}{2}\left(v_{o2}(0) + \tau_0 v_{o2}'(0)\right) e^{+\frac{t}{\tau_0}} \tag{8.12}$$

$$v_{o1}(t) = \frac{1}{2}\left(v_{o1}(0) - \tau_0 v_{o1}'(0)\right) e^{-\frac{t}{\tau_0}} + \frac{1}{2}\left(v_{o1}(0) + \tau_0 v_{o1}'(0)\right) e^{+\frac{t}{\tau_0}} \tag{8.13}$$

For large values of t, the time-domain expressions (8.12) and (8.13) can be simplified as follows.

$$v_{o1}(t) \approx \frac{1}{2}\left(v_{o1}(0) + \tau_0 v_{o1}'(0)\right) e^{+\frac{t}{\tau_0}}$$

$$v_{o2}(t) \approx \frac{1}{2}\left(v_{o2}(0) + \tau_0 v_{o2}'(0)\right) e^{+\frac{t}{\tau_0}} \tag{8.14}$$

Note that the magnitude of both output voltages increases exponentially with time. Depending on the polarity of the initial small perturbations ($dv_{o1}(0)$ and $dv_{o2}(0)$), the output voltages of both inverters will diverge from their initial value of V_{th} to either V_{OL} or V_{OH}. In fact, the polarity of the output-voltage perturbation dv_{o1} must always be opposite to that of dv_{o2}, because of the charge-conservation principle (8.5). Hence, the two output voltages always diverge into opposite directions, as expected.

$$v_{o1}: \quad V_{th} \rightarrow V_{OH} \text{ or } V_{OL}$$

$$v_{o2}: \quad V_{th} \rightarrow V_{OL} \text{ or } V_{OH} \tag{8.15}$$

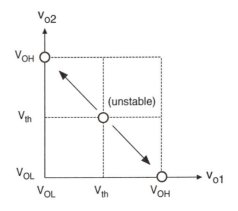

Figure 8.5.   Phase-plane representation of the bistable circuit behavior.

The phase-plane representation of this event, illustrated in Fig. 8.5, shows that the operating point ($v_{o1} = V_{th}, v_{o2} = V_{th}$) is unstable. The two operating points ($v_{o1} = V_{OL}, v_{o2} = V_{OH}$) and ($v_{o1} = V_{OH}, v_{o2} = V_{OL}$) can be shown to be stable, using small-signal models at the corresponding operating points.

In addition to the time-domain analysis presented here, an interesting observation can be made for the two-inverter bistable element: While the bistable circuit is settling from its unstable operating point into one of its stable operating points, we can envision a signal traveling the loop consisting of the two cascaded inverters several times (Fig. 8.6). The time-domain behavior of the output voltage v_{o1} during this period is approximated as follows.

$$\frac{v_{o1}(t)}{v_{o1}(0)} = e^{+\frac{t}{\tau_0}} \tag{8.16}$$

If during a time interval T, the signal travels the loop n times, then this is equivalent to the same signal propagating along a cascaded inverter chain, which consists of $2n$ inverters. Expressing the loop gain (combined voltage gain of two cascaded inverters) with A, we obtain

$$A^n = e^{+\frac{T}{\tau_0}} \qquad (8.17)$$

This expression describes the time-domain behavior of the diverging process until it reaches stable points, in terms of loop gains, as depicted in Fig. 8.6.

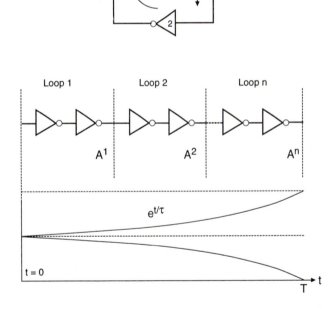

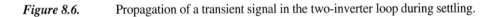

Figure 8.6.   Propagation of a transient signal in the two-inverter loop during settling.

8.3. The SR Latch Circuit

The bistable element consisting of two cross-coupled inverters (Fig. 8.2) has two stable operating modes, or states. The circuit preserves its state (either one of the two possible modes) as long as the power supply voltage is provided; hence, the circuit can perform a simple memory function of *holding* its state. However, the simple two-inverter circuit examined above has no provision for allowing its state to be changed externally from one stable operating mode to the other. To allow such a change of state, we must add simple switches to the bistable element, which can be used to force or trigger the circuit from one operating point to the other. Figure 8.7 shows the circuit structure of the simple CMOS SR latch, which has two such triggering inputs, S and R. In the literature, the SR latch is also called an SR flip-flop, since two stable states can be switched back and forth. The circuit consists of two CMOS NOR2 gates. One of the input terminals of each NOR gate is used to cross-couple to the output of the other NOR gate, while the second input enables triggering of the circuit.

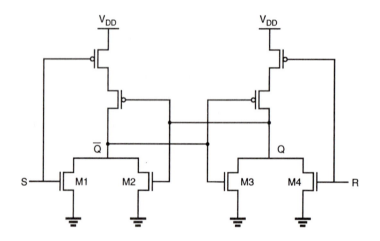

Figure 8.7.   CMOS SR latch circuit based on NOR2 gates.

The SR latch circuit has two complementary outputs, Q and $\overline{Q}$. By definition, the latch is said to be in its *set* state when Q is equal to logic "1" and $\overline{Q}$ is equal to logic "0." Conversely, the latch is in its *reset* state when the output Q is equal to logic "0" and $\overline{Q}$ is equal to "1." The gate-level schematic of the SR latch consisting of two NOR2 gates, and the corresponding block diagram representation are shown in Fig. 8.8. It can easily be seen that when both input signals are equal to logic "0," the SR latch will operate exactly like the simple cross-coupled bistable element examined earlier, i.e., it will preserve (hold) either one of its two stable operating points (states) as determined by the previous inputs.

If the *set input* (S) is equal to logic "1" and the *reset input* is equal to logic "0," then the output node Q will be forced to logic "1" while the output node $\overline{Q}$ is forced to logic "0." This means that the SR latch will be *set*, regardless of its previous state.

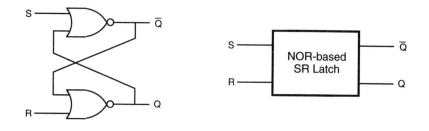

Figure 8.8.   Gate-level schematic and block diagram of the NOR-based SR latch.

Similarly, if S is equal to "0" and R is equal to "1," then the output node Q will be forced to "0" while $\overline{Q}$ is forced to "1." Thus, with this input combination, the latch is *reset*, regardless of its previously held state. Finally, consider the case in which both of the inputs S and R are equal to logic "1." In this case, both output nodes will be forced to logic "0," which conflicts with the complementarity of Q and $\overline{Q}$. Therefore, this input combination is not permitted during normal operation and is considered to be a *not-allowed* condition. The truth table of the NOR-based SR latch is summarized in the following.

S	R	Q_{n+1}	$\overline{Q}_{n+1}$	Operation
0	0	Q_n	$\overline{Q}_n$	hold
1	0	1	0	set
0	1	0	1	reset
1	1	0	0	not allowed

The operation of the CMOS SR latch circuit shown in Fig. 8.7 can be examined in more detail by considering the operating modes of the four nMOS transistors, M1, M2, M3, and M4. If the set input (S) is equal to V_{OH} and the reset input (R) is equal to V_{OL}, both of the parallel-connected transistors M1 and M2 will be on. Consequently, the voltage on node $\overline{Q}$ will assume a logic-low level of $V_{OL} = 0$. At the same time, both M3 and M4 are turned off, which results in a logic-high voltage V_{OH} at node Q. If the reset input (R) is equal to V_{OH} and the set input (S) is equal to V_{OL}, the situation will be reversed (M1 and M2 turned off and M3 and M4 turned on).

When both of the input voltages are equal to V_{OL}, on the other hand, there are two possibilities. Depending on the previous state of the SR latch, either M2 or M3 will be on, while both of the trigger transistors M1 and M4 are off. This will generate a logic-low level of $V_{OL} = 0$ at one of the output nodes, while the complementary output node is at V_{OH}. The static operation modes and voltage levels of the NOR-based CMOS SR latch circuit are summarized in the following table. For simplicity, the operating modes of the complementary pMOS transistors are not explicitly listed here.

S	R	Q_{n+1}	$\overline{Q_{n+1}}$	Operation
V_{OH}	V_{OL}	V_{OH}	V_{OL}	M1 and M2 on, M3 and M4 off
V_{OL}	V_{OH}	V_{OL}	V_{OH}	M1 and M2 off, M3 and M4 on
V_{OL}	V_{OL}	V_{OH}	V_{OL}	M1 and M4 off, M2 on, or
V_{OL}	V_{OL}	V_{OL}	V_{OH}	M1 and M4 off, M3 on

For the transient analysis of the SR latch circuit, we have to consider an event which results in a state change, i.e., either an initially reset latch being set by applying a *set* signal, or an initially set latch being reset by applying the *reset* signal. In either case, we note that both of the output nodes undergo simultaneous voltage transitions. While one output is rising from its logic-low level to logic-high, the other output node is falling from its initial logic-high level to logic-low. Thus, an interesting problem is to estimate the amount of time required for the simultaneous switching of the two output nodes. The exact solution of this problem obviously requires the simultaneous solution of two coupled differential equations, one each for each output node. The problem can, however, be simplified considerably if we assume that the two events described above take place in sequence rather than simultaneously. This assumption causes an overestimation of the switching time.

To calculate the switching times for both output nodes, we first have to find the total parasitic capacitance associated with each node. Simple inspection of the circuit shows that the total lumped capacitance at each output node can be approximated as follows.

$$C_Q = C_{gb,2} + C_{gb,5} + C_{db,3} + C_{db,4} + C_{db,7} + C_{sb,7} + C_{db,8}$$
$$C_{\overline{Q}} = C_{gb,3} + C_{gb,7} + C_{db,1} + C_{db,2} + C_{db,5} + C_{sb,5} + C_{db,6}$$

(8.18)

The circuit diagram of the SR latch is shown in Fig. 8.9 together with the lumped load capacitances at the nodes Q and $\overline{Q}$. Assuming that the latch is initially reset, and that

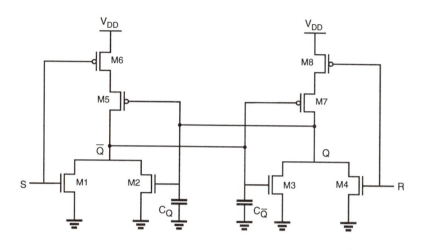

Figure 8.9.   Circuit diagram of the CMOS SR latch showing the lumped load capacitances at both output nodes.

a set operation is being performed by applying S = "1" and R = "0," the rise time associated with node Q can now be estimated as follows.

$$\tau_{rise,Q}(SR-latch)=\tau_{rise,Q}(NOR2)+\tau_{fall,\overline{Q}}(NOR2) \tag{8.19}$$

Note that the calculation of the switching time $\tau_{rise,Q}$ requires two separate calculations for the rise and fall times of the NOR2 gates. It is obvious that by considering the two events separately, i.e., first, one of the output node voltages $(\overline{Q})$ falling from high to low due to turn-on of M1, followed by the other node voltage (Q) rising from low to high due to turn-off of M3, we are bound to overestimate the actual switching time for the SR latch. Both M2 and M4 can be assumed to be off in this process, although M2 can be turned on as Q rises, thus actually shortening the $\overline{Q}$ node fall time. This approach, however, yields a simpler first-order prediction for the time delay, as opposed to the simultaneous solution of two coupled differential equations.

The NOR-based SR latch can also be implemented by using two cross-coupled depletion-load nMOS NOR2 gates, as shown in Fig. 8.10. From the logic point of view, the operation principle of the depletion-load nMOS NOR-based SR latch is identical to that of the CMOS SR latch. In terms of power dissipation and noise margins, however, the CMOS circuit implementation offers a better alternative, since both of the CMOS NOR2 gates dissipate virtually no static power for preserving a state, and since the output voltages can exhibit a full swing between 0 and V_{DD}.

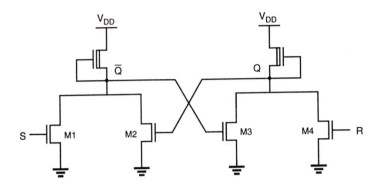

Figure 8.10.   Depletion-load nMOS SR latch circuit based on NOR2 gates.

Now consider a different approach for building the basic SR latch circuit. Instead of using two NOR2 gates, we can use two NAND2 gates, as shown in Fig. 8.11. Here, one input of each NAND gate is used to cross-couple to the output of the other NAND gate, while the second input enables external triggering.

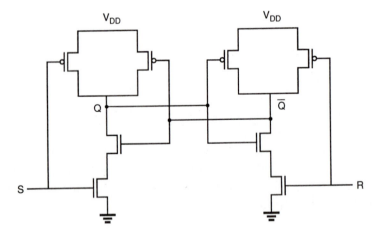

Figure 8.11.   CMOS SR latch circuit based on NAND2 gates.

A close inspection of the NAND-based SR latch circuit reveals that in order to hold (preserve) a state, both of the external trigger inputs must be equal to logic "1." The operating point or the state of the circuit can be changed only by pulling the *set* input to logic zero or by pulling the *reset* input to zero. We can observe that if S is equal to "0" and R is equal to "1," the output Q attains a logic "1" value and the complementary output $\overline{Q}$ becomes logic "0." Thus, in order to *set* the NAND SR latch, a logic "0" must be applied to the *set* (S) input. Similarly, in order to *reset* the latch, a logic "0" must be applied to the *reset* (R) input. The conclusion is that the NAND-based SR latch responds to *active low* input signals, as opposed to the NOR-based SR latch, which responds to *active high* inputs. Note that if both input signals are equal to logic "0," both output nodes assume a logic-high level, which is not allowed because it violates the complementarity of the two outputs.

The gate-level schematic and the corresponding block diagram representation of the NAND-based SR latch circuit are shown in Fig. 8.12. The small circles at the S and R input terminals indicate that the circuit responds to *active low* input signals. The truth table of the NAND SR latch is also shown in the following. The same approach used in the timing analysis of NOR-based SR latches can be applied to NAND-based SR latches.

The NAND-based SR latch can also be implemented by using two cross-coupled depletion-load NAND2 gates, as shown in Fig. 8.13. While the operation principle is identical to that of the CMOS NAND SR latch (Fig. 8.11) from the logic point of view, the CMOS circuit implementation again offers a better alternative in terms of static power dissipation and noise margins.

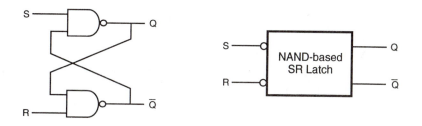

S	R	Q_{n+1}	$\overline{Q_{n+1}}$	Operation
0	0	1	1	*not allowed*
0	1	1	0	*set*
1	0	0	1	*reset*
1	1	Q_n	$\overline{Q_n}$	*hold*

Figure 8.12.   Gate-level schematic and block diagram of the NAND-based SR latch.

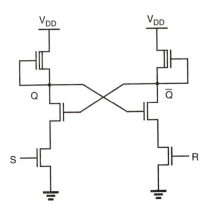

Figure 8.13.   Depletion-load nMOS NAND-based SR latch circuit.

8.4. Clocked Latch and Flip-Flop Circuits

Clocked SR Latch

All of the SR latch circuits examined in the previous section are essentially asynchronous sequential circuits, which will respond to the changes occurring in input signals at a circuit-delay-dependent time point during their operation. To facilitate synchronous operation, the circuit response can be controlled simply by adding a gating clock signal to the circuit, so that the outputs will respond to the input levels only during the active period of a clock pulse. For simple reference, the clock pulse will be assumed to be a periodic square waveform, which is applied simultaneously to all clocked logic gates in the system.

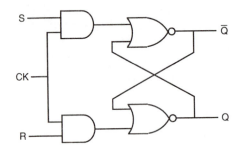
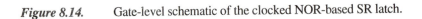

Figure 8.14.   Gate-level schematic of the clocked NOR-based SR latch.

The gate-level schematic of a clocked NOR-based SR latch is shown in Fig. 8.14. It can be seen that if the clock (CK) is equal to logic "0," the input signals have no influence upon the circuit response. The outputs of the two AND gates will remain at logic "0," which forces the SR latch to hold its current state regardless of the S and R input signals. When the clock input goes to logic "1," the logic levels applied to the S and R inputs are permitted to reach the SR latch, and possibly change its state. Note that as in the non-clocked SR latch, the input combination S = R = "1" is not allowed in the clocked SR latch. With both inputs S and R at logic "1," the occurrence of a clock pulse causes both outputs to go momentarily to zero. When the clock pulse is removed, i.e., when it becomes "0," the state of the latch is indeterminate. It can eventually settle into either state, depending on slight delay differences between the output signals.

To illustrate the operation of the clocked SR latch, a sample sequence of CK, S, and R waveforms, and the corresponding output waveform Q are shown in Fig. 8.15. Note that the circuit is strictly *level-sensitive* during active clock phases, i.e., any changes occurring in the S and R input voltages when the CK level is equal to "1" will be reflected onto the circuit outputs. Consequently, even a narrow spike or glitch occurring during an active clock phase can set or reset the latch, if the loop delay is shorter than the pulse width.

Figure 8.16 shows a CMOS implementation of the clocked NOR-based SR latch circuit, using two simple AOI gates. Notice that the AOI-based implementation of the circuit results in a very small transistor count, compared with the alternative circuit realization consisting of two AND2 and two NOR2 gates.

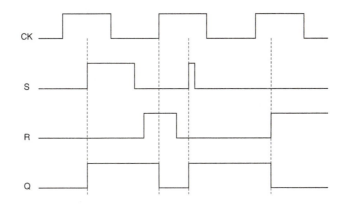

Figure 8.15. Sample input and output waveforms illustrating the operation of the clocked NOR-based SR latch circuit.

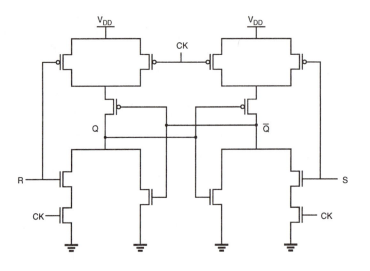

Figure 8.16.   AOI-based implementation of the clocked NOR-based SR latch circuit.

The NAND-based SR latch can also be implemented with gating clock input, as shown in Fig. 8.17. It must be noted, however, that both input signals S and R as well as the clock signal CK are *active low* in this case. This means that changes in the input signal levels will be ignored when the clock is equal to logic "1," and that inputs will influence the outputs only when the clock is active, i.e., CK = "0." For the circuit implementation of this clocked NAND-based SR latch, we can use a simple OAI structure, which is essentially analogous to the AOI-based realization of the clocked NOR SR latch circuit.

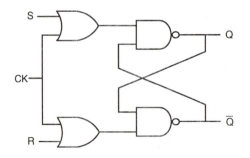

Figure 8.17.   Gate-level schematic of the clocked NAND-based SR latch circuit, with active low inputs.

A different implementation of the clocked NAND-based SR latch is shown in Fig. 8.18. Here, both input signals and the CK signal are *active high*, i.e., the latch output Q will be set when CK = "1," S = "1," and R = "0." Similarly, the latch will be reset when CK = "1," S = "0," and R = "1." The latch preserves its state as long as the clock signal is inactive, i.e., when CK = "0." The drawback of this implementation is that the transistor count is higher than the *active low* version shown in Fig. 8.17.

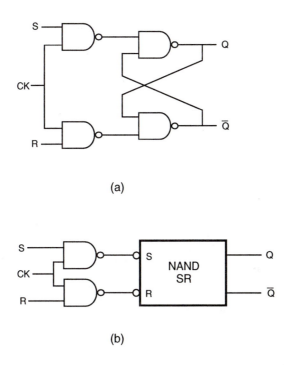

(a)

(b)

Figure 8.18.   (a) Gate-level schematic of the clocked NAND-based SR latch circuit, with active high inputs. (b) Partial block diagram representation of the same circuit.

Clocked JK Latch

All simple and clocked SR latch circuits examined to this point suffer from the common problem of having a not-allowed input combination, i.e., their state becomes indeterminate when both inputs S and R are activated at the same time. This problem can be overcome by adding two feedback lines from the outputs to the inputs, as shown in Fig. 8.19. The resulting circuit is called a JK latch. Figure 8.19 shows an all-NAND implementation of the JK latch with active high inputs, and the corresponding block diagram representation. The JK latch is commonly called a JK flip-flop.

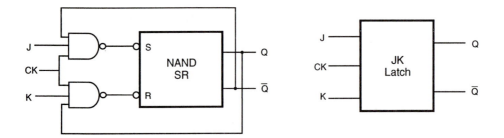

Figure 8.19.   Gate-level schematic of the clocked JK latch circuit.

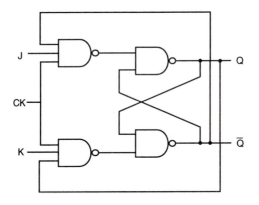

Figure 8.20.   All-NAND implementation of the clocked JK latch circuit.

The J and K inputs in this circuit correspond to the set and reset inputs of the basic SR latch. When the clock is active, the latch can be set with the input combination (J = "1," K = "0"), and it can be reset with the input combination (J = "0," K = "1"). If both inputs are equal to logic "0," the latch preserves its current state. If, on the other hand, both inputs are equal to "1" during the active clock phase, the latch simply switches its state due to feedback. In other words, the JK latch does not have a not-allowed input combination. As in the other clocked latch circuits, the JK latch will hold its current state when the clock is inactive (CK = "0"). The operation of the clocked JK latch is summarized in the truth table on the following page.

Figure 8.21 shows an alternative, NOR-based implementation of the clocked JK latch, and CMOS realization of this circuit. Note that the AOI-based circuit structure results in a relatively low transistor count, and consequently, a more compact circuit compared to the all-NAND realization shown in Fig. 8.20.

J	K	Q_n	$\overline{Q_n}$	S	R	Q_{n+1}	$\overline{Q_{n+1}}$	Operation
0	0	0	1	1	1	0	1	hold
		1	0	1	1	1	0	
0	1	0	1	1	1	0	1	reset
		1	0	1	0	0	1	
1	0	0	1	0	1	1	0	set
		1	0	1	1	1	0	
1	1	0	1	0	1	1	0	toggle
		1	0	1	0	0	1	

Detailed truth table of the JK latch circuit.

While there is no not-allowed input combination for the JK latch, there is still a potential problem. If both inputs are equal to logic "1" during the active phase of the clock pulse, the output of the circuit will oscillate (toggle) continuously until either the clock becomes inactive (goes to zero), or one of the input signals goes to zero. To prevent this undesirable timing problem, the clock pulse width must be made smaller

(a)

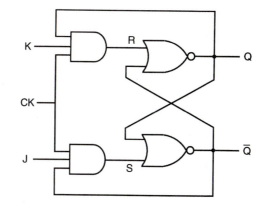

Figure 8.21. (a) Gate-level schematic of the NOR-based clocked JK latch circuit.

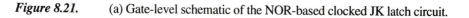

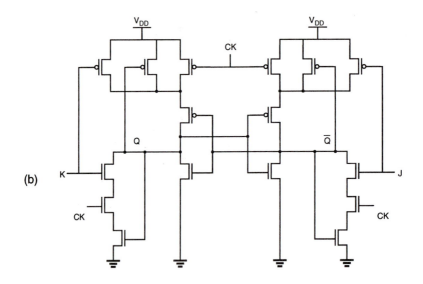

Figure 8.21. (continued)   (b) CMOS AOI realization of the JK latch.

than the input-to-output propagation delay of the JK latch circuit. This restriction dictates that the clock signal must go low before the output level has an opportunity to switch again, which prevents uncontrolled oscillation of the output. However, note that this clock constraint is difficult to implement for most practical applications.

Assuming that the clock timing constraint described above is satisfied, the output of the JK latch will toggle (change its state) only once for each clock pulse, if both inputs are equal to logic "1" (Fig. 8.22). A circuit which is operated exclusively in this mode is called a *toggle switch*.

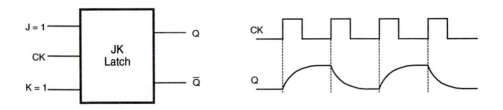

Figure 8.22.   Operation of the JK latch as a toggle switch.

Master-Slave Flip-Flop

Most of the timing limitations encountered in the previously examined clocked latch circuits can be prevented by using two latch stages in a cascaded configuration. The key operation principle is that the two cascaded stages are activated with opposite clock phases. This configuration is called the *master-slave flip-flop*. Our definition of flip-flop is designed to distinguish it from latches discussed previously, although they are mostly used interchangeably in the literature.

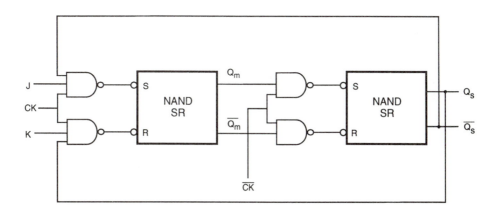

Figure 8.23.   Master-slave flip-flop consisting of NAND-based JK latches.

The input latch in Fig. 8.23, called the "master," is activated when the clock pulse is high. During this phase, the inputs J and K allow data to be entered into the flip-flop, and the first-stage outputs are set according to the primary inputs. When the clock pulse goes to zero, the master latch becomes inactive and the second-stage latch, called the "slave," becomes active. The output levels of the flip-flop circuit are determined during this second phase, based on the master-stage outputs set in the previous phase.

Since the master and the slave stages are effectively decoupled from each other with the opposite clocking scheme, the circuit is never *transparent*, i.e., a change occurring in the primary inputs is never reflected directly to the outputs. This very important property clearly separates the master-slave flip-flop from all of the latch circuits examined earlier in this section. Figure 8.24 shows a sample set of input and output waveforms associated with the JK master-slave flip-flop, which can help the reader to study the basic operation principles.

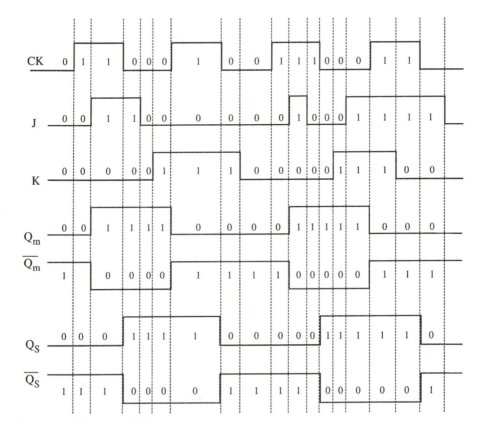

Figure 8.24.   Sample input and output waveforms of the master-slave flip-flop circuit.

Since the master and the slave stages are decoupled from each other, the circuit allows for toggling when $J = K =$ "1," but it eliminates the possibility of uncontrolled oscillations since only one stage is active at any given time. A NOR-based alternative realization for the master-slave flip-flop circuit is shown in Fig. 8.25.

Figure 8.24 also shows that the master-slave flip-flop circuit examined here has the potential problem of "one's catching." When the clock pulse is high, a narrow spike or glitch in one of the inputs, for instance the glitch in the J line (or K line), may set (or reset) the master latch and thus cause an unwanted state transition, which will then be propagated into the slave stage during the following phase. This problem can be eliminated to a large extent by building an edge-triggered master-slave flip-flop, which will be examined in the following section.

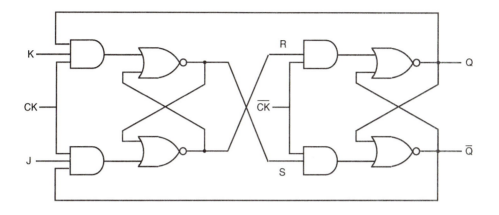

Figure 8.25.   NOR-based realization of the JK master-slave flip-flop.

8.5. CMOS D-Latch and Edge-Triggered Flip-Flop

With the widespread use of CMOS circuit techniques in digital integrated circuit design, a large selection of CMOS-based sequential circuits have also gained popularity and prominence, especially in VLSI design. Throughout this chapter, we have seen examples showing that virtually all of the latch and flip-flop circuits can be implemented with CMOS gates, and that their design is quite straightforward. However, direct CMOS implementations of conventional circuits such as the clocked JK latch or the JK master-slave flip-flop tend to require a large number of transistors.

In this section, we will see that specific versions of sequential circuits built primarily with CMOS transmission gates are generally simpler and require fewer transistors than the circuits designed with conventional structuring. As an introduction to the issue, let us first consider the simple D-latch circuit shown in Fig. 8.26. The gate-level representation of the D-latch is simply obtained by modifying the clocked NOR-based SR latch circuit. Here, the circuit has a single input D, which is directly connected to the S input of the latch. The input variable D is also inverted and connected to the R input of the latch. It can be seen from the gate-level schematic that the output Q assumes the value of the input D when the clock is active, i.e., for CK = "1." When the clock signal goes to zero, the output will simply preserve its state. Thus, the CK input acts as an enable signal which allows data to be accepted into the D-latch.

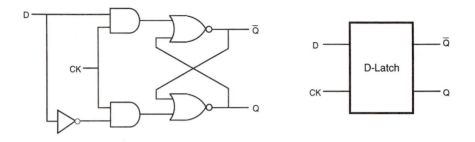

Figure 8.26.   Gate-level schematic and the block diagram view of the D-latch.

The D-latch finds many applications in digital circuit design, primarily for temporary storage of data or as a delay element. In the following, we will examine its simple CMOS implementation. Consider the circuit diagram given in Fig. 8.27, which shows a basic two-inverter loop and two CMOS transmission gate (TG) switches.

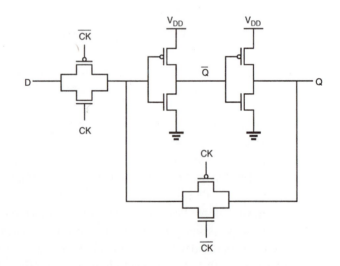

Figure 8.27.   CMOS implementation of the D-latch (version 1).

The TG at the input is activated by the CK signal, whereas the TG in the inverter loop is activated by the inverse of the CK signal, $\overline{CK}$. Thus, the input signal is accepted (latched) into the circuit when the clock is high, and this information is preserved as the state of the inverter loop when the clock is low. The operation of the CMOS D-latch circuit can be better visualized by replacing the CMOS transmission gates with simple switches, as shown in Fig. 8.28. A timing diagram accompanying this figure shows the time intervals during which the input and the output signals should be valid (unshaded).

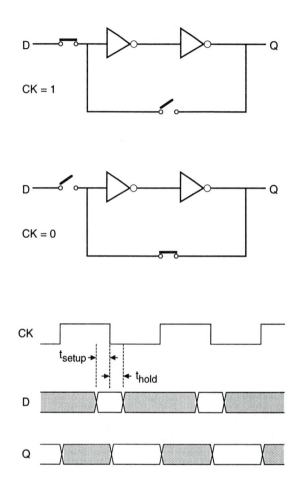

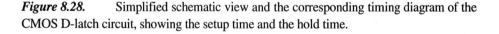

Figure 8.28.   Simplified schematic view and the corresponding timing diagram of the CMOS D-latch circuit, showing the setup time and the hold time.

Note that the valid D input must be stable for a short time before (*setup time*, t_{setup}) and after (*hold time*, t_{hold}) the negative clock transition, during which the input switch opens and the loop switch closes. Once the inverter loop is completed by closing the loop switch, the output will preserve its valid level. In the D-latch design, the requirements for setup time and hold time should be met carefully. Any violation of such specifications can cause *metastability* problems which lead to seemingly chaotic transient behavior, and can result in an unpredictable state after the transitional period.

The D-latch shown in Fig. 8.27 is not an edge-triggered storage element because the output changes according to the input, i.e., the latch is transparent, while the clock is high. The transparency property makes the application of this D-latch unsuitable for counters and some data storage implementations.

Figure 8.29 shows a different version of the CMOS D-latch. The circuit contains two tristate inverters, driven by the clock signal and its inverse, respectively. Although the circuit appears to be quite different from that shown in Fig. 8.27, the basic operation principle of the circuit is the same as that shown in Fig. 8.28. The first tri-state inverter acts as the input switch, accepting the input signal when the clock is high. At this time, the second tristate inverter is at its high-impedance state, and the output Q is following the input signal. When the clock goes low, the input buffer becomes inactive, and the second tristate inverter completes the two-inverter loop, which preserves its state until the next clock pulse.

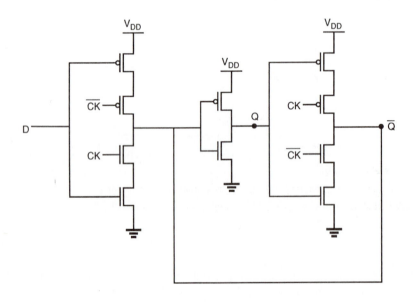

Figure 8.29.   CMOS implementation of the D-latch (version 2).

Finally, consider the two-stage master-slave flip-flop circuit shown in Fig. 8.30, which is constructed by simply cascading two D-latch circuits. The first stage (master) is driven by the clock signal, while the second stage (slave) is driven by the inverted clock signal. Thus, the master stage is positive level-sensitive, while the slave stage is negative level-sensitive.

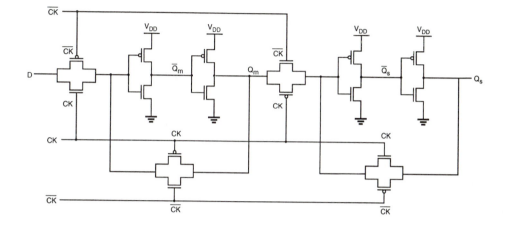

Figure 8.30.   CMOS edge-triggered master-slave D flip-flop.

When the clock is high, the master stage follows the D input while the slave stage holds the previous value. When the clock changes from logic "1" to logic "0," the master latch ceases to sample the input and stores the D value at the time of the clock transition. At the same time, the slave latch becomes transparent, passing the stored master value Q_m to the output of the slave stage, Q_s. The input cannot affect the output because the master stage is disconnected from the D input. When the clock changes again from logic "0" to "1," the slave latch locks in the master latch output and the master stage starts sampling the input again. Thus, this circuit is a negative edge-triggered D flip-flop by virtue of the fact that it samples the input at the falling edge of the clock pulse.

Another implementation of edge-triggered D flip-flop is shown in Fig. 8.31, which consists of 6 NAND3 gates. This D flip-flop is positive edge-triggered as illustrated in the waveform chart in Fig. 8.32. Initially all the signal values except S are 0, i.e., (S , R , CK , D) = (1 , 0 , 0 , 0), and Q = 0. In the second phase, both D and R switch to 1, i.e., (S , R , CK , D) = (1 , 0 , 1 , 1), but no change in Q occurs and the Q value remains at 0. However, in the third phase, if CK goes to high, i.e., (S , R , CK , D) = (1 , 1 , 1 , 1), the output of gate 2 switches to 0, which in turn sets the output of the last stage SR latch to 1. Thus, the output of this D flip-flop switches to 1 at the positive-going edge of the clock signal, CK. However, as can be observed in the ninth phase of the waveform diagram chart, the Q output is not affected by the negative-going edge of CK, nor by other signal changes.

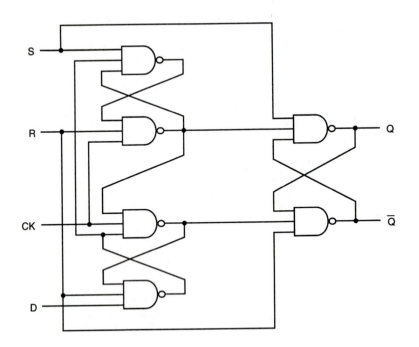

Figure 8.31.   NAND3-based positive edge-triggered D flip-flop circuit.

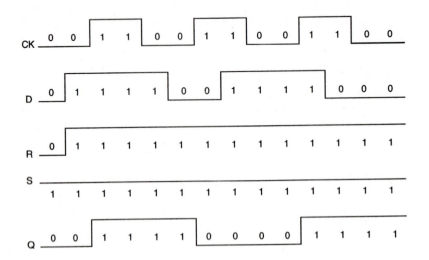

Figure 8.32.   Timing diagram of the positive edge-triggered D flip-flop.

Example 8.1.

In this example, we will examine the Schmitt trigger circuit, which is a very useful regenerative circuit. The Schmitt trigger has an inverter-like voltage transfer characteristic, but with two different logic threshold voltages for increasing and for decreasing input signals. With this unique property, the circuit can be utilized for the detection of low-to-high and high-to-low switching events in noisy environments.

The circuit diagram of a CMOS Schmitt trigger and the typical features of its voltage transfer characteristic are shown below. Also listed here is the corresponding SPICE circuit input file. In the following, we will first calculate the important points in the VTC, and then compare our results with SPICE simulation.

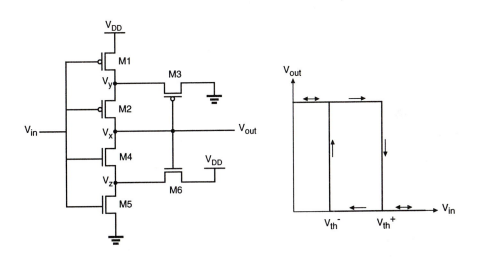

```
CMOS Schmitt Trigger DC analysis
vdd 5 0 dc 5V
vin 1 0 dc 1v
m5 2 1 0 0 mn l=1u w=1u
m4 3 1 2 0 mn l=1u w=2.5u
m6 5 3 2 0 mn l=1u w=3u
m1 4 1 5 5 mp l=1u w=1u
m2 3 1 4 5 mp l=1u w=2.5u
m3 0 3 4 5 mp l=1u w=3u
.model mn nmos vto=1 gamma=0.4 kp=2.5e-5
.model mp pmos vto=-1 gamma=0.4 kp=1.0e-5
.dc vin 0 5 0.1
.print dc v(3)
.end
```

We start our step-by-step analysis by considering a positive input sweep, i.e., assuming that the input voltage is increasing from 0 to V_{DD}.

i) At $V_{in} = 0$ V:
    M1 and M2 are turned on, then

$$V_x = V_y = V_{DD} = 5 \text{ V}$$

At the same time, M4 and M5 are turned off. M3 is off; M6 is on and operates in the saturation region. Calculating the threshold voltage of M6 with $2\phi_F = -0.6$ V,

$$V_z = V_{DD} - V_{T,6} = 3.5 \text{ V}$$

ii) At $V_{in} = V_{T0,n} = 1.0$ V:
    M5 starts to turn on, M4 is still off.

$$V_x = 5 \text{ V}$$

iii) At $V_{in} = 2.0$ V:
    Assume M4 is off, while both M5 and M6 operate in the saturation region.

$$\frac{1}{2}k'\left(\frac{W}{L}\right)_5 \left(V_{in} - V_{T0,n}\right)^2 = \frac{1}{2}k'\left(\frac{W}{L}\right)_6 \left(V_{DD} - V_z - V_{T,6}\right)^2$$

$$(2-1)^2 = 3\left(5 - V_z - \left[1 + 0.4\left(\sqrt{0.6+V_z} - \sqrt{0.6}\right)\right]\right)^2$$

Solving this equation for V_z, we find that there is only one physically reasonable root.

$$V_z = 2.976 \text{ V}$$

Now, we check our assumption made above, i.e., M4 is indeed turned off:

$$V_{GS,4} = 2 - 2.976 = -0.976 < V_{T0,n} = 1$$

iv) At $V_{in} = 3.5$ V:

V_z continues to decrease. Assuming M5 in linear region and M6 in saturation, we arrive at the following current equation.

$$\frac{1}{2}k'\left(\frac{W}{L}\right)_5\left[2\left(V_{in}-V_{T0,n}\right)V_z-V_z^2\right]=\frac{1}{2}k'\left(\frac{W}{L}\right)_6\left(V_{DD}-V_z-V_{T,6}\right)^2$$

$$\left[2(3.5-1.0)V_z-V_z^2\right]=3\left(5-V_z-\left[1+0.4\left(\sqrt{0.6+V_z}-\sqrt{0.6}\right)\right]\right)^2$$

Solving this equation for V_z, we obtain, $V_z = 2.2$ V. Now determine the gate-to-source voltage of M4 as

$$V_{GS,4} = 3.5 - 2.2 = 1.3 > V_{T0,n} = 1$$

It is seen that at this point, M4 is already on. Thus, the analysis above, which is based on the assumption that M4 is not conducting, can no longer be valid. At this input voltage, node x is being pulled down toward "0." This can also be seen clearly from the simulation results. We conclude that the upper logic threshold voltage V_{th}^+ is approximately equal to 3.5 V.

Next, we consider a negative input sweep, i.e., assume that the input voltage is decreasing from V_{DD} to 0.

i)   At $V_{in} = 5.0$ V:

M4 and M5 are on, so that the output voltage is $V_x = 0$ V. The pMOS transistors M1 and M2 are off, and M3 is in saturation, thus,

$$\frac{1}{2}k'\left(\frac{W}{L}\right)_3\left(0-V_y-V_{T,3}\right)^2=0$$

$$V_y = -V_{T,3} = -\left[V_{T0,p}-0.4\left(\sqrt{0.6+V_{DD}-V_y}-\sqrt{0.6}\right)\right]$$

$$V_y = 1.5\ [V]$$

ii)   At $V_{in} = 4.0$ V:

M1 is at the edge of turning on, M2 is off, and M3 is in saturation. The output voltage is still unchanged.

iii)   At $V_{in} = 3.0$ V:
M1 is on and in saturation region. M3 is also in saturation, thus,

$$\frac{1}{2}k'\left(\frac{W}{L}\right)_1 \left(V_{in} - V_{DD} - V_{T0,p}\right)^2 = \frac{1}{2}k'\left(\frac{W}{L}\right)_3 \left(0 - V_y - V_{T,3}\right)^2$$

$$[3 - 5 - (-1)]^2 = 3\left(0 - V_y - \left[-1 - 0.4\left(\sqrt{0.6 + 5 - V_y} - \sqrt{0.6}\right)\right]\right)^2$$

The solution of this equation yields:

$$V_{in} = 2.02 \text{ V}$$

Now we determine the gate-to-source voltage of M2 as

$$V_{GS,2} = 3.0 - 2.02 = 0.98 > V_{T0,p} = -1$$

which indicates that M2 is still turned off at this point.

iv)   At $V_{in} = 1.5$ V:
       If M2 is still off, M1 is in the linear region, and M3 is in the saturation region:

$$\frac{1}{2}k'\left(\frac{W}{L}\right)_1 \left(2\left(V_{in} - V_{DD} - V_{T0,p}\right)\left(V_y - V_{DD}\right) - \left(V_y - V_{DD}\right)^2\right)$$

$$= \frac{1}{2}k'\left(\frac{W}{L}\right)_3 \left(0 - V_y - V_{T,3}\right)^2$$

$$2(1.5 - 5 + 1)\left(V_y - 5\right) - \left(V_y - 5\right)^2$$

$$= 3\left(-V_y - \left[-1 - 0.4\left(\sqrt{0.6 + 5 - V_y} - \sqrt{0.6}\right)\right]\right)^2$$

Solving this quadratic equation yields

$$V_y = 2.79 \text{ V}$$

It can be shown that at this point, the pMOS transistor M2 is already turned on. Consequently, the output voltage is being pulled up to V_{DD}. We conclude that the lower logic threshold voltage V_{th}^- is approximately equal to 1.5 V.

The SPICE simulation results are plotted for both increasing and decreasing input voltages.

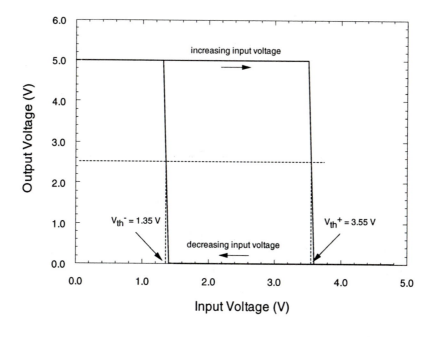

References

1.   C. Mead and L. Conway, *Introduction to VLSI Systems*, Reading, MA: Addison-Wesley Publishing Company, Inc., 1980.

2.   F. J. Hill and G. R. Peterson, *Computer-Aided Logical Design with Emphasis on VLSI*, fourth edition, John Wiley & Sons, Inc., New York, 1993.

Exercise Problems

8.1   Figure P8.1 shows a schematic for a positive edge-triggered D flip-flop. Use a layout editor (e.g., Magic) to design a layout of the circuit. Use CMOS technology, and assume that you have n-type substrate. On the printout of your layout, clearly indicate the location of each logic gate in the figure below. Also, calculate the parasitic capacitances of your layout.

- $W_n = 4\ \mu m$ and $W_p = 8\ \mu m$ for all gates
- $L_M = 2\ \mu m$
- $L_D = 0.25\ \mu m$
- $V_{T0,n} = 1\ V$
- $V_{T0,p} = -1\ V$
- $k'_n = 40\ \mu A/V^2$
- $k'_p = 25\ \mu A/V^2$
- $t_{ox} = 20\ nm$

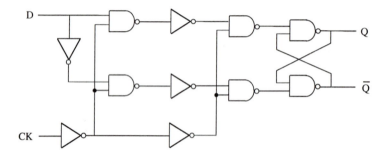

Figure P8.1

8.2   For the layout in Problem 8.1, find the minimum setup time (t_{setup}) and hold time (t_{hold}) for the flip-flop using SPICE simulation. This will require you to obtain four plots.

    (a)   A plot of the output using the minimum setup time t_{setup}
    (b)   A plot of the output using a setup time of $0.8t_{setup}$
    (c)   A plot of the output using the minimum hold time t_{hold}
    (d)   A plot of the output using a hold time of $0.8t_{hold}$

8.3   We have discussed the features of the CMOS Schmitt trigger in Example 8.1. It has been pointed out that a useful application lies in the receiver circuit design to filter out noises. However, in terms of speed performance it delays the switching activity. In view of speed alone, it would be useful to reverse the switching directions. In particular, we want to have the negative-going (high-to-low transition) edge to occur at an input voltage smaller than the typical inverter's saturation voltage and also the positive-going (low-to-high transition) edge to occur at an input voltage larger than the inverter's saturation voltage. Complete the circuit connection in Fig. P8.3 to realize such a circuit block for the assembly of the following components. Justify your answer by using SPICE circuit analysis. You can use some approximation technique to simulate the circuit. For instance, the different VTC curves can be simulated by using inverters of different β ratios. To be more specific, for larger saturation voltage you can use an inverter with a strong pull-up transistor, and for smaller saturation voltage use an inverter with a strong pull-down transistor.

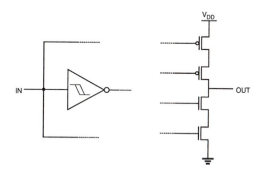

Figure P8.3

8.4 Consider the monostable multivibrator circuit drawn in Fig. P8.4. Calculate the output pulse width.

- $V_T(\text{dep}) = -2$ V
- $V_T(\text{enh}) = 1$ V
- $k' = 20 \ \mu\text{A/V}^2$
- $\gamma = 0$

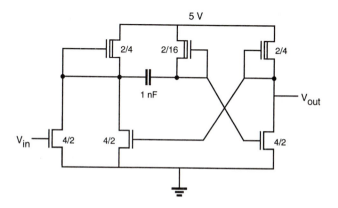

Figure P8.4

8.5   Shown in Fig. P8.5 is an nMOS Schmitt trigger. Draw the voltage transfer characteristic. Include values for all important points on the graph. Use the parameters in Problem 8.4 and $\lambda = 0$. W/L ratios for the transistors are given below.

	M1	M2	M3	M4
W/L	1	0.5	10	1

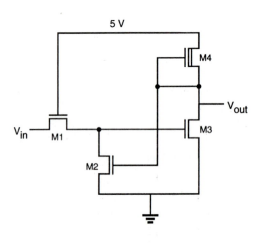

Figure P8.5

8.6   Design a circuit to implement the truth table shown in Fig. P8.6. A gate-level design is sufficient.

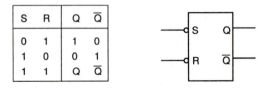

S	R	Q	$\overline{Q}$
0	1	1	0
1	0	0	1
1	1	Q	$\overline{Q}$

Figure P8.6

8.7   The circuit you have designed in Problem 8.6 is embedded in the larger circuit shown in Fig. P8.7. Complete the timing diagram for the output.

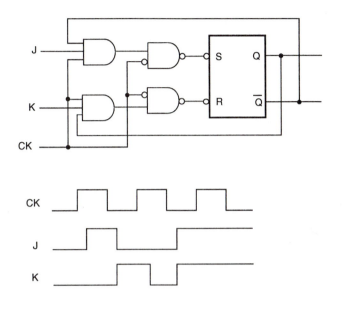

Figure P8.7

8.8 The voltage waveforms shown below are applied to the nMOS JK master-slave flip-flop shown in Figure 8.23 in Chapter 8. With the flip-flop initially reset, show the resulting waveforms at nodes Q_m (master flip-flop output) and Q_s (slave flip-flop output).

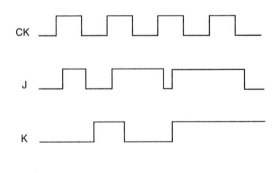

Figure P8.8

CHAPTER 9

DYNAMIC LOGIC CIRCUITS

9.1. Introduction

A wide range of static combinational and sequential logic circuits was introduced in the previous chapters. Static logic circuits allow versatile implementation of logic functions based on static, or steady-state, behavior of simple nMOS or CMOS structures. In other words, all valid output levels in static gates are associated with steady-state operating points of the circuit in question. Hence, a typical static logic gate generates its output corresponding to the applied input voltages after a certain time delay, and it can preserve its output level (or state) as long as the power supply is provided. This approach, however, may require a large number of transistors to implement a function, and may cause a considerable time delay.

In high-density, high-performance digital implementations where reduction of circuit delay and silicon area is a major objective, *dynamic logic circuits* offer several significant advantages over static logic circuits. The operation of all dynamic logic gates depends on temporary (transient) storage of charge in parasitic node capacitances, instead of relying on steady-state circuit behavior. This operational property necessitates periodic updating of internal node voltage levels, since stored charge in a capacitor cannot be retained indefinitely. Consequently, dynamic logic circuits require periodic clock signals for periodic charge refreshing of transfer, or timing discipline. The capability of temporarily storing a state, i.e., a voltage level, at a capacitive node allows us to implement very simple sequential circuits with memory functions. Also, the use of common clock signals throughout the system enables us to *synchronize* the operations of various circuit blocks. As a result, dynamic circuit techniques lend themselves well to synchronous logic design. Finally, the dynamic logic implementation of complex functions generally requires a smaller silicon area than does the static logic implementation. As for the power consumption which increases with the parasitic capacitances, the dynamic circuit implementation in a smaller area will, in many cases, consume less power than the static counterpart, despite its use of clock signals.

The following example presents the simple realization of a depletion-load nMOS dynamic D-latch circuit. This simple example illustrates most of the basic operational concepts involved in dynamic circuit design.

Example 9.1.

Consider the dynamic D-latch circuit shown below. The circuit consists of two cascaded depletion-load nMOS inverters and one enhancement-type nMOS pass transistor driving the input of the primary inverter stage.

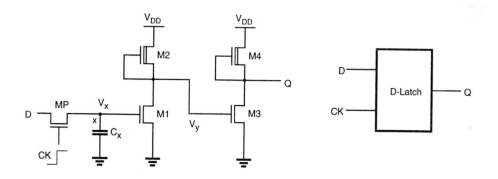

We will see that the parasitic input capacitance C_x of the primary inverter stage plays an important role in the dynamic operation of this circuit. The input pass transistor is being driven by the external periodic clock signal, as follows:

- When the clock is high (CK = 1), the pass transistor turns on. The capacitor C_x is either charged up, or charged down through the pass transistor MP, depending on the input (D) voltage level. The output (Q) assumes the same logic level as the input.

- When the clock is low (CK = 0), the pass transistor MP turns off, and the capacitor C_x is isolated from the input D. Since there is no current path from the intermediate node X to either V_{DD} or ground, the amount of charge stored in C_x during the previous cycle determines the output voltage level Q.

It can easily be seen that this circuit performs the function of a simple D-latch. In fact, the transistor count can be reduced by removing the last inverter stage if the latch output can be inverted. This option will be elaborated on in Section 9.2. The "hold" operation during the inactive clock cycle is accomplished by temporarily storing charge in the parasitic capacitance C_x. Correct operation of the circuit critically depends on how long a sufficient amount of charge can be retained at node X, before the output state changes due to charge leakage. Therefore, the capacitive intermediate node X is also called a *soft node*. The nature of the soft node makes the dynamic circuits more vulnerable to the so-called single-event upsets (SEUs) caused by α-particle or cosmic ray hits in integrated circuits.

The critical transistor parameters for this circuit are as follows.

$$\left(\frac{W}{L}\right)_{driver} = 2$$

$$\left(\frac{W}{L}\right)_{load} = 0.5$$

$$k_n' = \mu_n C_{ox} = 20 \ \mu A/V^2$$

$$V_{T,driver} = 1.0 \ V$$

$$V_{T,load} = -3.0 \ V$$

$$\gamma = 0.37 \ V^{1/2}$$

$$|2\phi_F| = 0.6 \ V$$

During the active clock phase (CK = 1), assume that the input is equal to logic "1," i.e., $V_{in} = V_{OH} = V_{DD}$. The pass transistor MP is conducting during this phase, and the parasitic intermediate node capacitance C_x is charged up to a logic-high level. It can be recalled that the nMOS pass transistor is a poor conductor for logic "1," and its output voltage will be lower by one threshold voltage. The output voltage V_y of the primary inverter will become very close to V_{OL}, which is lower than the threshold voltage of M3. The detailed analysis will be presented in Section 9.2. Consequently, the output level Q of the secondary inverter becomes a logic "1," $V_Q = V_{DD}$.

Next, the clock signal goes to zero, and the pass transistor turns off. Initially, the logic-high level at node X is preserved through charge storage in C_x. Thus, the output level Q also remains at logic "1." However, the voltage level V_x ultimately starts to drop because of charge leakage from the soft node.

We have to determine the minimum value of V_x which is necessary in order to keep the output node at logic "1," i.e., $V_Q = V_{DD}$.

First, note that to guarantee $V_Q = V_{DD}$, the driver transistor of the secondary inverter, M3, must be off. This means that the node voltage V_y must be smaller than the threshold voltage of M3, $V_y < 1.0$ V. When this condition is satisfied, it can be found that the driver transistor of the primary inverter, M1, operates in the linear region and the load transistor M2 operates in saturation. Thus,

$$\frac{k_n'}{2} \cdot 0.5 \cdot \left(0 - V_{T,load}\right)^2 = \frac{k_n'}{2} \cdot 2 \cdot \left(2\left(V_x - 1.0\right) V_y - V_y^2\right)$$

As the limiting case, assume that the voltage V_y is barely sufficient to keep M3 off, i.e., $V_y = 1.0$ V. Then, the threshold voltage of the load transistor M2 can be found as

$$V_{T,load} = -3.0 + 0.37 \left(\sqrt{0.6 + 1.0} - \sqrt{0.6}\right) = -2.84 \text{ V}$$

The corresponding voltage level V_x is determined from the current equation.

$$0.5 \cdot \left(0 + 2.84\right)^2 = 2\left(2\left(V_x - 1.0\right) 1.0 - 1.0^2\right)$$

$$V_x = \underline{\underline{2.51 \text{ V}}}$$

Thus, the voltage level at the intermediate node X can be allowed to drop from its logic-high level to 2.51 V due to charge leakage, before the output level is affected. To avoid an erroneous output, the charge stored in C_x must be restored, or refreshed, to its original level before V_x reaches 2.51 V.

This example shows that the simple dynamic-charge storage principle employed in the D-latch circuit is quite feasible for preserving an output state during the inactive clock phase, considering that the leakage currents responsible for draining the capacitance C_x are relatively small. In the following, we will examine the charge-up and charge-down events for the soft-node capacitance C_x in greater detail.

9.2. Basic Principles of Pass Transistor Circuits

The fundamental building block of nMOS dynamic logic circuits, consisting of an nMOS pass transistor and an inverter, is shown in Fig. 9.1. As already discussed in Example 9.1, the pass transistor MP is driven by the periodic clock signal and acts as an access switch to either charge up or charge down the parasitic capacitance C_x, depending on the input signal V_{in}. Thus, the two possible operations when the clock signal is active (CK = 1) are the logic "1" transfer (charging up the capacitance C_x to a logic-high level) and the logic "0" transfer (charging down the capacitance C_x to a logic-low level). In either case, the output of the depletion-load nMOS inverter obviously assumes a logic-low or a logic-high level, depending on the voltage V_x.

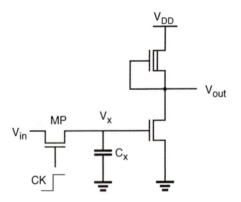

Figure 9.1.   The basic building block for nMOS dynamic logic, which consists of an nMOS pass transistor and an inverter.

Notice that the pass transistor MP provides the only current path to the intermediate capacitive node (soft node) X. When the clock signal becomes inactive (CK = 0), the pass transistor ceases to conduct and the charge stored in the parasitic capacitor C_x continues to determine the output level of the inverter. In the following, we will first examine the charge-up event.

Logic "1" Transfer

Assume that the soft node voltage is equal to 0 initially, i.e., $V_x(t=0) = 0$ V. A logic "1" level is applied to the input terminal, which corresponds to $V_{in} = V_{OH} = V_{DD}$. Now, the clock signal at the gate of the pass transistor goes from 0 to V_{DD} at $t=0$. It can be seen that the pass transistor MP starts to conduct as soon as the clock signal becomes active and that MP will operate in saturation throughout this cycle since it has $V_{DS} = V_{GS}$. Consequently, $V_{DS} > V_{GS} - V_{T,n}$.

The circuit to be analyzed for the logic "1" transfer event can be simplified into an equivalent circuit as shown in Fig. 9.2. Here, the depletion-load nMOS inverter is not to be considered, since it does not affect the charge-up event.

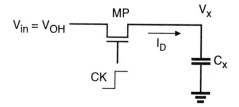

Figure 9.2. Equivalent circuit for the logic "1" transfer event.

The pass transistor MP operating in the saturation region starts to charge up the capacitor C_x, thus,

$$C_x \frac{dV_x}{dt} = \frac{k_n}{2} \left(V_{DD} - V_x - V_{T,n} \right)^2 \tag{9.1}$$

Note that the threshold voltage of the pass transistor is actually subject to substrate bias effect and therefore, depends on the voltage level V_x. To simplify our analysis, we will neglect the substrate bias effect at this point. Integrating (9.1), we obtain

$$\int_0^t dt = \frac{2C_x}{k_n} \int_0^{V_x} \frac{dV_x}{\left(V_{DD} - V_x - V_{T,n} \right)^2}$$

$$= \frac{2C_x}{k_n} \left(\frac{1}{\left(V_{DD} - V_x - V_{T,n} \right)} \right) \Bigg|_0^{V_x} \tag{9.2}$$

$$t = \frac{2\,C_x}{k_n}\left[\left(\frac{1}{V_{DD}-V_x-V_{T,n}}\right)-\left(\frac{1}{V_{DD}-V_{T,n}}\right)\right]$$ (9.3)

This equation can be solved for $V_x(t)$, as follows.

$$V_x(t) = \left(V_{DD}-V_{T,n}\right)\frac{\left(\dfrac{k_n\left(V_{DD}-V_{T,n}\right)}{2\,C_x}\right)t}{1+\left(\dfrac{k_n\left(V_{DD}-V_{T,n}\right)}{2\,C_x}\right)t}$$ (9.4)

The variation of the node voltage V_x according to (9.4) is plotted as a function of time in Fig. 9.3. The voltage rises from its initial value of 0 V and approaches a limit value for large t, but it cannot exceed its limit value of $V_{max}=(V_{DD}-V_{T,n})$. The pass transistor will turn off when $V_x=V_{max}$, since at this point, its gate-to-source voltage will be equal to its threshold voltage. Therefore, the voltage at node X can never attain the full power supply voltage level of V_{DD} during the logic "1" transfer.

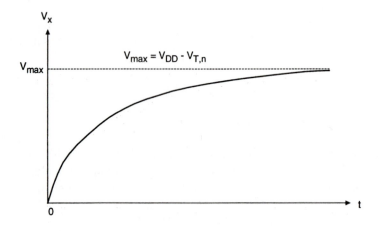

Figure 9.3.   Variation of V_x as a function of time during logic "1" transfer.

The actual value of the maximum possible voltage V_{max} at node X can be found by taking into account the substrate bias effect for MP.

$$V_{max} = V_x|_{t \to \infty} = V_{DD} - V_{T,n}$$

$$= V_{DD} - V_{T0,n} - \gamma \left(\sqrt{|2\phi_F| + V_{max}} - \sqrt{|2\phi_F|} \right)$$

(9.5)

Thus, the voltage V_x which is obtained at node X following a logic "1" transfer can be considerably lower than V_{DD}. Also note that the rise time of the voltage V_x will be *underestimated* if the zero-bias threshold voltage V_{T0} is used in (9.3). In that case, the actual charge-up time will be longer than predicted by (9.3), because the drain current of the nMOS transistor is decreased due to the substrate bias effect.

The fact that the node voltage V_x has an upper limit of $V_{max} = (V_{DD} - V_{T,n})$ has a significant implication for circuit design. As an example, consider the following case in which a logic "1" at the input node ($V_{in} = V_{DD}$) is being transferred through a chain of cascaded pass transistors (Fig. 9.4). For simple analysis, we assume that initially all internal node voltages, V_1 through V_4, are zero. The first pass transistor M1 operates in saturation with $V_{DS1} > V_{GS1} - V_{T,n1}$. Therefore, the voltage at node 1 cannot exceed the limit value $V_{max1} = (V_{DD} - V_{T,n1})$. Now, assuming that the pass transistors in this circuit are identical, the second pass transistor M2 operates at the *saturation boundary*. As a result, the voltage at node 2 will be equal to $V_{max2} = (V_{DD} - V_{T,n2})$. It can easily be seen that with $V_{T,n1} = V_{T,n2} = V_{T,n3} = ...$, the node voltage at the end of the pass transistor chain will become one threshold voltage lower than V_{DD}, regardless of the number of pass transistors in the chain. It can be observed that the steady-state internal node voltages in this circuit are always one threshold voltage below V_{DD}, regardless of the initial voltages.

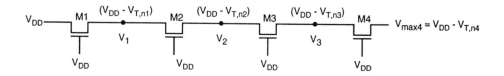

Figure 9.4.   Node voltages in a pass-transistor chain during the logic "1" transfer.

Now consider a different case in which the output of each pass transistor drives the gate of another pass transistor, as depicted in Fig. 9.5.

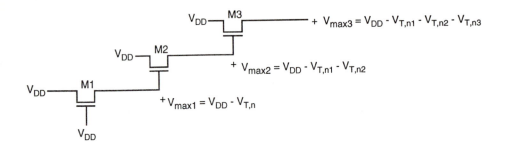

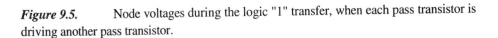

Figure 9.5.   Node voltages during the logic "1" transfer, when each pass transistor is driving another pass transistor.

Here, the output of the first pass transistor M1 can reach the limit $V_{max1} = (V_{DD} - V_{T,n1})$. This voltage drives the gate of the second pass transistor, which also operates in the saturation region. Its gate-to-source voltage cannot exceed $V_{Tn,2}$, hence, the upper limit for V_2 is found as $V_{max2} = V_{DD} - V_{T,n1} - V_{T,n2}$. It can be seen that in this case, each stage causes a significant loss of voltage level. The amount of voltage drop at each stage can be approximated more realistically by taking into account the corresponding substrate bias effect, which is different in all stages.

$$V_{T,n1} = V_{T0,n} - \gamma \left(\sqrt{|2 \phi_F| + V_{max1}} - \sqrt{|2 \phi_F|} \right)$$

$$V_{T,n2} = V_{T0,n} - \gamma \left(\sqrt{|2 \phi_F| + V_{max2}} - \sqrt{|2 \phi_F|} \right) \qquad (9.6)$$

$$\vdots$$

The preceding analysis helped us to examine important characteristics of the logic "1" transfer event. Next, we will examine the charge-down event, which is also called a logic "0" transfer.

Logic "0" Transfer

Assume that the soft-node voltage V_x is equal to a logic "1" level initially, i.e., $V_x(t=0) = V_{max} = (V_{DD} - V_{T,n})$. A logic "0" level is applied to the input terminal, which corresponds to $V_{in} = 0$ V. Now, the clock signal at the gate of the pass transistor goes from 0 to V_{DD} at $t = 0$. The pass transistor MP starts to conduct as soon as the clock signal becomes active, and the direction of drain current flow through MP will be opposite to that during the charge-up (logic "1" transfer) event. This means that the intermediate node X will now correspond to the drain terminal of MP and that the input node will correspond to its source terminal. With $V_{GS} = V_{DD}$ and $V_{DS} = V_{max}$, it can be seen that the pass transistor operates in the linear region throughout this cycle, since $V_{DS} < V_{GS} - V_{T,n}$.

The circuit to be analyzed for the logic "0" transfer event can be simplified into an equivalent circuit as shown in Fig. 9.6. As in the logic "1" transfer case, the depletion-load nMOS inverter does not affect this event.

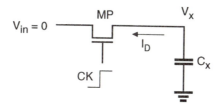

Figure 9.6.   Equivalent circuit for the logic "0" transfer event.

The pass transistor MP operating in the linear region discharges the parasitic capacitor C_x, as follows.

$$-C_x \frac{dV_x}{dt} = \frac{k_n}{2}\left(2\left(V_{DD} - V_{T,n}\right)V_x - V_x^2\right) \tag{9.7}$$

$$dt = -\frac{2C_x}{k_n} \cdot \frac{dV_x}{2\left(V_{DD} - V_{T,n}\right)V_x - V_x^2} \tag{9.8}$$

Note that the source voltage of the nMOS pass transistor is equal to 0 V during this event, hence there is no substrate bias effect which affects the threshold voltage of MP. Integrating both sides of (9.8) yields

$$\int_0^t dt = -\frac{2C_x}{k_n}\int_{V_{DD}-V_{T,n}}^{V_x}\left(\frac{\dfrac{1}{2\left(V_{DD}-V_{T,n}\right)}}{2\left(V_{DD}-V_{T,n}\right)-V_x} + \frac{\dfrac{1}{2\left(V_{DD}-V_{T,n}\right)}}{V_x}\right)dV_x \tag{9.9}$$

$$t = \frac{C_x}{k_n\left(V_{DD}-V_{T,n}\right)}\left[\ln\left(\frac{2\left(V_{DD}-V_{T,n}\right)-V_x}{V_x}\right)\right]\Bigg|_{V_{DD}-V_{T,n}}^{V_x} \tag{9.10}$$

Finally, the fall-time expression for the node voltage V_x can be obtained as

$$t = \frac{C_x}{k_n \left(V_{DD} - V_{T,n} \right)} \ln \left(\frac{2 \left(V_{DD} - V_{T,n} \right) - V_x}{V_x} \right) \tag{9.11}$$

The variation of the node voltage V_x according to (9.11) is plotted as a function of time in Fig. 9.7. It is seen that the voltage drops from its logic-high level of V_{max} to 0 V. Hence, unlike the charge-up case, the applied input voltage level (logic 0) can be transferred to the soft node without any modification during this event.

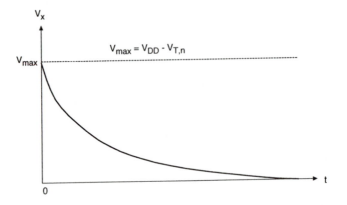

Figure 9.7.   Variation of V_x as a function of time during logic "0" transfer.

The fall time (τ_{fall}) for the soft-node voltage V_x can be calculated from (9.11) as follows. First, define the two time points $t_{90\%}$ and $t_{10\%}$ as the times at which the node voltage is equal to $0.9\,V_{max}$ and $0.1\,V_{max}$, respectively. These two time points can easily be found by using (9.11).

$$
\begin{aligned}
t_{90\%} &= \frac{C_x}{k_n \left(V_{DD} - V_{T,n} \right)} \ln \left(\frac{(2-0.9)\left(V_{DD} - V_{T,n} \right)}{0.9 \left(V_{DD} - V_{T,n} \right)} \right) \\
&= \frac{C_x}{k_n \left(V_{DD} - V_{T,n} \right)} \ln \left(\frac{1.1}{0.9} \right)
\end{aligned}
\tag{9.12}
$$

$$t_{10\%} = \frac{C_x}{k_n \left(V_{DD} - V_{T,n} \right)} \ln \left(\frac{1.9}{0.1} \right) \tag{9.13}$$

The fall time of the soft-node voltage V_x is by definition the difference between $t_{10\%}$ and $t_{90\%}$, which is found as

$$\tau_{fall} = t_{10\%} - t_{90\%}$$

$$= \frac{C_x}{k_n(V_{DD} - V_{T,n})}\left[\ln(19) - \ln(1.22)\right]$$

$$= 2.74\frac{C_x}{k_n(V_{DD} - V_{T,n})} \tag{9.14}$$

Until this point, we have examined the transient charge-up and charge-down events which are responsible for logic "1" transfer and logic "0" transfer during the active clock phase, i.e., when $CK = 1$. Now we will turn our attention to the storage of logic levels at the soft node X during the inactive clock cycle, i.e., when $CK = 0$.

Charge Storage and Charge Leakage

As already discussed qualitatively in the preceding section, the preservation of a correct logic level at the soft node during the inactive clock phase depends on preserving a sufficient amount of charge in C_x, despite the leakage currents. To analyze the events during the inactive clock phase in more detail, consider the scenario shown in Fig. 9.8 below. We will assume that a logic-high voltage level has been transferred to the soft node during the active clock phase and that now both the input voltage V_{in} and the clock are equal to 0 V. The charge stored in C_x will gradually leak away, primarily due to the leakage currents associated with the pass transistor. The gate current of the inverter driver transistor is negligible for all practical purposes.

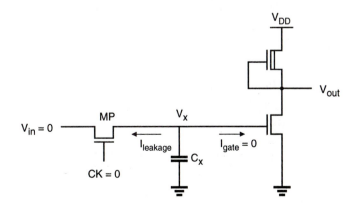

Figure 9.8.   Charge leakage from the soft node.

Figure 9.9 shows a simplified cross-section of the nMOS pass transistor, together with the lumped node capacitance C_x. We see that the leakage current responsible for draining the soft-node capacitance over time has two main components, namely, the subthreshold channel current and the reverse conduction current of the drain-substrate junction.

$$I_{leakage} = I_{subthreshold(MP)} + I_{reverse(MP)} \qquad (9.15)$$

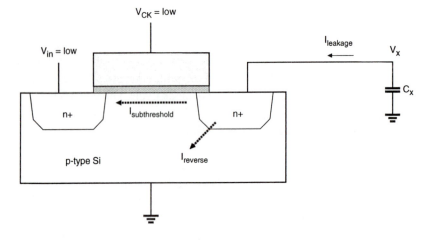

Figure 9.9. Simplified cross-section of the nMOS pass transistor, showing the leakage current components responsible for draining the soft-node capacitance C_x.

Note that a certain portion of the total soft-node capacitance C_x is due to the reverse biased drain-substrate junction, which is also a function of the soft-node voltage V_x. Other components of C_x, which are primarily due to oxide-related parasitics, can be considered constants. In our analysis, these constant capacitance components will be represented by C_{in} (Fig. 9.10). Thus, we have to express the total charge stored in the soft node as the sum of two main components, as follows.

$$Q = Q_j(V_x) + Q_{in} \quad \text{where} \quad Q_{in} = C_{in} \cdot V_x$$

$$(9.16)$$

$$C_{in} = C_{gb} + C_{poly} + C_{metal}$$

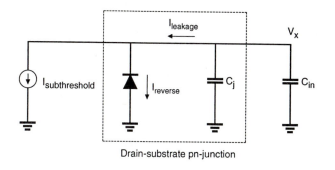

Figure 9.10.   Equivalent circuit used for analyzing the charge leakage process.

The total leakage current can be expressed as the time derivative of the total soft-node charge Q.

$$
\begin{aligned}
I_{leakage} &= \frac{dQ}{dt} \\
&= \frac{dQ_j(V_x)}{dt} + \frac{dQ_{in}}{dt} \\
&= \frac{dQ_j(V_x)}{dV_x}\frac{dV_x}{dt} + C_{in}\frac{dV_x}{dt}
\end{aligned}
\tag{9.17}
$$

where

$$
\frac{dQ_j(V_x)}{dt} = C_j(V_x) = \frac{A \cdot C_{j0}}{\sqrt{1+\dfrac{V_x}{\phi_0}}} = A \cdot \sqrt{\frac{q\,\varepsilon_{Si}\,N_A}{2(\phi_0+V_x)}}
\tag{9.18}
$$

according to (3.104), and

$$
\phi_0 = \frac{kT}{q}\ln\!\left(\frac{N_D \cdot N_A}{n_i^{\,2}}\right)
\tag{9.19}
$$

$$
C_{j0} = \sqrt{\frac{q\,\varepsilon_{Si}\,N_A\,N_D}{2(N_A+N_D)\phi_0}} \approx \sqrt{\frac{q\,\varepsilon_{Si}\,N_A}{2\,\phi_0}}
\tag{9.20}
$$

Also note that typically, the reverse conduction current $I_{reverse}$ of the drain-substrate pn-junction is significantly larger than the subthreshold channel current of the pass transistor. The reverse conduction current in turn has two main components, the constant reverse saturation current I_0, and the generation current I_{gen} which originates in the depletion region and is a function of the applied bias voltage V_x.

To estimate the actual charge leakage time from the soft node, we have to solve the differential equation given in (9.17), taking into account the voltage-dependent capacitance components and the nonlinear leakage currents. For a quick estimate of the worst-case leakage behavior, on the other hand, the problem can be further simplified.

Assume that the *minimum* combined soft-node capacitance is given as

$$C_{x,min} = C_{gb} + C_{poly} + C_{metal} + C_{db,min} \tag{9.21}$$

where $C_{db,min}$ represents the minimum junction capacitance, obtained under the bias condition $V_x = V_{max}$. Now we define the *worst-case holding time* (t_{hold}) as the shortest time required for the soft-node voltage to drop from its initial logic-high value to the logic threshold voltage due to leakage. Once the soft-node voltage reaches the logic threshold, the logic stage being driven by this node will lose its previously held state.

$$t_{hold} = \frac{\Delta Q_{critical,min}}{I_{leakage,max}} \tag{9.22}$$

where

$$\Delta Q_{critical,min} = C_{x,min} \left(V_{max} - \frac{V_{DD}}{2} \right) \tag{9.23}$$

The calculation of the worst-case leakage time can be simplified with this approximation, as will be shown in the following example.

Example 9.2.

Consider the soft-node structure shown below, which consists of the drain (or source, depending on current direction) terminal of the pass transistor, connected to the polysilicon gate of an nMOS driver transistor via a metal interconnect.

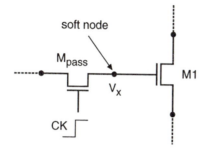

We will assume that the power supply voltage used in this circuit is $V_{DD} = 5$ V, and that the soft node has initially been charged up to its maximum voltage, V_{max}. In order to estimate the worst-case holding time, the total soft-node capacitance must be calculated first. The simplified mask layout of the structure is shown in the following. All dimensions are given in micrometers.

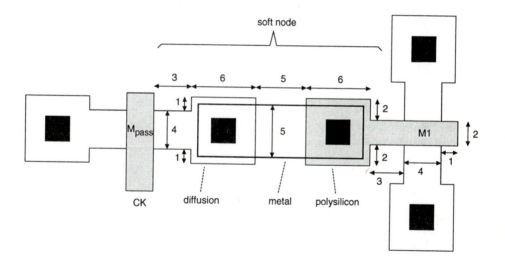

The critical material parameters to be used in this example are listed below.

$$V_{T0} = 0.8 \text{ V}$$

$$\gamma = 0.4 \text{ V}^{1/2}$$

$$|2\phi_F| = 0.6 \text{ V}$$

$$C_{ox} = 0.065 \text{ fF/}\mu\text{m}^2$$

$$C'_{metal} = 0.036 \text{ fF/}\mu\text{m}^2$$

$$C'_{poly} = 0.055 \text{ fF/}\mu\text{m}^2$$

$$C_{j0} = 0.095 \text{ fF/}\mu\text{m}^2$$

$$C_{j0sw} = 0.2 \text{ fF/}\mu\text{m}$$

First, we calculate the oxide-related (constant) parasitic capacitance components associated with the soft node.

$$
\begin{aligned}
C_{gb} &= C_{ox} \cdot W \cdot L_{mask} \\
&= 0.065 \text{ fF/}\mu\text{m}^2 \cdot (4 \ \mu\text{m} \times 2 \ \mu\text{m}) \\
&= 0.52 \text{ fF}
\end{aligned}
$$

$$
\begin{aligned}
C_{metal} &= 0.036 \text{ fF/}\mu\text{m}^2 \cdot (5 \ \mu\text{m} \times 5 \ \mu\text{m}) \\
&= 0.90 \text{ fF}
\end{aligned}
$$

$$
\begin{aligned}
C_{poly} &= 0.055 \text{ fF/}\mu\text{m}^2 \cdot (36 \ \mu\text{m}^2 + 8 \ \mu\text{m}^2) \\
&= 2.42 \text{ fF}
\end{aligned}
$$

Now, we have to calculate the parasitic junction capacitance associated with the drain-substrate pn-junction of the pass transistor. Using the zero-bias unit capacitance values given here, we obtain

$$
\begin{aligned}
C_{db,max} &= C_{bottom} + C_{sidewall} \\
&= A_{bottom} \cdot C_{j0} + P_{sidewall} \cdot C_{j0sw} \\
&= (36 \ \mu\text{m}^2 + 12 \ \mu\text{m}^2) \cdot 0.095 \text{ fF/}\mu\text{m}^2 + 30 \ \mu\text{m} \cdot 0.2 \text{ fF/}\mu\text{m} \\
&= 4.56 \text{ fF} + 6.0 \text{ fF} \\
&= 10.56 \text{ fF}
\end{aligned}
$$

The minimum value of the drain junction capacitance is achieved when the junction is biased (in reverse) with its maximum possible voltage, V_{max}. In order to calculate the minimum capacitance value, we first find V_{max} using (9.5), as follows.

$$V_{max} = 5.0 - 0.8 - 0.4\left(\sqrt{0.6 + V_{max}} - \sqrt{0.6}\right)$$

$$\Rightarrow V_{max} = 3.68 \text{ V}$$

Now, the minimum value of the drain junction capacitance can be calculated.

$$C_{db,min} = \frac{C_{bottom}}{\sqrt{1 + \dfrac{V_{x,max}}{\phi_0}}} + \frac{C_{sidewall}}{\sqrt{1 + \dfrac{V_{x,max}}{\phi_{0sw}}}}$$

$$= \frac{4.56 \text{ fF}}{\sqrt{1 + \dfrac{3.68}{0.88}}} + \frac{6.0 \text{ fF}}{\sqrt{1 + \dfrac{3.68}{0.95}}} = 4.71 \text{ fF}$$

The minimum value of the total soft-node capacitance is found by using (9.21).

$$C_{x,min} = C_{gb} + C_{metal} + C_{poly} + C_{db,min}$$
$$= 0.52 \text{ fF} + 0.90 \text{ fF} + 2.42 \text{ fF} + 4.71 \text{ fF}$$
$$= 8.55 \text{ fF}$$

The amount of the critical charge drop in the soft node, which will eventually cause a change of logic state, is

$$\Delta Q_{critical} = C_{x,min} \cdot \left(V_{x,max} - \frac{V_{DD}}{2}\right)$$
$$= 8.55 \text{ fF} \cdot (3.68 \text{ V} - 2.5 \text{ V})$$
$$= 10.09 \text{ fC}$$

assuming that the logic threshold voltage of the next gate is ($V_{DD}/2$). In this example, the maximum leakage current responsible for charge depletion is given from the junction diode characteristics and the pn-junction area as

$$I_{leakage} = I_{subthreshold} + I_{reverse} \approx I_{reverse} = 0.85 \text{ pA}$$

Finally, we calculate the worst-case (minimum) hold time for the soft node using the expression (9.22).

$$t_{hold,min} = \frac{\Delta Q_{critical}}{I_{leakage,max}}$$

$$= \frac{10.09 \text{ fC}}{0.85 \text{ pA}} = \underline{\underline{11.87 \text{ ms}}}$$

It is interesting to note that even with a very small soft-node capacitance of 8.55 fF, the worst-case hold time for this structure is relatively long, especially compared with the signal propagation delays encountered in nMOS or CMOS logic gates. This example proves the feasibility of the dynamic charge storage concept and shows that a logic state can be safely preserved in a soft node for long time periods.

9.3. Voltage Bootstrapping

In this section, we will briefly examine a very useful dynamic circuit technique for overcoming threshold voltage drops in digital circuits, which is called *voltage bootstrapping*. We have already seen that output voltage levels may suffer from threshold voltage drops in several circuit structures, such as pass transistor gates or enhancement-load inverters and logic gates. This situation is typically due to enhancement-type nMOS transistors being forced to operate exclusively in the saturation region: When an enhancement-type nMOS transistor biased in the saturation region does not conduct a drain current, then its gate-to-source voltage drop must be equal to its threshold voltage. The resulting threshold voltage drop at the source terminal may cause significant problems in terms of reduced logic swing and noise margins.

Dynamic voltage bootstrapping techniques offer a simple yet effective way to overcome threshold voltage drops which occur in most situations. Consider the circuit shown in Fig. 9.11, where the voltage V_x is equal to or smaller than the power supply voltage, $V_x \leq V_{DD}$. Consequently, the enhancement-type nMOS transistor M2 will operate in saturation.

When the input voltage V_{in} is low, i.e., when M1 and M2 are not conducting, the maximum value that the output voltage can attain is limited by

$$V_{out}(max) = V_x - V_{T2}(V_{out}) \tag{9.24}$$

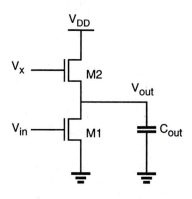

Figure 9.11.   Enhancement-type circuit in which the output node is weakly driven.

To overcome the threshold voltage drop and to obtain a full logic-high level (V_{DD}) at the output node, the voltage V_x must be increased. Now consider the circuit shown in Fig. 9.12, where a third transistor M3 has been added to the circuit. The two capacitors C_S and C_{boot} seen in the circuit diagram represent the capacitances which dynamically couple the voltage V_x to the ground and to the output, respectively. We will see that this circuit can produce a high V_x during switching, so that the threshold voltage drop can be overcome at the output node.

$$V_x \geq V_{DD} + V_{T2}\left(V_{out}\right) \tag{9.25}$$

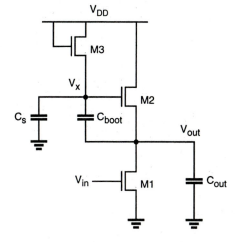

Figure 9.12.   Dynamic bootstrapping arrangement to boost V_x during switching.

Initially assume that the input voltage V_{in} is logic-high, so that M1 and M2 have a nonzero drain current and that the output voltage is low. At this point, M1 is in the linear region and M2 is in saturation. Since $I_{D3} = 0$, the initial condition for the voltage V_x can be found as

$$V_x = V_{DD} - V_{T3}(V_x) \tag{9.26}$$

Now, assume that the input switches from its logic-high level to 0 V at $t = 0$. As a result, the driver transistor M1 will turn off and the output voltage V_{out} will start to rise. This change in the output voltage level will now be coupled to V_x through the bootstrap capacitor, C_{boot}. Let i_{Cboot} represent the transient current flowing through the capacitor C_{boot} during this charge-up event, and let i_{Cs} be the current through C_S. Assuming that the two current components are approximately equal, we obtain

$$i_{Cs} \approx i_{Cboot} \Leftrightarrow C_S \frac{dV_x}{dt} \approx C_{boot} \frac{d(V_{out} - V_x)}{dt} \tag{9.27}$$

Reorganizing (9.27) yields the following equation.

$$(C_S + C_{boot}) \frac{dV_x}{dt} \approx C_{boot} \frac{dV_{out}}{dt} \tag{9.28}$$

$$\frac{dV_x}{dt} \approx \frac{C_{boot}}{(C_S + C_{boot})} \cdot \frac{dV_{out}}{dt} \tag{9.29}$$

It can be seen from (9.29) that the increase in the output voltage V_{out} during this switching event will generate a proportional increase in the voltage level V_x. Integrating both sides of (9.29), we obtain the final value of V_x at the end of the voltage transition required to achieve $V_{out} = V_{DD}$.

$$\int_{V_{DD} - V_{T3}}^{V_x} dV_x = \frac{C_{boot}}{(C_S + C_{boot})} \cdot \int_{V_{OL}}^{V_{DD}} dV_{out} \tag{9.30}$$

$$V_x = (V_{DD} - V_{T3}) + \frac{C_{boot}}{(C_S + C_{boot})} (V_{DD} - V_{OL}) \tag{9.31}$$

If the capacitor C_{boot} is much larger than C_S ($C_{boot} \gg C_S$), the maximum value of V_x can be approximated as

$$V_x(max) = 2V_{DD} - V_{T3} - V_{OL} \tag{9.32}$$

which proves that voltage bootstrapping can significantly boost the voltage level V_x. Now remember that in order to overcome the threshold voltage drop at the output, the minimum required voltage level V_x is

$$V_x(min) = V_{DD} + V_{T2}\big|_{V_{out} = V_{DD}}$$

$$= \left(V_{DD} - V_{T3}(V_x)\right) + \frac{C_{boot}}{\left(C_S + C_{boot}\right)}\left(V_{DD} - V_{OL}\right) \tag{9.33}$$

This equation can be rearranged to give the required capacitance ratio, as follows.

$$\frac{C_{boot}}{\left(C_S + C_{boot}\right)} = \frac{V_{T2}\big|_{V_{out} = V_{DD}} + V_{T3}\big|_{V_x}}{\left(V_{DD} - V_{OL}\right)} \tag{9.34}$$

$$\frac{C_{boot}}{C_S} = \frac{V_{T2}\big|_{V_{out} = V_{DD}} + V_{T3}\big|_{V_x}}{V_{DD} - V_{OL} - V_{T2}\big|_{V_{out} = V_{DD}} - V_{T3}\big|_{V_x}} \tag{9.35}$$

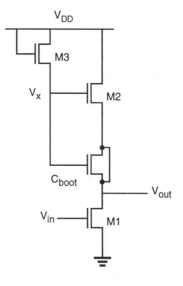

Figure 9.13.   Realization of the bootstrapping capacitor with a dummy MOS device.

Note that C_S is essentially the sum of the parasitic source-to-substrate capacitance of M3 and the gate-to-substrate capacitance of M2. To obtain a sufficiently large bootstrap capacitance C_{boot} in comparison to C_S, an extra "dummy" transistor is typically added to the circuit, as shown in Fig. 9.13.

Since its drain and source terminals are connected together, the dummy transistor simply acts as an MOS capacitor between V_x and V_{out}. Although this circuit arrangement contains two additional transistors to achieve voltage bootstrapping, the resulting circuit-performance improvement is usually well worth the extra silicon area used for the bootstrapping devices.

Example 9.3.

The transient operation of the simple bootstrap circuit shown in Fig. 9.13 is simulated using SPICE in the following. It is assumed that the output node is loaded with an external capacitance of 10 fF, in addition to the parasitic device capacitances. To provide the needed bootstrap capacitance C_{boot}, a dummy nMOS device with channel length $L = 5\,\mu m$ and channel width $W = 50\,\mu m$ is used. Transistor M1 has a (W/L) ratio of 2, while M2 and M3 each have a (W/L) ratio of 1.

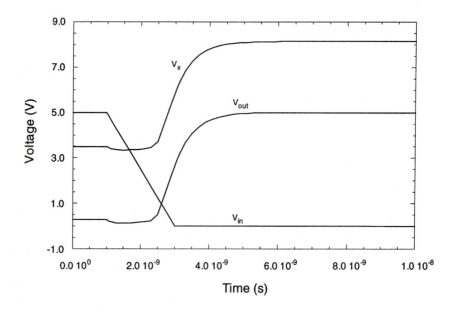

9.4. Synchronous Dynamic Circuit Techniques

Having examined the basic concepts associated with temporary storage of logic levels in capacitive circuit nodes, we now turn our attention to digital circuit design techniques which take advantage of this simple yet effective principle. In the following, we will investigate different examples of synchronous dynamic circuits implemented using depletion-load nMOS, enhancement-load nMOS, and CMOS building blocks.

Dynamic Pass Transistor Circuits

Consider the generalized view of a multi-stage synchronous circuit shown in Fig. 9.14. The circuit consists of cascaded combinational logic stages, which are interconnected through nMOS pass transistors. All inputs of each combinational logic block are driven by a single clock signal. Individual input capacitances are not shown in this figure for simplicity, but the operation of the circuit obviously depends on temporary charge storage in the parasitic input capacitances.

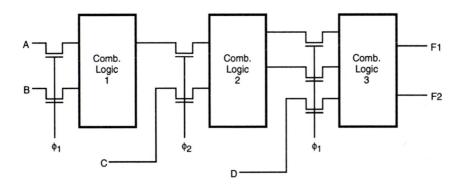

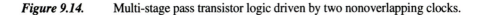

Figure 9.14.   Multi-stage pass transistor logic driven by two nonoverlapping clocks.

To drive the pass transistors in this system, two nonoverlapping clock signals, ϕ_1 and ϕ_2, are used. The nonoverlapping property of the two clock signals guarantees that at any given time point, only one of the two clock signals can be active, as illustrated in Fig. 9.15. When clock ϕ_1 is active, the input levels of Stage 1 (and also of Stage 3) are applied through the pass transistors, while the input capacitances of Stage 2 retain their previously set logic levels. During the next phase, when clock ϕ_2 is active, the input levels of Stage 2 will be applied through the pass transistors, while the input capacitances of Stage 1 and Stage 3 retain their logic levels. This allows us to incorporate the simple dynamic memory function at each stage input, and at the same time, to facilitate synchronous operation by controlling the signal flow in the circuit using the two periodic clock signals. This signal timing scheme is also called *two-phase clocking* and is one of the most widely used timing strategies.

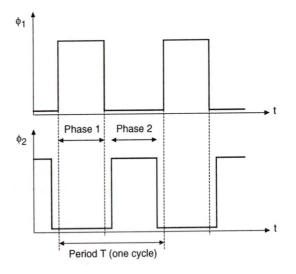

Figure 9.15.　　　　Nonoverlapping clock signals used for two-phase synchronous operation.

By introducing the two-phase clocking scheme, we have not made any specific assumptions about the internal structure of the combinational logic stages. It will be seen that depletion-load nMOS, enhancement-load nMOS, or CMOS logic circuits can be used for implementing the combinational logic. Figure 9.16 shows a depletion-load dynamic shift register circuit, in which the input data are inverted once and transferred, or *shifted* into the next stage during each clock phase.

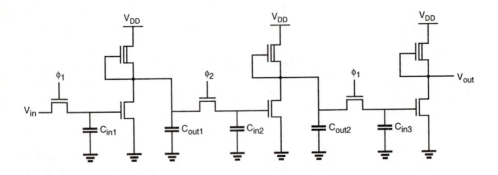

Figure 9.16.　　　　Three stages of a depletion-load nMOS dynamic shift register circuit driven with two-phase clocking.

The operation of the shift register circuit is as follows. During the active phase of ϕ_1, the input voltage level V_{in} is transferred into the input capacitance C_{in1}. Thus, the valid output voltage level of the first stage is determined as the inverse of the current input during this cycle. When ϕ_2 becomes active during the next phase, the output voltage level of the first stage is transferred into the second stage input capacitance C_{in2}, and the valid output voltage level of the second stage is determined. During the active ϕ_2 phase, the first-stage input capacitance continues to retain its previous level via charge storage. When ϕ_1 becomes active again, the original data bit *written* into the register during the previous cycle is transferred into the third stage, and the first stage can now accept the next data bit.

In this circuit, the maximum clock frequency is determined by the signal propagation delay through one inverter stage. One half-period of the clock signal must be long enough to allow the input capacitance C_{in} to charge up or down, and the logic level to propagate to the output by charging C_{out}. Also notice that the logic-high input level of each inverter stage in this circuit is one threshold voltage lower than the power supply voltage level.

The same operation principle used in the simple shift register circuit can easily be extended to synchronous complex logic. Figures 9.17 and 9.18 show a two-stage circuit example implemented using depletion-load nMOS complex logic gates.

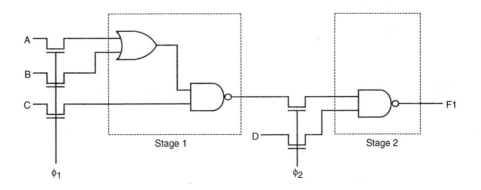

Figure 9.17.   A two-stage synchronous complex logic circuit example.

In a complex logic circuit such as the one shown in Fig. 9.18, we see that the signal propagation delay of each stage may be different. Thus, in order to guarantee that correct logic levels are propagated during each active clock cycle, the half-period length of the clock signal must be longer than the largest single-stage signal propagation delay found in the circuit.

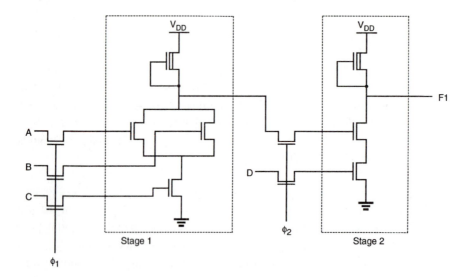

Figure 9.18.   Depletion-load nMOS implementation of synchronous complex logic.

Now consider a different implementation of the simple shift register circuit, using enhancement-load nMOS inverters. One important difference is that, instead of biasing the load transistors with a constant gate voltage, we apply the clock signal to the gate of the load transistor as well. It can be shown that the power dissipation and the silicon area can be reduced significantly by using this dynamic (clocked) load approach. Two variants of the dynamic enhancement-load shift register will be examined in the following, both of which are driven by two non-overlapping clock signals. Figure 9.19 shows the first implementation, where in each stage the input pass transistor and the load transistor are driven by opposite clock phases, ϕ_1 and ϕ_2.

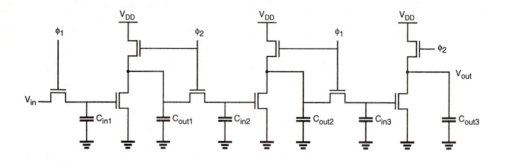

Figure 9.19.   Enhancement-load dynamic shift register (ratioed logic).

When ϕ_1 is active, the input voltage level V_{in} is transferred into the first-stage input capacitance C_{in1} through the pass transistor. In this phase, the enhancement-type nMOS load transistor of the first-stage inverter is not active yet. During the next phase (active ϕ_2), the load transistor is turned on. Since the input logic level is still being preserved in C_{in1}, the output of the first inverter stage attains its valid logic level. At the same time, the input pass transistor of the second stage is also turned on, which allows this newly determined output level to be transferred into the input capacitance C_{in2} of the second stage. When clock ϕ_1 becomes active again, the valid output level across C_{out2} is determined, and transferred into C_{in3}. Also, a new input level can be accepted (*pipelined*) into C_{in1} during this phase.

In this circuit, the valid low-output voltage level V_{OL} of each stage is strictly determined by the driver-to-load ratio, since the output pass transistor (input pass transistor of next stage) turns on in phase with the load transistor. Therefore, this circuit arrangement is also called *ratioed dynamic logic*. The basic operation principle can obviously be extended to arbitrary complex logic, as shown in Fig. 9.20. Since the power supply current flows only when the load devices are activated by the clock signal, the overall power consumption of dynamic enhancement-load logic is generally lower than for depletion-load nMOS logic.

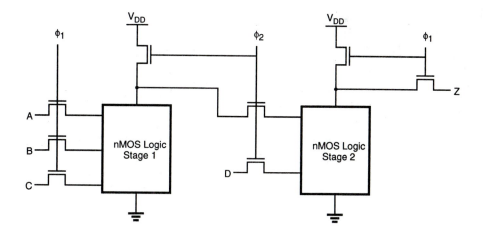

Figure 9.20.   General circuit structure of ratioed synchronous dynamic logic.

Next, consider the second dynamic enhancement-load shift register implementation where, in each stage, the input pass transistor and the load transistor are driven by the same clock phase (Fig. 9.21).

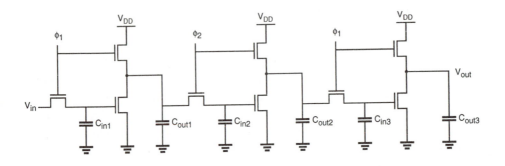

Figure 9.21. Enhancement-load dynamic shift register (ratioless logic).

When ϕ_1 is active, the input voltage level V_{in} is transferred into the first-stage input capacitance C_{in1} through the pass transistor. Note that at the same time, the enhancement-type nMOS load transistor of the first-stage inverter is active. Therefore, the output of the first inverter stage attains its valid logic level. During the next phase (active ϕ_2), the input pass transistor of the next stage is turned on, and the logic level is transferred onto the next stage. Here, we have to consider two cases, as follows.

If the output level across C_{out1} is logic-high at the end of the active ϕ_1 phase, this voltage level is transferred to C_{in2} via charge sharing over the pass transistor during the active ϕ_2 phase. Note that the logic-high level at the output node is subject to threshold voltage drop, i.e., it is one threshold voltage lower than the power supply voltage. To correctly transfer a logic-high level after charge sharing, the ratio of the capacitors (C_{out}/C_{in}) must be made large enough during circuit design.

If, on the other hand, the output level of the first stage is logic-low at the end of the active ϕ_1 phase, then the output capacitor C_{out1} will be completely drained to a voltage of $V_{OL}=0$ V when ϕ_1 turns off. This can be achieved because a logic-high level is being stored in the input capacitance C_{in1} in this case, which forces the driver transistor to remain in conduction. Obviously, the logic-low level of $V_{OL}=0$ V is also transferred into the next stage via the pass transistor during the active ϕ_2 phase.

When clock ϕ_1 becomes active again, the valid output level across C_{out2} is determined and transferred into C_{in3}. Also, a new input level can be accepted into C_{in1} during this phase. Since the valid logic-low level of $V_{OL} = 0$ V can be achieved regardless of the driver-to-load ratio, this circuit arrangement is called *ratioless dynamic logic*. The basic operation principle can be extended to arbitrary complex logic, as shown in Fig. 9.22.

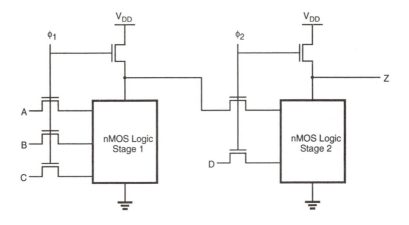

Figure 9.22.   General circuit structure of ratioless synchronous dynamic logic.

CMOS Transmission Gate Logic

The basic two-phase synchronous logic circuit principle, in which individual logic blocks are cascaded via clock-controlled switches, can easily be adopted to CMOS structures as well. Here, static CMOS gates are used for implementing the logic blocks, and CMOS transmission gates are used for transferring the output levels of one stage to the inputs of the next stage (Fig. 9.23). Notice that each transmission gate is actually controlled by the clock signal *and* its complement. As a result, two-phase clocking in CMOS transmission gate logic requires that a total of four clock signals are generated and routed throughout the circuit.

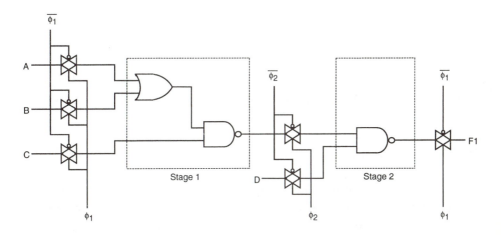

Figure 9.23.   Typical example of dynamic CMOS transmission gate logic.

As in the nMOS-based dynamic circuit structures, the operation of CMOS dynamic logic relies on charge storage in the parasitic input capacitances during the inactive clock cycles. To illustrate the basic operation principles, the fundamental building block of a dynamic CMOS transmission gate shift register is shown in Fig. 9.24. It consists of a CMOS inverter, which is driven by a CMOS transmission gate. During the active clock phase (CK = 1), the input voltage V_{in} is transferred onto the parasitic input capacitance C_x via the transmission gate. Note that the low on-resistance of the CMOS transmission gate usually results in a smaller transfer time compared to those for nMOS-only switches. Also, there is no threshold voltage drop across the CMOS transmission gate. When the clock signal becomes inactive, the CMOS transmission gate turns off and the voltage level across C_x can be preserved until the next cycle.

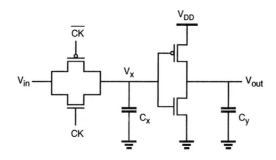

Figure 9.24.   Basic building block of a CMOS transmission gate dynamic shift register.

Figure 9.25 shows a single-phase CMOS shift register, which is built by cascading identical units as in Fig. 9.24 and by driving each stage alternately with the clock signal and its complement.

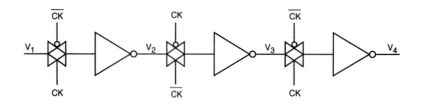

Figure 9.25.   Single-phase CMOS transmission gate dynamic shift register.

Ideally, the transmission gates of the first- and odd-numbered stages would conduct during the active clock phase (when CK = 1), while the transmission gates of the second and even-numbered stages are off, so that the cascaded inverter stages in the chain are alternately isolated. This would ensure that inputs are permitted in alternating half cycles. In practice, however, the clock signal and its complement do not constitute a truly nonoverlapping signal pair, since the clock voltage waveform has finite rise and fall times. Also, the clock skew between CK and $\overline{CK}$ may be unavoidable because one of the signals is generated by inverting the other. Therefore, true two-phase clocking with two nonoverlapping clock signals (ϕ_1 and ϕ_2) *and* their complements is usually preferred over single-phase clocking in dynamic CMOS transmission gate logic.

Dynamic CMOS Logic (Precharge-Evaluate Logic)

In the following, we will introduce a dynamic CMOS circuit technique which allows us to significantly reduce the number of transistors used to implement any logic function. The circuit operation is based on first *precharging* the output node capacitance and subsequently, *evaluating* the output level according to the applied inputs. Both of these operations are scheduled by a single clock signal, which drives one nMOS and one pMOS transistor in each dynamic stage. A dynamic CMOS logic gate which implements the function $F = \overline{(A_1 A_2 A_3 + B_1 B_2)}$ is shown in Fig. 9.26.

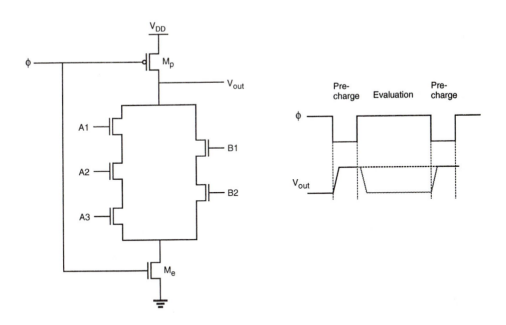

Figure 9.26.   Dynamic CMOS logic gate implementing a complex Boolean function.

When the clock signal is low (precharge phase), the pMOS precharge transistor M_p is conducting, while the complementary nMOS transistor M_e is off. The parasitic output capacitance of the circuit is charged up through the conducting pMOS transistor to a logic-high level of $V_{out} = V_{DD}$. The input voltages are also applied during this phase, but they have no influence yet upon the output level since M_e is turned off.

When the clock signal becomes high (evaluate phase), the precharge transistor M_p turns off and M_e turns on. The output node voltage may now remain at the logic-high level or drop to a logic low, depending on the input voltage levels. If the input signals create a conducting path between the output node and the ground, the output capacitance will discharge toward $V_{OL} = 0$ V. The final discharged output level depends on the time span of the evaluation phase. Otherwise, V_{out} remains at V_{DD}.

The operation of the single-stage dynamic CMOS logic gate is quite straightforward. For practical multi-stage applications, however, the dynamic CMOS gate presents a significant problem. To examine this fundamental limitation, consider the two-stage cascaded structure shown in Fig. 9.27. Here, the output of the first dynamic CMOS stage drives one of the inputs of the second dynamic CMOS stage, which is assumed to be a two-input NAND gate for simplicity.

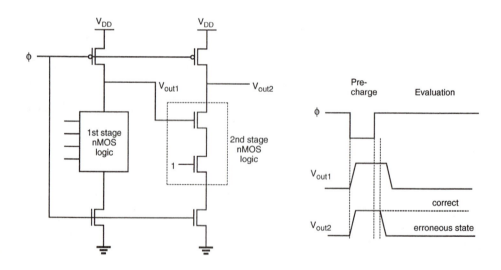

Figure 9.27.   Illustration of the cascading problem in dynamic CMOS logic.

During the precharge phase, both output voltages V_{out1} and V_{out2} are pulled up by the respective pMOS precharge devices. Also, the external inputs are applied during this phase. The input variables of the first stage are assumed to be such that the output V_{out1} will drop to logic "0" during the evaluation phase. On the other hand, the external input of the second-stage NAND2 gate is assumed to be a logic "1," as shown in Fig. 9.27. When the evaluation phase begins, both output voltages V_{out1} and V_{out2} are logic-high. The output of the first stage (V_{out1}) eventually drops to its correct logic level after a certain time delay. However, since the evaluation in the second stage is done concurrently, starting with the high value of V_{out1} at the beginning of the evaluation phase, the output voltage V_{out2} at the end of the evaluation phase will be *erroneously* low. Although the first stage output subsequently assumes its correct output value once the stored charge is drained, the correction of the second-stage output is not possible.

This example illustrates that dynamic CMOS logic stages driven by the same clock signal cannot be cascaded directly. This severe limitation seems to undermine all the other advantages of dynamic CMOS logic, such as low power dissipation, large noise margins, and low transistor count. Alternative clocking schemes and circuit structures must be developed to overcome this problem. In fact, the search for viable circuit alternatives has spawned a large array of high-performance dynamic CMOS circuit techniques, some of which will be examined in the following section.

9.5. High-Performance Dynamic CMOS Circuits

The circuits presented here are variants of the basic dynamic CMOS logic gate structure. We will see that they are designed to take full advantage of the obvious benefits of dynamic operation and at the same time, to allow unrestricted cascading of multiple stages. The ultimate goal is to achieve reliable, high-speed, compact circuits using the least complicated clocking scheme possible.

Domino CMOS Logic

Consider the generalized circuit diagram of a domino CMOS logic gate shown in Fig. 9.28. A dynamic CMOS logic stage, such as the one shown in Fig. 9.26, is cascaded with a static CMOS inverter stage. The addition of the inverter allows us to operate a number of such structures in cascade, as explained in the following.

During the precharge phase (when CK = 0), the output node of the dynamic CMOS stage is precharged to a high logic level, and the output of the CMOS inverter (buffer) becomes low. When the clock signal rises at the beginning of the evaluation phase, there are two possibilities: The output node of the dynamic CMOS stage is either discharged to a low level through the nMOS circuitry (1 to 0 transition), or it remains high. Consequently, the inverter output voltage can also make at most one transition during the evaluation phase, from 0 to 1. Regardless of the input voltages applied to the dynamic CMOS stage, it is not possible for the buffer output to make a 1 to 0 transition during the evaluation phase.

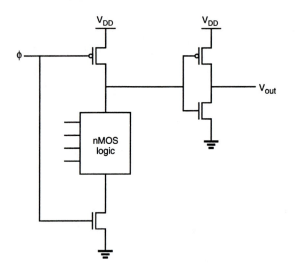

Figure 9.28.   Generalized circuit diagram of a domino CMOS logic gate.

Remember that the problem in cascading conventional dynamic CMOS stages occurs when one or more inputs of a stage make a 1 to 0 transition *during* the evaluation phase, as illustrated in Fig. 9.27. On the other hand, if we build a system by cascading domino CMOS logic gates as shown in Fig. 9.29, all input transistors in subsequent logic blocks will be turned off during the precharge phase, since all buffer outputs are equal to 0. During the evaluation phase, each buffer output can make at

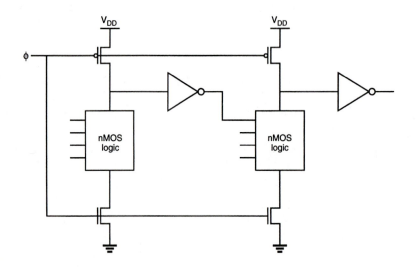

Figure 9.29.   Cascaded domino CMOS logic gates.

most one transition (from 0 to 1), and thus each input of all subsequent logic stages can also make at most one (0 to 1) transition. In a cascade structure consisting of several such stages, the evaluation of each stage ripples the next stage evaluation, similar to a chain of dominos falling one after the other. The structure is hence called *domino CMOS logic*.

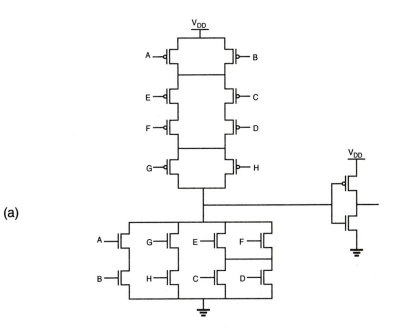

(a)

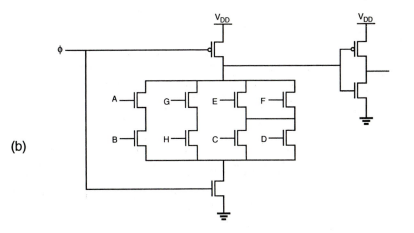

(b)

Figure 9.30.   (a) An 8-input complex logic gate, realized using conventional CMOS logic and (b) domino CMOS logic.

Domino CMOS logic gates allow a significant reduction in the number of transistors required to realize any complex Boolean function. The implementation of the 8-input Boolean function, $Z = AB + (C + D)(C + D) + GH$, using standard CMOS and domino CMOS, is shown in Fig. 9.30, where the reduction of circuit complexity is obvious. The distribution of the clock signal within the system is quite straightforward, since a single clock can be used to precharge and evaluate any number of cascaded stages, as long as the signal propagation delay from the first stage to the last stage does not exceed the time span of the evaluation phase. Also, conventional static CMOS logic gates can be used together with domino CMOS gates in a cascaded configuration (Fig. 9.31). The limitation is that the number of inverting static logic stages in cascade must be even, so that the inputs of the next domino CMOS stage experience only 0 to 1 transitions during the evaluation.

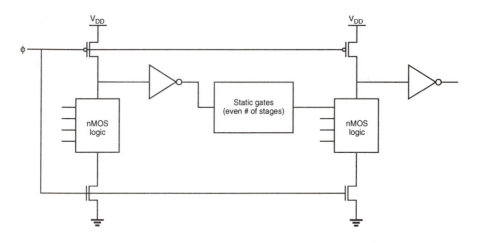

Figure 9.31.   Cascading domino CMOS logic gates with static CMOS logic gates.

There are also some other limitations associated with domino CMOS logic gates. First, only non-inverting structures can be implemented using domino CMOS. If necessary, inversion must be carried out using conventional CMOS logic. Also, charge sharing between the dynamic stage output node and the intermediate nodes of the nMOS logic block during the evaluation phase may cause erroneous outputs, as will be explained in the following.

Consider the domino CMOS logic gate shown in Fig. 9.32, in which the intermediate node capacitance C_2 is comparable in size to the output node capacitance C_1. We will assume that all inputs are low initially, and that the intermediate node voltage across C_2 has an initial value of 0 V. During the precharge phase, the output node capacitance C_1 is charged up to its logic-high level of V_{DD} through the pMOS transistor. In the next phase, the clock signal becomes high and the evaluation begins.

If the input signal of the uppermost nMOS transistor switches from low to high during this evaluation phase, as shown in Fig. 9.32, the charge initially stored in the output capacitance C_1 will now be shared by C_2, leading to the so-called *charge-sharing* phenomenon. The output node voltage after charge sharing becomes $V_{DD}/(1 + C_2/C_1)$. For example, if $C_1 = C_2$, the output voltage becomes $V_{DD}/2$ in the evaluation phase. Unless its logic threshold voltage is less than $V_{DD}/2$, the output voltage of the *following inverter* will then inadvertently switch high, which is a logic error. Thus, it is important to have C_2 much smaller than C_1.

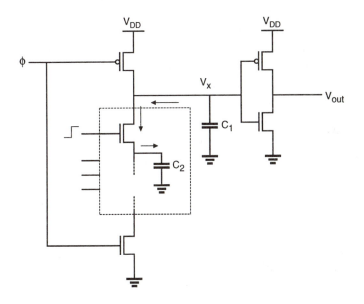

Figure 9.32.   Charge sharing between the output capacitance C_1 and an intermediate node capacitance C_2 during the evaluation cycle may reduce the output voltage level.

Several measures can be taken in order to prevent erroneous output levels due to charge sharing in domino CMOS gates. One simple solution is to add a weak pMOS pull-up device (with a small (W/L) ratio) to the dynamic CMOS stage output, which essentially forces a high output level unless there is a strong pull-down path between the output and the ground (Fig. 9.33). It can be observed that the weak pMOS transistor will be turned on only when the precharge node voltage is kept high. Otherwise, it will be turned off as V_{out} becomes high.

Another solution is to use separate pMOS transistors to precharge all intermediate nodes in the nMOS pull-down tree which have a large parasitic capacitance. The precharging of all high-capacitance nodes within the circuit effectively eliminates all potential charge-sharing problems during evaluation. However, it can also cause additional delay time since the nMOS logic tree now has to drain a larger charge in order to pull down the node voltage V_x. Another way of preventing logic errors due to charge sharing is to make the logic threshold voltage of the inverter smaller, such that the final stage output is not affected by lowering of V_x due to charge sharing. It should be noted that this design approach would trade off the pull-up speed (weaker pMOS transistor) for lower sensitivity to the charge-sharing problem.

The use of multiple precharge transistors also enables us to use the precharged intermediate nodes as resources for additional outputs. Thus, additional logic functions can be realized by tapping the internal nodes of the dynamic CMOS stage, as illustrated by two series-connected logic blocks in Fig. 9.34. The resulting multiple-output domino CMOS logic gate allows us to simultaneously realize several complex functions using a small number of transistors.

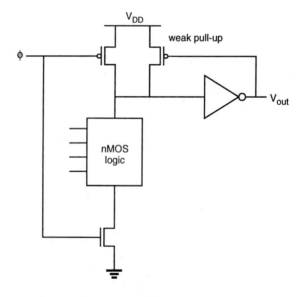

Figure 9.33.   A weak pMOS pull-up device in a feedback loop can be used to prevent the loss of output voltage level due to charge sharing.

Figure 9.35 shows the realization of four Boolean functions of nine variables, using a single domino CMOS logic gate. The four functions to be realized are listed in the following.

$$C_1 = G_1 + P_1 C_0$$
$$C_2 = G_2 + P_2 G_1 + P_2 P_1 C_0$$
$$C_3 = G_3 + P_3 G_2 + P_3 P_2 G_1 + P_3 P_2 P_1 C_0$$
$$C_4 = G_4 + P_4 G_3 + P_4 P_3 G_2 + P_4 P_3 P_2 G_1 + P_4 P_3 P_2 P_1 C_0$$

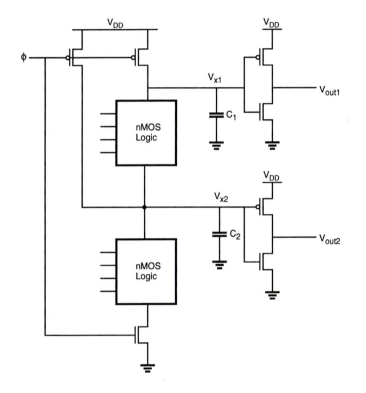

Figure 9.34.   Precharging of internal nodes to prevent charge sharing also allows implementation of multi-output domino CMOS structures.

It can be shown that the functions C_1 through C_4 are the four carry terms to be used in a four-stage carry-lookahead adder, where the variables G_i and P_i are defined as

$$G_i = A_i \cdot B_i$$
$$P_i = A_i + B_i$$

and A_i and B_i are the input bits associated with the i_{th} stage. Hence, this circuit is also known as the Manchester carry chain. The generation of the four carry terms using four separate standard CMOS logic gates or four separate single-output domino CMOS circuits, on the other hand, would require a larger number of transistors and consequently a much larger silicon area. Variants of the multiple-output dynamic CMOS circuit shown in Fig. 9.35 are used widely in high-performance adder structures.

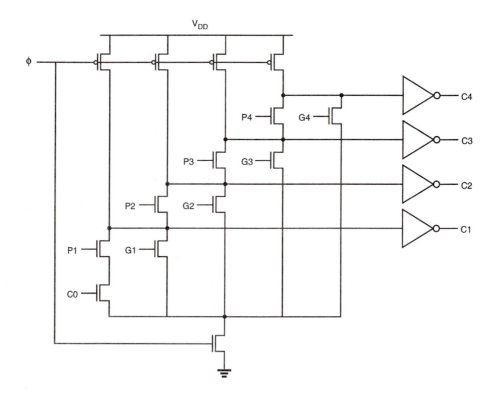

Figure 9.35.   Example of a multiple-output domino CMOS gate realizing four functions.

The transient performance of domino CMOS logic gates can be improved by adjusting the nMOS transistor sizes in the pull-down path, with the objective to reduce the discharge time. Shoji has shown that a graded sizing of nMOS transistors in series structures, where the nMOS transistor closest to the output node also has the smallest (W/L) ratio, yields the best performance. The domino CMOS circuit diagram and the corresponding stick-diagram layout of an optimized example are shown in Fig. 9.36. The fact that a graded reduction of transistor sizes from bottom to top ultimately leads to a better transient performance may seem counterintuitive. But, this effect can be explained by observing the RC delay of the combined pull-down path consisting of series-connected nMOS transistors.

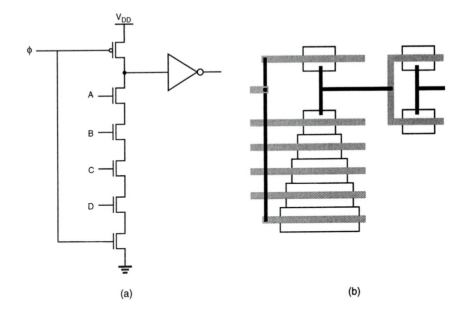

Figure 9.36. (a) Four-input domino CMOS NAND gate and (b) the corresponding stick-diagram layout to show the graded scaling of nMOS transistor sizes for improving the transient performance.

Consider first the nMOS transistor closest to the output node. If the (W/L) ratio of this transistor is reduced by a certain factor, two effects occur. First, the current-driving capability will decrease, i.e., the equivalent resistance of the nMOS transistor will increase. Second, the parasitic drain capacitance associated with this transistor will decrease. If the length of the nMOS chain is sufficiently long, the increase in resistance has little influence upon the combined RC delay time, whereas a reduction of the capacitance significantly decreases the delay.

In fact, by applying Elmore's RC delay formula to series-connected nMOS structures, one can determine if a reduction of nMOS transistor sizes will improve the transient performance. Let C_L represent the precharge-node load capacitance of the domino CMOS gate, and let C_1 represent the parasitic drain capacitance of the nMOS transistor closest to the precharge node. It is assumed that the inverter transistor sizes are fixed and that the pull-down chain contains N series-connected nMOS transistors. Shoji has shown that if the following condition

$$C_L < (N-1)\frac{C_1}{2}$$

(9.36)

is satisfied, the overall delay time can be reduced by decreasing the size of the nMOS transistor closest to the output node. This result can be iteratively applied to the other transistors in the pull-down chain, which leads to graded sizing of all nMOS devices. On the other hand, if the inverter gate transistors are allowed to be optimized along with the series-connected transistors, even shorter delays can be achieved with smaller chip area.

Example 9.4.

Consider the domino CMOS NAND2 gate shown below, where $C_1 = C_2 = 0.05$ pF. First, the operation of the circuit with only one pMOS precharge transistor will be examined. Since the two capacitances C_1 and C_2 are assumed to be equal to each other, we expect that the charge-sharing phenomenon will cause erroneous output values, as explained above, unless specific measures are taken to prevent it.

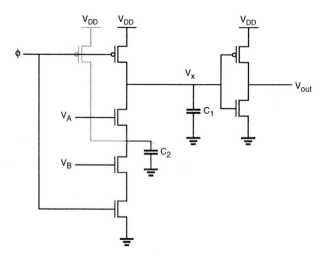

The SPICE-circuit input file of the domino CMOS NAND2 gate is listed below.

```
Domino CMOS with charge sharing
vdd 10 0 dc 5V
vin 1 0 dc pulse( 5 0 1ns 0.1ns 0.1ns 10ns 22ns)
vb 4 0 dc pulse ( 0 5 35ns 0.1ns 0.1ns 11ns 22n)
va 5 0 dc pulse( 0 5 12ns 0.1ns 0.1ns 11ns 22ns)
m1 2 1 0 0 mn l=5u w=10u
m2 3 4 2 0 mn l=5u w=10u
m3 6 5 3 0 mn l=5u w=10u
```

```
m4 6 1 10 10 mp l=5u w=25u
m6 7 6 0 0 mn l=5u w=10u
m7 7 6 10 10 mp l=5u w=100u
cload 6 0 0.05p
cs 3 0 0.05p
cout 7 0 0.1p
.model mn nmos vto=1 gamma=0.4 kp=2.5e-5
.model mp pmos vto=-1 gamma=0.4 kp=1.0e-5
.tran 0.1ns 50ns
.print tran v(1) v(6) v(7) v(3)
.end
```

The transient simulation of this circuit shows that the precharge node voltage V_x drops to about 2.5 V during the evaluation phase, due to charge sharing. As a result, the inverter output voltage erroneously switches to logic-high level during the first evaluation phase.

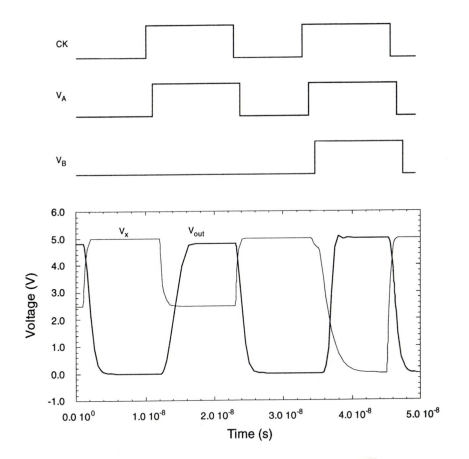

Now consider the case where an additional pMOS precharge transistor is connected between the power supply voltage V_{DD} and the intermediate node, as indicated in the figure above. Both pMOS transistors conduct during the precharge phase, and charge up the node capacitances to the same voltage level. Consequently, charge sharing can no longer cause a logic error at the output node. The simulation results *with* the additional pMOS precharge transistor are plotted below, showing that the output node voltage is pulled up to logic "1" only when both inputs are equal to logic "1."

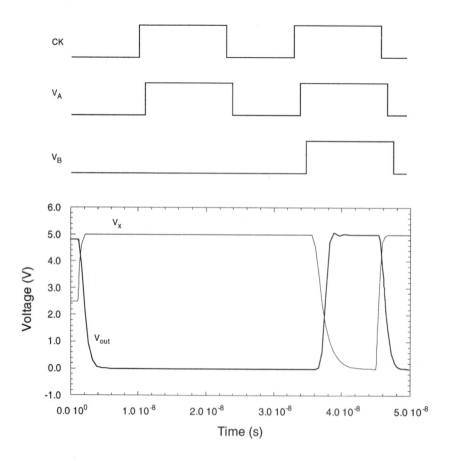

It must be emphasized that there is a speed penalty for adding another pMOS precharge transistor to the circuit. Simulation results indicate that the pull-down delay of the node voltage V_x is actually increased by about 1 ns (approx. 25 percent) as a result of the additional parasitic capacitance which is due to the precharge device.

NORA CMOS Logic (NP-Domino Logic)

In domino CMOS logic gates, all logic operations are performed by the nMOS transistors acting as pull-down networks, while the role of pMOS transistors is limited to precharging the dynamic nodes. As an alternative and a complement to nMOS-based domino CMOS logic, we can construct dynamic logic stages using pMOS transistors as well. Consider the circuit shown in Fig. 9.37, with alternating nMOS and pMOS logic stages.

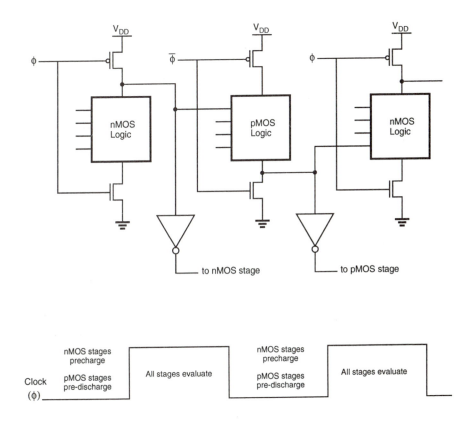

Figure 9.37. NORA CMOS logic consisting of alternating nMOS and pMOS stages, and the scheduling of precharge/evaluation phases.

Note that the precharge-and-evaluate timing of nMOS logic stages is accomplished by the clock signal ϕ, whereas the pMOS logic stages are controlled by the inverted clock signal, $\overline{\phi}$. The operation of the NORA CMOS circuit is as follows: When the clock signal is low, the output nodes of nMOS logic blocks are precharged to V_{DD}

through the pMOS precharge transistors, whereas the output nodes of pMOS logic blocks are pre-discharged to 0 V through the nMOS discharge transistors, driven by $\overline{\phi}$. When the clock signal makes a low-to-high transition (note that the inverted clock signal $\overline{\phi}$ makes a high-to-low transition simultaneously), all cascaded nMOS and pMOS logic stages evaluate one after the other, much like the domino CMOS examined earlier. A simple NORA CMOS circuit example is shown in Fig. 9.38.

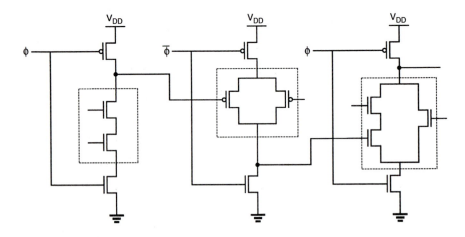

Figure 9.38.   NORA CMOS logic circuit example.

The advantage of NORA CMOS logic is that a static CMOS inverter is not required at the output of every dynamic logic stage. Instead, direct coupling of logic blocks is feasible by alternating nMOS and pMOS logic blocks. NORA logic is also compatible with domino CMOS logic. Outputs of NORA nMOS logic blocks can be inverted, and then applied to the input of a domino CMOS block, which is also driven by the clock signal ϕ. Similarly, the buffered output of a domino CMOS stage can be applied directly to the input of a NORA nMOS stage.

The second important advantage of NORA CMOS logic is that it allows pipelined system architecture. Consider the circuit shown in Fig. 9.39(a), which consists of an nMOS-pMOS logic sequence similar to the one shown in Fig. 9.37, and a clocked CMOS (C^2MOS) output buffer. It can easily be seen that all stages of this circuit perform the precharge-discharge operation when the clock is low, and all stages of the circuit evaluate output levels when the clock is high. Therefore, we will call this circuit a ϕ-section, meaning that evaluation occurs during active ϕ.

Now consider the circuit shown in Fig. 9.39(b), which is essentially the same circuit shown in Fig. 9.39(a), with the only difference being that the signals ϕ and $\overline{\phi}$ have been interchanged. In this circuit, all logic stages perform the precharge-discharge operation when the clock is high, and all stages evaluate output levels when the clock is low. Therefore, we will call this circuit a $\overline{\phi}$-section, meaning that evaluation occurs during active $\overline{\phi}$.

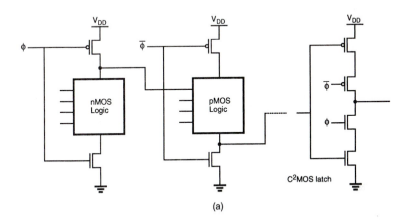

(a)

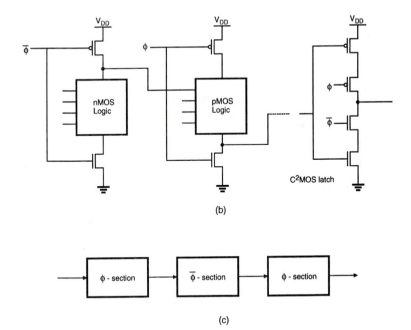

(b)

(c)

Figure 9.39. (a) NORA CMOS ϕ-section; evaluation occurs during $\phi = 1$. (b) NORA CMOS $\overline{\phi}$-section; evaluation occurs during $\phi = 0$. (c) A pipelined NORA CMOS system.

A pipelined system can be constructed by simply cascading alternating ϕ- and $\overline{\phi}$-sections, as shown in Fig. 9.39(c). Note that each of the sections may consist of several logic stages, and that all logic stages in one section are evaluated during the same clock cycle. When the clock is low, the ϕ-sections in the pipelined system undergo the precharge cycle, while the $\overline{\phi}$-sections undergo evaluation. When the clock signal changes from low to high, the ϕ-sections start the evaluation cycle, while the $\overline{\phi}$-sections undergo precharge. Thus, consecutive sets of input data can be processed in alternating sections of the pipelined system.

As in all dynamic CMOS structures, NORA CMOS logic gates also suffer from charge sharing and leakage. To overcome the dynamic charge sharing and soft-node leakage problems in NORA CMOS structures, a circuit technique called Zipper CMOS can be used.

Zipper CMOS Circuits

The basic circuit architecture of Zipper CMOS is essentially identical to NORA CMOS, with the exception of the clock signals. The Zipper CMOS clock scheme requires the generation of slightly different clock signals for the precharge (discharge) transistors and for the pull-down (pull-up) transistors. In particular, the clock signals which drive the pMOS precharge and nMOS discharge transistors allow these transistors to remain in weak conduction or near cut-off during the evaluation phase, thus compensating for the charge leakage and charge-sharing problems. The generalized circuit diagram and the clock signals of the Zipper CMOS architecture are shown in Fig. 9.40.

True Single-Phase Clock Dynamic CMOS

The dynamic circuit technique to be presented in the following is distinctly different from the NORA CMOS circuit architecture in that it uses only one clock signal which is never inverted. Since the inverted clock signal $\overline{\phi}$ is not used anywhere in the system, no clock skew problem exists. Consequently, higher clock frequencies can be reached for dynamic pipelined operation.

Consider the circuit diagram shown in Fig. 9.41. The circuit consists of alternating stages called n-blocks and p-blocks, and each block is being driven by the same clock signal ϕ. An n-block is constructed by cascading a dynamic nMOS stage and a dynamic latch, while a p-block is constructed by cascading a dynamic pMOS stage and a dynamic latch.

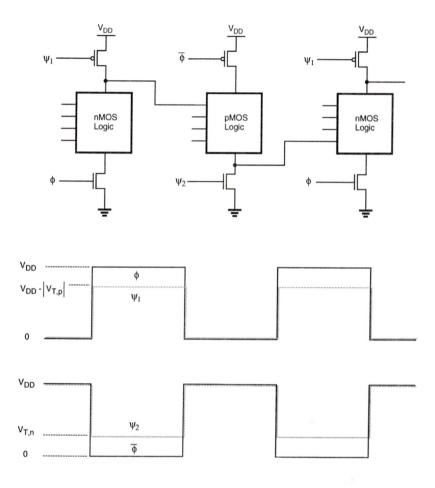

Figure 9.40.   General circuit structure and the clock signals of Zipper CMOS.

When the clock signal is low, the output node of the n-block is being precharged to V_{DD} by the pMOS precharge transistor. When the clock signal switches from low to high, the logic stage output is evaluated and the output latch generates a valid output level. On the other hand, we can see by inspection that the p-block pre-discharges when the clock is high, and evaluates when the clock is low. This means that a cascade-connection of alternating n-block and p-block circuits as shown in Fig. 9.41 will allow pipelined operation using a single clock signal. Compared to NORA CMOS, we need two extra transistors per stage, but the ability to operate with a true single-phase clock signal offers very attractive possibilities from the system-design point of view.

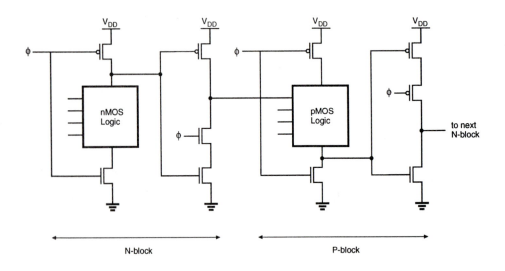

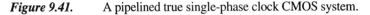

Figure 9.41. A pipelined true single-phase clock CMOS system.

References

1.   R. H. Krambeck, C. M. Lee, and H.-F. S. Law, "High-speed compact circuits with CMOS," *IEEE Journal of Solid-State Circuits*, vol. SC-17, no. 3, pp. 614-619, June 1982.

2.   N. F. Goncalves and H. De Man, "NORA: A racefree dynamic CMOS technique for pipelined logic structures," *IEEE Journal of Solid-State Circuits*, vol. SC-18, no. 3, pp. 261-266, June 1983.

3.   C. M. Lee and E. W. Szeto, "Zipper CMOS," *IEEE Circuits and Devices Magazine*, pp. 10-16, May 1986.

4.   I. S. Hwang and A. L. Fisher, "Ultrafast compact 32-bit CMOS adders in multiple-output domino logic," *IEEE Journal of Solid-State Circuits*, vol. 24, no. 2, pp. 358-369, June 1982.

5.   J. Yuan and C. Svensson, "High-speed CMOS circuit technique," *IEEE Journal of Solid-State Circuits*, vol. 24, no.1, pp. 62-70, February 1989.

6.   M. Shoji, "FET scaling in domino CMOS gates," *IEEE Journal of Solid-State Circuits*, vol. SC-20, no. 5, pp. 1067-1071, October 1985.

7.   H. Y. Chen and S. M. Kang, "iCOACH: A circuit optimization aid for CMOS high-performance circuits," *Integration, the VLSI Journal*, 10 (1991), pp. 185-212.

Exercise Problems

9.1   Consider the CMOS circuit shown in Fig. P9.1 that was designed to drive a total capacitive load of $C_L = 0.2$ pF. For the n-channel devices, assume zero bias threshold voltage $V_{T0} = 1.0$ V and transconductance $k'_n = 50$ μA/V^2. For the p-channel devices, assume $V_{T0} = -1.0$ V and $k'_p = 25$ μA/V^2. Use the *W/L* ratios for each device as shown in the figure. The initial voltage across the load capacitor C_L is 0 V. The waveform at input E is identical to 0 V for all time. For the clock CK and the rest of the inputs, the waveforms are also shown in the figure. Sketch the voltage waveform across the load capacitor C_L and provide clear marking of the 50% crossings along the time axis in nanoseconds for both rise and fall transitions. Hint: The n-channel transistor group can be approximated by a single equivalent n-channel transistor and either an average-current method or a state-equation method may be used to compute the required delay times.

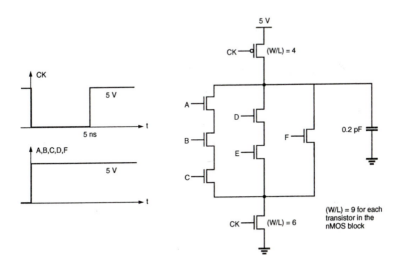

Figure P9.1

9.2   In logic design, complex gates such as AOI or OAI gates are often used to combine the functions of several gates into a single gate, thus reducing the chip area and parasitics of the circuit. Let us consider an OAI432 whose logic function is $Z = \overline{(A+B+C+D)(E+F+G)(H+I)}$. Assume that only A, E, H inputs are high and other inputs are low. The device parameters are:

- $C_{jsw} = 250$ pF/m
- $C_{j0} = 80$ μF/m^2
- $C_{ox} = 350$ μF/m^2
- $L_D = 0.5$ μm
- $K_{eq} = 1.0$ for worst-case capacitance
- $C_{metal} = 2.0$ pF/cm
- $C_{poly} = 2.2$ pF/cm
- $R_{poly} = 25$ Ω/sq.
- $R_{metal} = 0.2$ Ω/sq.
- Polysilicon line width $= 2$ μm, metal line width $= 2$ μm
- $k'_n = 20$ μA/V^2
- $k'_p = 10$ μA/V^2
- $|V_{T0}| = 1.0$ V

(a)   Draw a full CMOS circuit diagram for this OAI432.

(b)   Draw a domino CMOS circuit diagram which implements $\overline{Z}$.

(c)   Draw an equivalent circuit for this case by using equivalent transistor sizes with W/L=30/2 (both for nMOS and pMOS).

(d)   By assuming that the total parasitic capacitances at the precharging node and the output nodes are 0.2 pF, calculate the delay from the end of precharging (clock signal is a rectangular pulse) to the time point at which output voltage reaches 2.5 V. For simple analysis, neglect the body effect. For approximate solution of this problem, first calculate the delay for the precharging node to fall to 2.5 V. Then use a rectangular pulse at that time point which falls from 5 V to 0 V to calculate the delay of the inverter gate. The total delay can be found by adding the two delay components. (A more accurate method is to approximate the precharging node voltage by a falling ramp function.)

9.3   Discuss the charge-sharing problems in VLSI circuits. Explain various circuit techniques used in domino CMOS circuits for solving charge-sharing problems. State as many as you know.

9.4 Bootstrapping circuits are used to increase the gate voltages of transistors such that the drain node voltage can be pulled high (despite the threshold voltage drop). The circuit shown in Fig. P9.4 can be used for such purposes and can indeed increase the node voltage at X well beyond $V_{DD} = 5$ V. Determine the maximum achievable voltage at node X using the following parameters:

- $V_{T0} = 1.0$ V
- $\gamma = 0.3$ V$^{1/2}$
- $|2\phi_F| = 0.6$ V
- $V_{OL} = 0.2$ V
- $C_{sub} = 12$ fF
- $C_{boot} = 20$ fF

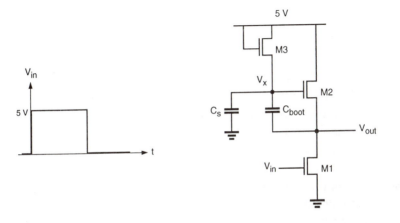

Figure P9.4

9.5 A CMOS circuit is shown in Fig. P9.5. Suppose that the precharge transistor was chosen such that node X is guaranteed to be charged to V_{DD}. All nMOS transistors have $W/L = 20$. Determine how long it takes for the node voltage at X to decrease to $0.8\, V_{DD}$ after the clock signal pulse goes to high (with zero rise time) when input voltages at A, B, D are 5 V and the input voltage at C is zero. Assume the following parameter values:

- $\gamma = 0.0$ V$^{1/2}$
- $V_{T0} = 1.0$ V
- $k_n' = 10$ µA/V^2

Hint: The nMOS transistor tree between node X and the ground can be approximated by an equivalent transistor with an effective W/L.

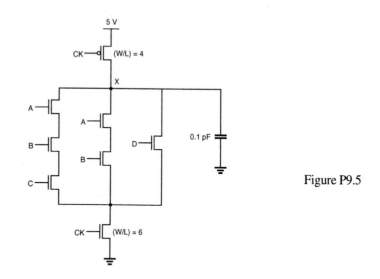

Figure P9.5

9.6   The inputs to a domino logic gate are always LO during the precharge phase ($\phi=$LO) and may undergo a LO-to-HI transition during the evaluation phase ($\phi=$HI). Consider the domino 3-input AND gate shown in Fig. P9.6 below. If, during the evaluation phase, $A=$HI, $B=$LO, and $C=$LO, charge sharing will cause the voltage at the input of the inverter to drop. Given that the switching threshold of the inverter is 3 V, calculate the maximum ratio C_p/C_L necessary to ensure that charge sharing does not corrupt the value of F for the given case.

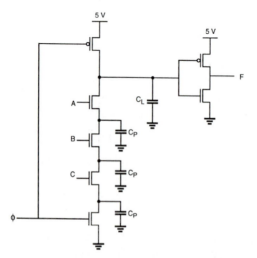

Figure P9.6

9.7   Consider the following CMOS logic circuit, which is a simple domino circuit. Node X is connected to a CMOS inverter so that the output of the inverter can be directly fed to the next stage of the domino circuit.

   (a)   Explain how the voltage level at node X, after it is precharged to 5 V, can be affected by the charge sharing between node X and node Y if their node capacitances are the same. Express the final voltage at node X in terms of the initial voltage at node Y when the charge sharing is completed, following the full precharge operation when the gate terminal of transistor M2 is fixed at 0 V.

   (b)   Determine the ratio between device transconductance parameters, k_p and k_n, of the inverter to prevent any logic error due to charge sharing between nodes X and Y under all circumstances. Assume that the magnitudes of threshold voltages in the inverter are equal to 1.0 V. The use of Level 1 transistor current equations is deemed adequate.

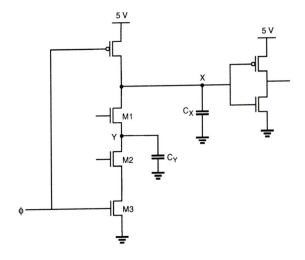

Figure P9.7

9.8   Consider the domino CMOS circuit shown in Fig. P9.6. Using the input voltage waveforms illustrated in Fig. P9.8 determine the output voltage waveform.

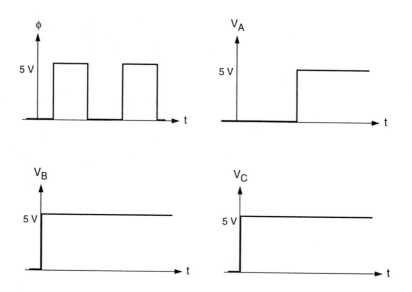

Figure P9.8

CHAPTER 10

SEMICONDUCTOR MEMORIES

10.1. Introduction

Semiconductor memory arrays capable of storing large quantities of digital information are essential to all digital systems. The amount of memory required in a particular system depends on the type of application, but, in general, the number of transistors utilized for the information (data) storage function is much larger than the number of transistors used in logic operations and for other purposes. The ever-increasing demand for larger data storage capacity has driven the fabrication technology and memory development towards more compact design rules and, consequently, toward higher data storage densities. Thus, the maximum realizable data storage capacity of single-chip semiconductor memory arrays approximately doubles every two years. On-chip memory arrays have become widely used subsystems in many VLSI circuits, and commercially available single-chip read/write memory capacity has reached 64 megabits. This trend toward higher memory density and larger storage capacity will continue to push the leading edge of digital system design.

The area efficiency of the memory array, i.e., the number of stored data bits per unit area, is one of the key design criteria that determine the overall storage capacity and, hence, the memory *cost per bit*. Another important issue is the memory access time, i.e., the time required to store and/or retrieve a particular data bit in the memory array. The access time determines the memory *speed*, which is an important performance criterion of the memory array. Finally, the static and dynamic *power consumption* of the memory array is a significar: factor to be considered in design, because of the increasing importance of low-power applications. In the following, we will investigate different types of MOS memory arrays and discuss in detail the issues of area, speed, and power consumption for each circuit type.

Memory circuits are generally classified according to the type of data storage and the type of data access. *Read-Only Memory* (ROM) circuits allow, as the name implies, only the retrieval of previously stored data, and do not permit modifications of the stored information contents during normal operation. ROMs are *non-volatile* memories, i.e., the data storage function is not lost even when the power supply voltage is off. Depending on the type of data storage (data write) method, ROMs are classified as mask-programmed ROMs, Programmable ROMs (PROM), Erasable PROMs (EPROM), and Electrically Erasable PROMs (EEPROM).

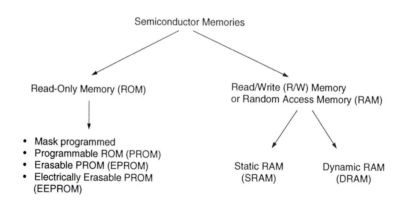

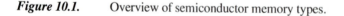

Figure 10.1.   Overview of semiconductor memory types.

Read-write (R/W) memory circuits, on the other hand, must permit the modification (writing) of data bits stored in the memory array, as well as their retrieval (reading) on demand. This requires that the data storage function be *volatile*, i.e., the stored data are lost when the power supply voltage is turned off. The read-write memory circuit is commonly called *Random Access Memory* (RAM), mostly due to historical reasons. Compared to sequential-access memories such as magnetic tapes, any cell in the R/W memory array can be accessed with nearly equal access time. Based on the operation type of individual data storage cells, RAMs are classified into two main categories, the *Static* RAMs (SRAM) and the *Dynamic* RAMs (DRAM). Figure 10.1 shows an overview of the different memory types and their classifications.

A typical memory array organization is shown in Fig. 10.2. The data storage structure, or *core*, consists of individual memory cells arranged in an array of horizontal rows and vertical columns. Each *cell* is capable of storing one bit of binary information. Also, each memory cell shares a common connection with the other cells in the same row, and another common connection with the other cells in the same column. In this structure, there are 2^N rows, also called *word lines*, and 2^M columns, also called *bit lines*. Thus, the total number of memory cells in this array is $2^M \times 2^N$.

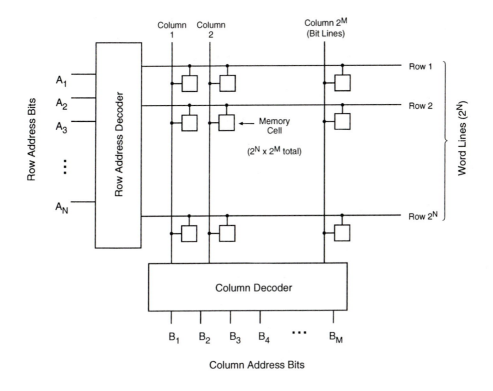

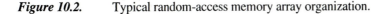

Figure 10.2.   Typical random-access memory array organization.

To access a particular memory cell, i.e., a particular data bit in this array, the corresponding bit line *and* the corresponding word line must be activated (selected). The row and column selection operations are accomplished by row and column *decoders*, respectively. The row decoder circuit selects one out of 2^N word lines according to an N-bit row address, while the column decoder circuit selects one out of 2^M bit lines according to an M-bit column address. Once a memory cell or a group of memory cells are selected in this fashion, a data read and/or a data write operation may be performed on the selected single bit or multiple bits on a particular row. The column decoder circuit serves the double duties of selecting the particular columns and routing the corresponding data content in a selected row to the output.

We can see from this simple discussion that individual memory cells can be accessed for data read and/or data write operations in random order, independent of their physical locations in the memory array. Thus, the array organization examined here is called a *Random Access Memory* (RAM) structure. Notice that this organization can be used for both read-write memory arrays and read-only memory arrays. In the following sections, however, we will use the acronym RAM specifically for read-write memories, because it is the universally accepted abbreviation for this particular type of memory array.

10.2. Read-Only Memory (ROM) Circuits

The read-only memory array can also be seen as a simple combinational Boolean network which produces a specified output value for each input combination, i.e., for each address. Thus, storing binary information at a particular address location can be achieved by the presence or absence of a data path from the selected row (word line) to the selected column (bit line), which is equivalent to the presence or absence of a device at that particular location. In the following, we will examine two different implementations for MOS ROM arrays. Consider first the 4 x 4 memory array shown in Fig. 10.3. Here, each column consists of a pseudo-nMOS NOR gate driven by some of the row signals, i.e., the word lines.

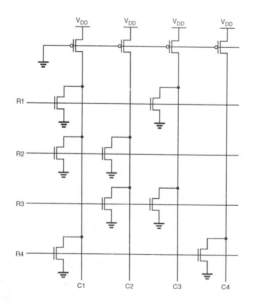

R1	R2	R3	R4	C1	C2	C3	C4
1	0	0	0	0	1	0	1
0	1	0	0	0	0	1	1
0	0	1	0	1	0	0	1
0	0	0	1	0	1	1	0

Figure 10.3.   Example of a 4 x 4 bit NOR-based ROM array .

As described in the previous section, only one word line is activated (selected) at a time by raising its voltage to V_{DD}, while all other rows are held at a low voltage level. If an active transistor exists at the crosspoint of a column and the selected row, the column voltage is pulled down to the logic low level by that transistor. If no active transistor exists at the crosspoint, the column voltage is pulled high by the pMOS load device. Thus, a logic "1"-bit is stored as the absence of an active transistor, while a logic "0"-bit is stored as the presence of an active transistor at the crosspoint. To reduce static power consumption, the pMOS load transistors in the ROM array shown in Fig. 10.3 can also be driven by a periodic precharge signal, resulting in a *dynamic* ROM.

In actual ROM layout, the array can be initially manufactured with nMOS transistors at every row-column intersection. The "1"-bits are then realized by omitting the drain or source connection, or the gate electrode of the corresponding nMOS transistors in the final metallization step. Figure 10.4 shows four nMOS transistors in a NOR ROM array, forming the intersection of two metal bit lines and two polysilicon word lines. To save silicon area, the transistors in every two adjacent rows are arranged to share a common ground line, also routed in n-type diffusion. To store a "0"-bit at a particular address location, the drain diffusion of the corresponding transistor must be connected to the metal bit line via a metal-to-diffusion contact. Omission of this contact, on the other hand, results in a stored "1"-bit.

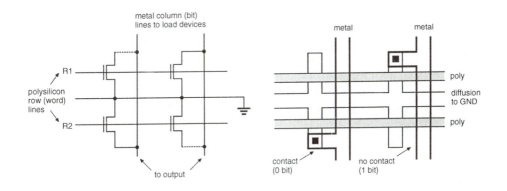

Figure 10.4.   Layout example of a NOR ROM array.

Figure 10.5 shows a larger portion of the ROM array, except for the pMOS load transistors connected to the metal columns. Here, the 4 x 4 bit ROM array shown in Fig. 10.3 is realized using the contact-mask programming methodology described above. Note that only 8 of the 16 nMOS transistors fabricated in this structure are actually connected to the bit lines via metal-to-diffusion contacts. In reality, the metal column lines are laid out directly on top of diffusion columns to reduce the horizontal dimension of the ROM array.

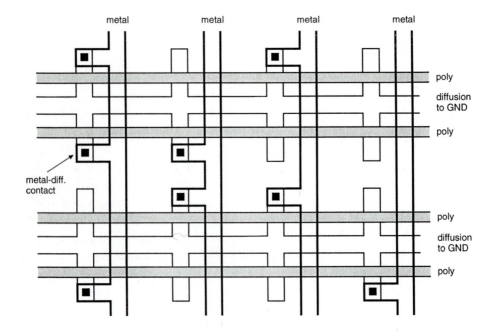

Figure 10.5.   Layout of the 4 x 4 bit NOR ROM array example shown in Fig. 10.3.

A different NOR ROM layout implementation is based on deactivation of the nMOS transistors by raising their threshold voltages through channel implants. Figure 10.6 shows the circuit diagram of a NOR ROM array in which every two rows of nMOS transistors share a common ground connection, and every drain diffusion contact to the metal bit line is shared by two adjacent transistors. In this case, all nMOS transistors are already connected to the column lines (bit lines); therefore, storing a "1"-bit at a particular location by omitting the corresponding drain contact is not possible. Instead, the nMOS transistor corresponding to the stored "1"-bit can be deactivated, i.e., permanently turned off, by raising its threshold voltage *above* the V_{OH} level through a selective channel implant during fabrication.

The alternative layout of the 4 x 4 bit ROM array example (Fig. 10.3), which is based on implant-mask programming, is shown in Fig. 10.7. Note that in this case, each threshold voltage implant signifies a stored "1"-bit, and all other (non-implanted) transistors correspond to stored "0"-bits. Since each diffusion-to-metal contact in this structure is shared by two adjacent transistors, the implant-mask ROM layout can yield a higher core density, i.e., a smaller silicon area per stored bit, compared to the contact-mask ROM layout.

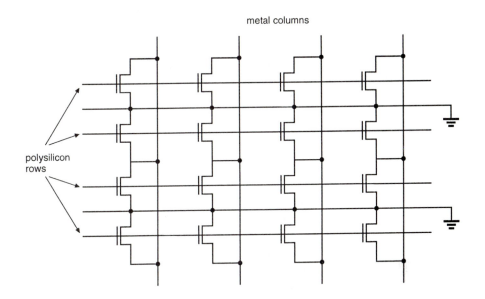

metal columns

polysilicon
rows

Figure 10.6. Arrangement of the nMOS transistors in the implant-mask programmable NOR ROM array. Every metal-to-diffusion contact is shared by two adjacent devices.

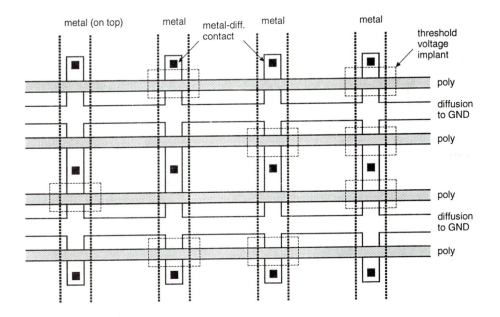
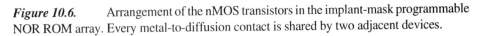

metal (on top)   metal   metal-diff.   metal   metal
                           contact

                                                                    threshold
                                                                    voltage
                                                                    implant

                                                                    poly

                                                                    diffusion
                                                                    to GND

                                                                    poly

                                                                    poly

                                                                    diffusion
                                                                    to GND

                                                                    poly

Figure 10.7. Layout of the 4 x 4 bit NOR ROM array example shown in Fig. 10.3. The threshold voltages of "1"-bit transistors are raised above V_{DD} through implant.

Next, we will examine a significantly different ROM array design, which is also called a NAND ROM (Fig. 10.8). Here, each bit line consists of a depletion-load NAND gate, driven by some of the row signals, i.e., the word lines. In normal operation, all word lines are held at the logic-high voltage level except for the selected line, which is pulled *down* to logic-low level. If a transistor exists at the crosspoint of a column and the selected row, that transistor is turned off and the column voltage is pulled high by the load device. On the other hand, if no transistor exists (shorted) at that particular crosspoint, the column voltage is pulled low by the other nMOS transistors in the multi-input NAND structure. Thus, a logic "1"-bit is stored by the presence of a transistor that can be deactivated, while a logic "0"-bit is stored by a shorted or normally on transistor at the crosspoint.

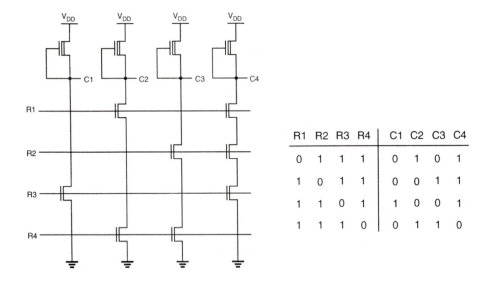

R1	R2	R3	R4	C1	C2	C3	C4
0	1	1	1	0	1	0	1
1	0	1	1	0	0	1	1
1	1	0	1	1	0	0	1
1	1	1	0	0	1	1	0

Figure 10.8.   A 4 x 4 bit NAND-based ROM array.

As in the NOR ROM case, the NAND-based ROM array can be fabricated initially with a transistor connection present at every row-column intersection. A "0"-bit is then stored by lowering the threshold voltage of the corresponding nMOS transistor at the crosspoint through a channel implant, so that the transistor remains on regardless of the gate voltage (i.e., the nMOS transistor at the intersection becomes a depletion-type device). The availability of this process step is also the reason why depletion-type nMOS load transistors are used instead of pMOS loads in the example shown above. Figure 10.9 shows a sample 4 x 4 bit layout of the implant-mask NAND ROM array. Here, vertical columns of n-type diffusion intersect at regular intervals with horizontal rows of polysilicon, which results in an nMOS transistor at each intersection point. The transistors with threshold voltage implants operate as normally-on depletion devices, thereby providing a continuous current path regardless of the gate voltage level. Since this structure has no contacts embedded in the array, it is much more compact than the NOR ROM array. However, the access time is usually slower than the NOR ROM, due to multiple series-connected nMOS transistors in each column. An alternative layout method for the NAND ROM array is not to place the nMOS transistors at "0"-bit locations, as in the case of the PLA (Programmable Logic Array) layout generation. In this case, the missing transistor is simply replaced by a metal line, instead of using a threshold voltage implant at that location.

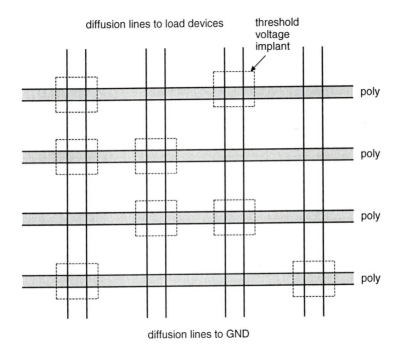

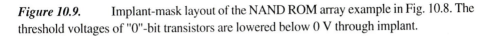

Figure 10.9.   Implant-mask layout of the NAND ROM array example in Fig. 10.8. The threshold voltages of "0"-bit transistors are lowered below 0 V through implant.

Design of Row and Column Decoders

Now we will turn our attention to the circuit structures of row and column address decoders, which select a particular memory location in the array, based on the binary row and column addresses. A row decoder designed to drive a NOR ROM array must, by definition, select one of the 2^N word lines by raising its voltage to V_{OH}. As an example, consider the simple row address decoder shown in Fig. 10.10, which decodes a two-bit row address and selects one out of four word lines by raising its level.

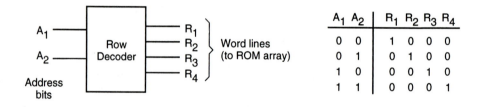

A₁ A₂	R₁ R₂ R₃ R₄
0 0	1 0 0 0
0 1	0 1 0 0
1 0	0 0 1 0
1 1	0 0 0 1

Figure 10.10. Row address decoder example for 2 address bits and 4 word lines.

A most straightforward implementation of this decoder is another NOR array, consisting of 4 rows (outputs) and 4 columns (two address bits and their complements). Note that this NOR-based decoder array can be built just like the NOR ROM array, using the same selective programming approach (Fig. 10.11). The ROM array and its row decoder can thus be fabricated as two adjacent NOR arrays, as shown in Fig. 10.12.

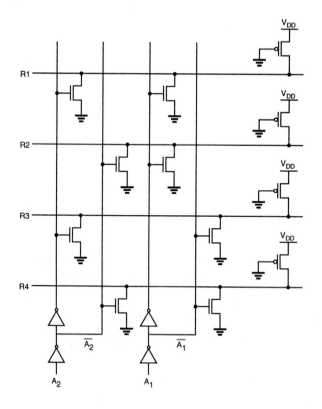

Figure 10.11. NOR-based row decoder circuit for 2 address bits and 4 word lines.

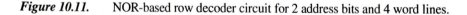

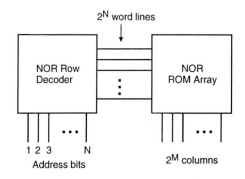

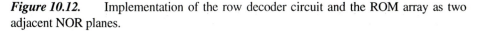

Figure 10.12.   Implementation of the row decoder circuit and the ROM array as two adjacent NOR planes.

A row decoder designed to drive a NAND ROM, on the other hand, must lower the voltage level of the selected row to logic "0" while keeping all other rows at a logic-high level. This function can be implemented by using an N-input NAND gate for each of the row outputs. The truth table of a simple address decoder for four rows and the double NAND-array implementation of the decoder and the ROM are shown in Fig. 10.13. As in the NOR ROM case, the row address decoder of the NAND ROM array can thus be realized using the same layout strategy as the memory array itself.

A_1	A_2	R_1	R_2	R_3	R_4
0	0	0	1	1	1
0	1	1	0	1	1
1	0	1	1	0	1
1	1	1	1	1	0

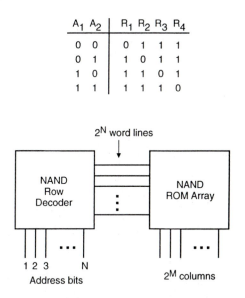

Figure 10.13.   Truth table of a row decoder for the NAND ROM array, and implementation of the row decoder circuit and the ROM array as two adjacent NAND planes.

The column decoder circuitry is designed to select one out of 2^M bit lines (columns) of the ROM array according to an M-bit column address, and to route the data content of the selected bit line to the data output. A straightforward but costly approach would be to connect an nMOS pass transistor to each bit-line (column) output, and to selectively drive one out of 2^M pass transistors by using a NOR-based column address decoder, as shown in Fig. 10.14. In this arrangement, only one nMOS pass transistor is turned on at a time, depending on the column address bits applied to the decoder inputs. The conducting pass transistor routes the selected column signal to the data output. Similarly, a number of columns can be chosen at a time, and the selected columns can be routed to a *parallel* data output port.

Note that the number of transistors required for this column decoder implementation is $2^M(M+1)$, i.e., 2^M pass transistors for each bit line and $M\,2^M$ transistors for the decoder circuit. This number can quickly become excessive for large M, i.e., for a large number of bit lines.

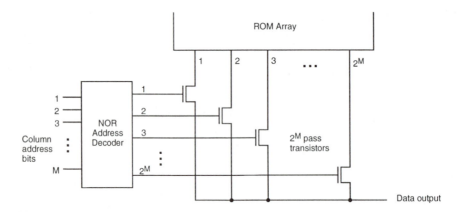
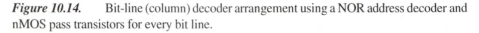

Figure 10.14. Bit-line (column) decoder arrangement using a NOR address decoder and nMOS pass transistors for every bit line.

An alternative approach to design of the column decoder circuit is to build a binary selection tree consisting of consecutive stages, as shown in Fig. 10.15. In this case, the pass transistor network is used to select one out of every two bit lines at each stage (level), whereas the column address bits drive the gates of the nMOS pass transistors. Notice that a NOR address decoder is not needed for this decoder tree structure, thereby reducing the number of transistors significantly although it requires M additional inverters ($2M$ transistors) for complementing column address bits. The example shown in Fig. 10.15 is a column decoder tree for eight bit lines, which requires three column address bits (and their complements) to select one of the eight columns.

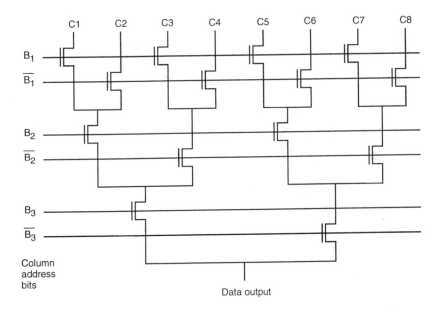

Figure 10.15. Column decoder circuit for eight bit lines, implemented as a binary tree decoder which is driven directly by the three column address bits.

One drawback of the decoder tree approach is that the number of series-connected nMOS pass transistors in the data path is equal to the number of column address bits, M. This situation can cause a long data access time, since the decoder delay time depends on the equivalent series resistance of the decoder branch that directs the column data to the output. To overcome this constraint, column address decoders can be built as a combination of the two structures presented here, i.e., consisting of relatively shallow, partial tree decoders and of additional selection circuits similar to that shown in Fig. 10.14.

Example 10.1.

In the following example, the design of a 32-kbit NOR ROM array will be considered, and the relevant design issues related to access time analysis will be examined.

A 32-kbit ROM array consists of $2^{15} = 32,768$ individual memory cells, which are arranged into a certain number of the rows and columns, as already explained in the beginning of this chapter. Note that the *sum* of row address bits and the column address bits must be equal to 15 in a 32-kbit array; the actual number of rows and columns in the memory array may be determined according to this and other constraints, as will be demonstrated in the following.

Assume that the ROM array under consideration has 7 row address bits and 8 column address bits, which results in a memory array of 128 rows and 256 columns. A section of the memory cell layout is shown in Fig. 10.16, in which the programming is done by an implant-mask to adjust the threshold voltages of those transistors which are to be left inactive (cf. Fig. 10.7). To provide a compact layout, the drain contact shown here is actually shared by two adjacent transistors, and the metal bit line (column) runs directly on top of the diffusion columns shown in this example. To simplify the mask view, the metal column line is not shown in Fig. 10.16.

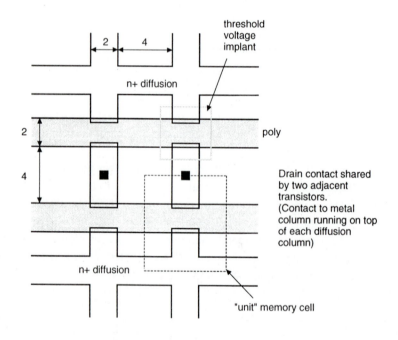

Figure 10.16.   Simplified mask layout of the implant-mask programmable NOR ROM array considered in the example. All dimensions are in micrometers ($W = 2\,\mu m$, $L = 1.5\,\mu m$).

Other relevant parameters of the structure are:

$$\mu_n C_{ox} = 20\ \mu A/V^2$$
$$C_{ox} = 3.47\ \mu F/cm^2$$
$$\text{Poly sheet resistance} = 20\ \Omega/\text{square}$$

First, we calculate the row resistance and row capacitance per bit, i.e., per memory cell. It is assumed that the row capacitance of each memory cell is dictated primarily by the thin oxide capacitance of the nMOS transistor, and that the polysilicon capacitance outside the active region is negligible. The row resistance associated with each memory cell, on the other hand, is calculated by summing the number of polysilicon squares (in this case, two) over the unit cell.

$$C_{row} = C_{ox} \cdot W \cdot L = 10.4 \text{ fF/bit}$$

$$R_{row} = (\text{\# of squares}) \times (\text{Poly sheet resistance}) = 60 \text{ } \Omega/\text{bit}$$

We note that each polysilicon row (word line) in this memory array is actually a distributed RC transmission line. Figure 10.17 depicts a few components of this structure, which ultimately affects the row access time of the memory array in consideration. It can be seen that once the row is selected by the row address decoder, i.e., the row voltage is forced to a logic-high level at one end of the word line, the gate voltage of the last (256th) transistor in that row will rise $last$ because of the RC line delay. The propagation delay time of the gate voltage of the 256th transistor in the row determines the row access time, t_{row} (Fig. 10.18).

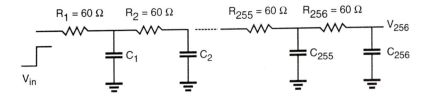

Figure 10.17. RC transmission line representation of the polysilicon word line.

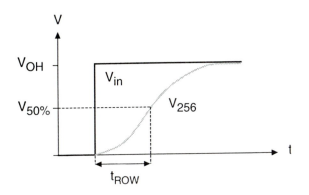

Figure 10.18. Definition of the row access time in a memory array with 256 columns.

By neglecting the signal propagation delay associated with the row address decoder circuit, and assuming that the row (word line) is driven by an ideal step voltage waveform, the row access time can be approximated by using the following empirical formula,

$$t_{row} \approx 0.38 \cdot R_T \cdot C_T = 15.53 \text{ ns}$$

where

$$R_T = \sum_{all\,columns} R_i = 15.36 \text{ k}\Omega$$

$$C_T = \sum_{all\,columns} C_i = 2.66 \text{ pF}$$

A more accurate RC delay value can be calculated by using the *Elmore time constant* for RC ladder circuits, as follows.

$$t_{row} = \sum_{k=1}^{256} R_{jk} C_k = 20.52 \text{ ns} \quad \text{where} \quad R_{jk} = \sum_{j=1}^{k} R_j$$

The row access time t_{row} is the time delay associated with selecting and activating one of the 128 word lines in this ROM array.

To calculate the column access time, we need to consider one of the 128-input NOR gates, which represent the bit lines in this ROM structure. The pseudo-nMOS NOR gate corresponding to each column is designed by using a pMOS load transistor, in which the (*W/L*) ratio is (4/1.5) (Fig. 10.19).

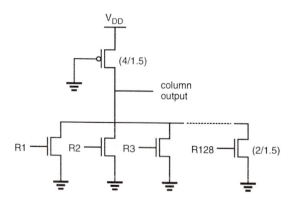

Figure 10.19.   128-input NOR gate representation of a column (bit line) in the ROM array.

The combined column capacitance loading the output node of the 128-input NOR gate can be approximated by adding up the parasitic capacitances of each driver transistor.

$$C_{column} = 128 \times \left(C_{gd,driver} + C_{db,driver} \right) \approx 1.5 \text{ pF}$$

where

$$C_{gd,driver} + C_{db,driver} = 0.0118 \text{ pF/word line}$$

Since only one word line (row) is activated at a time by the row address decoder, the NOR gate representing the column can actually be reduced to the inverter shown in Fig. 10.20. To calculate the column access time, we must consider the worst-case signal propagation delay τ_{PHL} (for falling output voltage) for this inverter. By using the propagation delay formula (6.22) in Chapter 6, the worst-case column access time can thus be calculated as $t_{column} = 18$ ns. Note that the propagation delay for rising output signals τ_{PLH} is not considered here because the bit line (column) is precharged high *before* each row access operation.

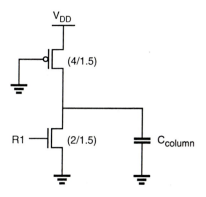

Figure 10.20. Equivalent inverter circuit representing the bit line (column). Note that only one word line (row) is activated at a time.

The total access time for this ROM array consisting of 128 rows and 256 columns is found by adding the row and column access times, as $t_{access} = 38.5$ ns.

At this point, we may also consider a different arrangement for the ROM array, which consists of 256 rows and 128 columns, i.e., using 8 row address bits and 7 column address bits. Since the number of columns in this arrangement is half of the number of columns in the previous example, the total row resistance and row capacitance values will be approximately half of those obtained for the 256-column arrangement. Consequently, the row access time for this structure will be *one-fourth* of the row access time found for the 256-column ROM array, i.e., approximately 5 ns. The number of rows in the new arrangement, on the other hand, is 256, which results in a column capacitance approximately *twice* that in the previous case. Thus, the column access time of the 256-by-128 memory array will be twice as large, approximately 36 ns. In conclusion, we find that the total access time of the 128-by-256 array examined initially is shorter than that of the 256-by-128 array.

10.3. Static Read-Write Memory (SRAM) Circuits

As already explained in Section 10.1, read-write (R/W) memory circuits are designed to permit the modification (writing) of data bits to be stored in the memory array, as well as their retrieval (reading) on demand. The memory circuit is said to be *static* if the stored data can be retained indefinitely (as long as a sufficient power supply voltage is provided), without any need for a periodic refresh operation. We will examine the circuit structure and the operation of simple SRAM cells, as well as the peripheral circuits designed to read and write the data.

The data storage cell, i.e., the 1-bit memory cell in static RAM arrays, invariably consists of a simple latch circuit with two stable operating points (states). Depending on the preserved state of the two-inverter latch circuit, the data being held in the memory cell will be interpreted either as a logic "0" or as a logic "1." To access (read and write) the data contained in the memory cell via the bit line, we need at least one switch, which is controlled by the corresponding word line, i.e., the row address selection signal (Fig. 10.21(a)). Usually, two complementary access switches consisting of nMOS pass transistors are implemented to connect the 1-bit SRAM cell to the complementary bit lines (columns). This can be likened to turning the car steering wheel with both left and right hands in complementary directions.

Figure 10.21(b) shows the generic structure of the MOS static RAM cell, consisting of two cross-coupled inverters and two access transistors. The load devices may be polysilicon resistors, depletion-type nMOS transistors, or pMOS transistors, depending on the type of the memory cell. The pass gates acting as data access switches are enhancement-type nMOS transistors.

The use of resistive-load inverters with undoped polysilicon resistors in the latch structure typically results in a significantly more compact cell size, compared with the other alternatives (Fig. 10.21(c)). This is true since the resistors can be stacked on top of the cell (using double-polysilicon technology), thereby reducing the cell size to four transistors, as opposed to the six-transistor cell topologies. If multiple polysilicon layers are available, one layer can be used for the gates of the enhancement-type nMOS transistors, while another level is used for load resistors and interconnects.

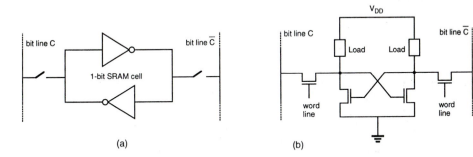

(a)

(b)

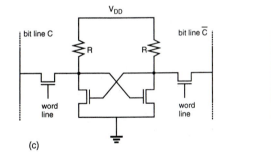

(c)

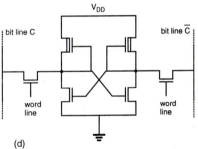

(d)

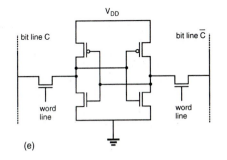

(e)

Figure 10.21. Various configurations of the static RAM cell. (a) Symbolic representation of the two-inverter latch circuit with access switches. (b) Generic circuit topology of the MOS static RAM cell. (c) Resistive-load SRAM cell. (d) Depletion-load nMOS SRAM cell. (e) Full CMOS SRAM cell.

In order to attain acceptable noise margins and output pull-up times for the resistive-load inverter, the value of the load resistor has to be kept relatively low, as already examined in Section 5.2. On the other hand, a high-valued load resistor is required in order to reduce the amount of standby current being drawn by each memory cell. Thus, there is a trade-off between the high resistance required for low power and the requirement to provide wider noise margins and high speed. The power consumption issue will be addressed later in more detail.

The six-transistor depletion-load nMOS SRAM cell shown in Fig. 10.21(d) can be easily implemented with one polysilicon and one metal layer, and the cell size tends to be relatively small, especially with the use of buried metal-diffusion contacts. The static characteristics and the noise margins of this memory cell are typically better than those of the resistive-load cell. The static power consumption of the depletion-load SRAM cell, however, makes it an unsuitable candidate for high-density SRAM arrays.

The full CMOS SRAM cell shown in Fig. 10.21(e) achieves the lowest static power dissipation among the various circuit configurations presented here. In addition, the CMOS cell offers superior noise margins and switching speed as well. The comparative advantages and disadvantages of the CMOS static RAM cell will be investigated in depth later in this section.

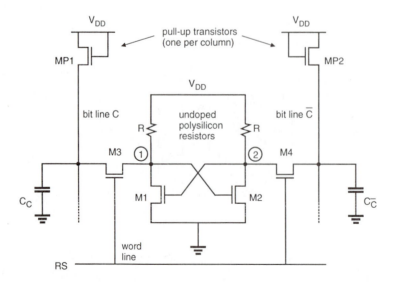

Figure 10.22. Basic structure of the resistive-load SRAM cell, shown with the column pull-up transistors.

SRAM Operation Principles

Figure 10.22 shows a typical four-transistor resistive-load SRAM cell widely used in high-density memory arrays, consisting of a pair of cross-coupled inverters. The two stable operating points of this basic latch circuit are used to store a one-bit piece of information; hence, this pair of cross-coupled inverters make up the central component of the SRAM cell. To perform read and write operations, we use two nMOS pass transistors, both of which are driven by the row select signal, RS. Note that the SRAM cell shown in Fig. 10.22 is accessed via two bit lines or columns, instead of one. This complementary column arrangement allows a more reliable operation.

When the word line (RS) is not selected, i.e., when the voltage level of line RS is equal to logic "0," the pass transistors M3 and M4 are turned off. The simple latch circuit consisting of two cross-connected inverters preserves one of its two stable operating points; hence, data is being *held*. At this point, consider the two columns, C and $\overline{C}$. If all word lines in the SRAM array are inactive, the relatively large column capacitances are charged-up by the column pull-up transistors, MP1 and MP2. Since both transistors operate in saturation, the steady-state voltage level V_C for both columns is determined by the following relationship.

$$V_{DD} - V_C = V_{T0} + \gamma \left(\sqrt{|2\phi_F| + V_C} - \sqrt{|2\phi_F|} \right) \tag{10.1}$$

Assuming $V_{DD} = 5$ V, $V_{T0} = 1$ V, $|2\phi_F| = 0.6$ V, and $\gamma = 0.4$ V$^{1/2}$, this voltage level is found to be approximately equal to 3.5 V. Note that the voltage levels of the two complementary bit lines (columns) are equal during this phase.

Now assume that we select the memory cell by raising its word line voltage to logic "1," hence, the pass transistors M3 and M4 are turned on. Once the memory cell is selected, four basic operations may be performed on this cell.

a) Write "1" operation: The voltage level of column $\overline{C}$ is forced to logic-low by the data-write circuitry. The driver transistor M1 turns off. The voltage V_1 attains a logic-high level, while V_2 goes low.

b) Read "1" operation: The voltage of column C retains its precharge level while the voltage of column $\overline{C}$ is pulled down by M2 and M4. The data-read circuitry detects the small voltage difference ($V_C > V_{\overline{C}}$) and amplifies it as a logic "1" data output.

c) Write "0" operation: The voltage level of column C is forced to logic-low by the data-write circuitry. The driver transistor M2 turns off. The voltage V_2 attains a logic-high level, while V_1 goes low.

d) Read "0" operation: The voltage of column $\overline{C}$ retains its precharge level while the voltage of column C is pulled down by M1 and M3. The data-read circuitry detects the small voltage difference $(V_C < V_{\overline{C}})$ and amplifies it as a logic "0" data output.

Typical voltage waveforms associated with the word line RS and the two pseudo-complementary bit lines are shown qualitatively in Fig. 10.23. Note that the voltage difference between the two columns during a read operation may be only a few hundred millivolts, which must be detected by the data-read circuitry. The reason for this is that the two nMOS transistors in series (e.g., M1 and M3 for read "0") pulling down the column during the read phase can not discharge the large column capacitance quickly.

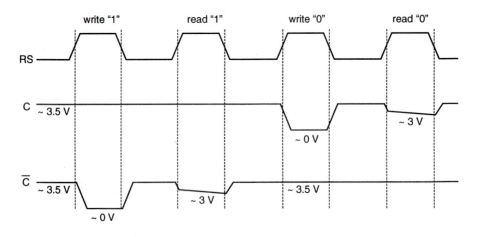

Figure 10.23.   Typical row and column voltage waveforms of the resistive-load SRAM shown in Fig. 10.22 during read and write.

Power Consumption

To estimate the standby power consumption of the static read-write memory cell, assume that a "1"-bit is stored in the cell. This means that the driver transistor M1 is turned off, while the driver transistor M2 is conducting, resulting in $V_1 = V_{OH}$ and $V_2 = V_{OL}$. In this circuit, one of the load resistors will always conduct a non-zero current and, consequently, consume steady-state power. The amount of the standby power consumption is ultimately determined by the value of the load resistor.

Large resistance values and a smaller cell size can be achieved by using lightly-doped or undoped polysilicon for the load resistors, which has a typical sheet resistivity of 10 MΩ per square or higher. The added process complexity for implementing resistive-load SRAM cells using undoped poly resistors is usually worth the advantage of low-power operation, which is manifested by a standby current of less than 10 nA per cell. With a load resistance value of $R = 100$ MΩ, for example, the standby power dissipation of the resistive SRAM cell shown in Fig. 10.22 becomes

$$P_{standby} = \frac{\left(V_{DD} - V_{OL}\right)^2}{R} < 0.25 \text{ μW/cell} \tag{10.2}$$

If we consider that a typical memory array contains a large number of memory cells, each consuming a non-zero standby power, the significance of the power consumption problem in static memory arrays becomes obvious.

Full CMOS SRAM Cell

A low-power SRAM cell may be designed simply by using cross-coupled CMOS inverters instead of the resistive-load nMOS inverters. In this case, the stand-by power consumption of the memory cell will be limited to the relatively small leakage currents of both CMOS inverters. The possible drawback of using CMOS SRAM cells, on the other hand, is that the cell area tends to increase in order to accommodate the n-well for the pMOS transistors and the polysilicon contacts.

The circuit structure of the full CMOS static RAM cell is shown in Fig. 10.24, along with the pMOS column pull-up transistors on the complementary bit lines. The basic operation principle of the CMOS SRAM cell is identical to that of the resistive-load nMOS cell examined earlier. The most important advantage of this circuit topology is that the static power dissipation is even smaller; essentially, it is limited by the leakage current of the pMOS transistors. A CMOS memory cell thus draws current from the power supply only during a switching transition. The low standby power consumption has certainly been a driving force for the increasing prominence of high-density CMOS SRAMs.

Other advantages of CMOS SRAM cells include high noise immunity due to larger noise margins, and the ability to operate at lower power supply voltages than,

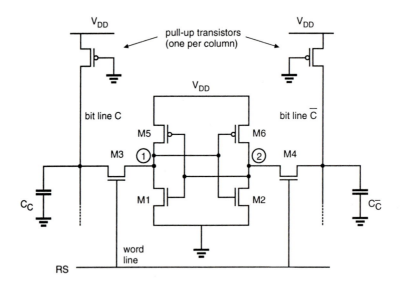

Figure 10.24.   Circuit topology of the CMOS SRAM cell.

for example, the resistive-load SRAM cells. The major disadvantages of CMOS memories historically were larger cell size, the added complexity of the CMOS process, and the tendency to exhibit "latch-up" phenomena. With the widespread use of multi-layer polysilicon and multi-layer metal processes, however, the area disadvantage of the CMOS SRAM cell has been reduced significantly in recent years. Considering the undisputable advantages of CMOS for low-power and low-voltage operation, the added process complexity and the required latch-up prevention measures do not present a substantial barrier against the implementation of CMOS cells in high density SRAM arrays. Figure 10.25 compares typical layouts of the four-transistor resistive-load SRAM cell and the six-transistor full CMOS SRAM cell.

Note that unlike the nMOS column pull-up devices used in resistive-load SRAM, the pMOS column pull-up transistors shown in Fig. 10.24 allow the column voltages to reach full V_{DD} level. To further reduce power consumption, these transistors can also be driven by a periodic precharge signal, which activates the pull-up devices to charge-up column capacitances.

CMOS SRAM Cell Design Strategy

To determine the (W/L) ratios of the transistors in a typical CMOS SRAM cell as shown in Fig. 10.24, a number of design criteria must be taken into consideration. The two basic requirements which dictate the (W/L) ratios are: (a) the data-read

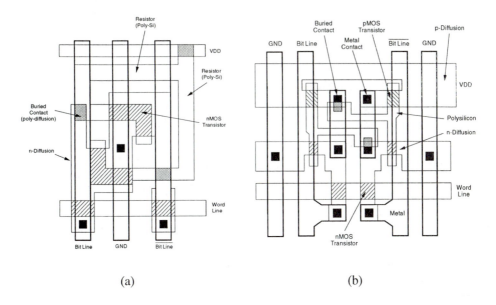

Figure 10.25. Layouts of (a) the resistive-load SRAM cell and (b) the CMOS SRAM cell.

operation should not destroy the stored information in the SRAM cell, and (b) the cell should allow modification of the stored information during the data-write phase. Consider the data-read operation first, assuming that a logic "0" is stored in the cell. The voltage levels in the CMOS SRAM cell at the beginning of the "read" operation are depicted in Fig. 10.26. Here, the transistors M2 and M5 are turned off, while the

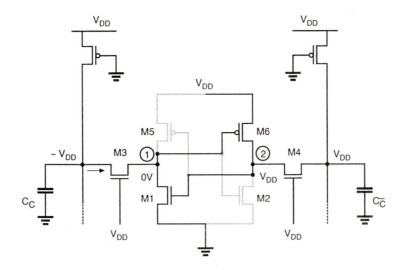

Figure 10.26. Voltage levels in the SRAM cell at the beginning of the "read" operation.

transistors M1 and M6 operate in the linear mode. Thus, the internal node voltages are $V_1 = 0$ and $V_2 = V_{DD}$ *before* the cell access (or pass) transistors M3 and M4 are turned on. The active transistors at the beginning of the data-read operation are highlighted in Fig. 10.26.

After the pass transistors M3 and M4 are turned on by the row selection circuitry, the voltage level of column $\overline{C}$ will not show any significant variation since no current will flow through M4. On the other half of the cell, however, M3 and M1 will conduct a nonzero current and the voltage level of column C will begin to drop slightly. Note that the column capacitance C_C is typically very large; therefore, the amount of decrease in the column voltage is limited to a few hundred millivolts during the read phase. The data-read circuitry to be examined later in this chapter is responsible for detecting this small voltage drop and amplifying it as a stored "0." While M1 and M3 are slowly discharging the column capacitance, the node voltage V_1 will increase from its initial value of 0 V. Especially if the (W/L) ratio of the access transistor M3 is large compared to the (W/L) ratio of M1, the node voltage V_1 may exceed the threshold voltage of M2 during this process, forcing an unintended change of the stored state. The key design issue for the data-read operation is then to guarantee that the voltage V_1 does not exceed the threshold voltage of M2, so that the transistor M2 remains turned off during the read phase, i.e.,

$$V_{1,max} \leq V_{T,2} \tag{10.3}$$

We can assume that after the access transistors are turned on, the column voltage V_C remains approximately equal to V_{DD}. Hence, M3 operates in saturation while M1 operates in the linear region.

$$\frac{k_{n,3}}{2}\left(V_{DD} - V_1 - V_{T,n}\right)^2 = \frac{k_{n,1}}{2}\left(2\left(V_{DD} - V_{T,n}\right)V_1 - V_1^2\right) \tag{10.4}$$

Combining this equation with (10.3) results in:

$$\frac{k_{n,3}}{k_{n,1}} = \frac{\left(\dfrac{W}{L}\right)_3}{\left(\dfrac{W}{L}\right)_1} < \frac{2\left(V_{DD} - 1.5\,V_{T,n}\right)V_{T,n}}{\left(V_{DD} - 2\,V_{T,n}\right)^2} \tag{10.5}$$

The upper limit of the aspect ratio found above is actually more conservative, since a portion of the drain current of M3 will also be used to charge-up the parasitic node capacitance of node (1). To summarize, the transistor M2 will remain in cut-off during the read "0" operation if condition (10.5) is satisfied. A symmetrical condition also dictates the aspect ratios of M2 and M4.

Now consider the write "0" operation, assuming that a logic "1" is stored in the SRAM cell initially. Figure 10.27 shows the voltage levels in the CMOS SRAM cell at the beginning of the data-write operation. The transistors M1 and M6 are turned off, while the transistors M2 and M5 operate in the linear mode. Thus, the internal node voltages are $V_1 = V_{DD}$ and $V_2 = 0$ V *before* the cell access (or pass) transistors M3 and M4 are turned on.

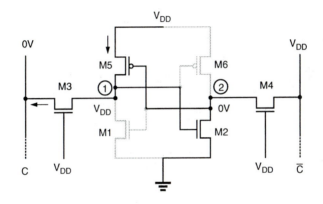

Figure 10.27. Voltage levels in the SRAM cell at the beginning of the "write" operation.

The column voltage V_C is forced to logic "0" level by the data-write circuitry; thus, we may assume that V_C is approximately equal to 0 V. Once the pass transistors M3 and M4 are turned on by the row selection circuitry, we expect that the node voltage V_2 remains *below* the threshold voltage of M1, since M2 and M4 are designed according to condition (10.5). Consequently, the voltage level at node (2) would not be sufficient to turn on M1. To change the stored information, i.e., to force V_1 to 0 V and V_2 to V_{DD}, the node voltage V_1 *must be reduced below* the threshold voltage of M2, so that M2 turns off first. When $V_1 = V_{T,n}$, the transistor M3 operates in the linear region while M5 operates in saturation.

$$\frac{k_{p,5}}{2}\left(0 - V_{DD} - V_{T,p}\right)^2 = \frac{k_{n,3}}{2}\left(2\left(V_{DD} - V_{T,n}\right)V_{T,n} - V_{T,n}^2\right) \qquad (10.6)$$

Rearranging this condition results in:

$$\frac{k_{p,5}}{k_{n,3}} < \frac{2\left(V_{DD} - 1.5\,V_{T,n}\right)V_{T,n}}{\left(V_{DD} + V_{T,p}\right)^2}$$

$$\frac{\left(\dfrac{W}{L}\right)_5}{\left(\dfrac{W}{L}\right)_3} < \frac{\mu_n}{\mu_p} \cdot \frac{2\left(V_{DD} - 1.5\,V_{T,n}\right)V_{T,n}}{\left(V_{DD} + V_{T,p}\right)^2} \qquad\qquad (10.7)$$

To summarize, the transistor M2 will be *forced* into cut-off mode during the write "0" operation if condition (10.7) is satisfied. This will guarantee that M1 subsequently turns on, changing the stored information. Note that a symmetrical condition also dictates the aspect ratios of M6 and M4.

SRAM Write Circuitry

As already discussed in the preceding section, a "write" operation is performed by forcing the voltage level of either column (bit line) to a logic-low level. To accomplish this task, a low-resistance, conducting path must be provided from each column to the ground, which can be selectively activated by the data-write signals. A simplified view of the SRAM "write" circuitry designed for this operation is shown in Fig. 10.28. Here, the nMOS transistors M1 and M2 are used to pull down the two column voltages, while the transistor M3 completes the conducting path to ground. Note that M3 is driven by the column address decoder circuitry, i.e., M3 turns on only when the corresponding column address is selected. The column pull-down transistors, on the other hand, are driven by two pseudo-complementary control signals, WB and $\overline{WB}$. The "write-enable" signal W (active low) and the data to be written ($DATA$) are used to generate the control signals, as shown in the following table.

The nMOS pull-down transistors M1 and M2, as well as the column selection transistor M3 must have sufficiently large (W/L) ratios so that the column voltages can be forced to almost 0 V level during a "write" operation. Also note that the data input circuitry consisting of two NOR2 gates can be shared by several columns, assuming that one column is activated, i.e., selected by the column address decoder at a time.

SRAM Read Circuitry

During the "data read" operation in the SRAM array, the voltage level on either one of the columns drops slightly after the pass transistors are turned on by the row address decoder circuit. In order to reduce the read access time, the "read" circuitry

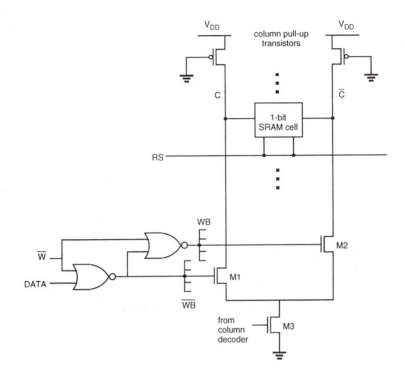

$\overline{W}$	DATA	WB	$\overline{WB}$	Operation
0	1	1	0	*M1 is off, M2 is on → $V_{\overline{C}}$ low*
0	0	0	1	*M1 is on, M2 is off → V_C low*
1	X	0	0	*M1 and M2 are off → both columns remain high*

Figure 10.28.   Data write circuitry associated with one column-pair in an SRAM array.

must detect a very small voltage difference between the two complementary columns, and amplify this difference to produce a valid logic output level. A simple source-coupled differential amplifier can be used for this task, as shown in Fig. 10.31. Here, the drain currents of the two complementary nMOS transistors are

$$I_{D,1} = \frac{k_n}{2}\left(V_C - V_x - V_{T,1}\right)^2 \tag{10.8}$$

$$I_{D,2} = \frac{k_n}{2}\left(V_{\overline{C}} - V_x - V_{T,2}\right)^2 \tag{10.9}$$

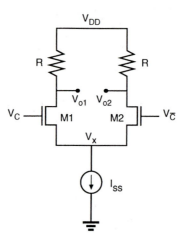

Figure 10.29.   Simple source-coupled differential amplifier circuit for "read" operation.

Small-signal analysis of the circuit yields the differential gain of this circuit as

$$\frac{\partial(V_{o1}-V_{o2})}{\partial(V_C-V_{\overline{C}})}=-R\cdot g_m \qquad (10.10)$$

where

$$g_m = \frac{\partial I_D}{\partial V_{GS}} = \sqrt{2\,k_n\,I_D}$$

The differential gain of the amplifier can be increased significantly by using active loads instead of resistors and by using *cascode* configuration, i.e., an intermediary common-gate stage between the common-source transistors and the load transistors. Finally, the differential output of the cascode stage must be converted into single-ended output, by using a level-shifter and buffer stage. Although the "data read" circuitry described above is capable of detecting small voltage differences between the two complementary bit lines (columns), other types of efficient sense amplifiers are also implemented with CMOS technology, as will be examined in the following.

The architecture of the output "read" circuitry is driven primarily by the constant demand for high access speed and high integration density. We must recognize first that the full CMOS SRAM cell has a natural speed advantage which is derived from using both active pull-up and active pull-down devices in the latch circuits. The CMOS rise and fall times are short and symmetrical, whereas the nMOS latch circuit has a short output fall time but a larger rise time because of its high-resistance load. Both the depletion-load nMOS SRAM cell and the resistive-load nMOS SRAM cell, hence, have slower average switching speed compared to that for the full CMOS cell.

Apart from the type of the SRAM cell, the *precharging* of bit lines also plays a significant role in the access time. In an unclocked SRAM array, data from the accessed cell develops a voltage difference on the bit lines. This voltage difference is then detected and amplified to drive the output buffer. When another cell on the same column is accessed next, one that contains data opposite to the data contained in the previously accessed cell, the output has to switch first to an equalized state and then to the opposite logic state. Since the capacitance on the bit lines is quite large, the time required for switching the differential from one state to the other becomes a significant portion of the overall access time. The access time penalty associated with this procedure can be substantially reduced by the *equalization* of bit lines prior to each new access. Equalization can be done when the memory array is deselected, i.e., between two access cycles.

Fast Sense Amplifiers

The availability of the CMOS technology for the manufacturing of high-density SRAM arrays, either with full-CMOS or resistive-load nMOS memory cells, also allows us to implement efficient sense amplifier structures with pMOS current mirrors. Figure 10.30 shows a single-stage differential current-mirror amplifier, which is typically used as a front-end (input stage) in operational amplifiers.

In this circuit, the gates of the two nMOS transistors M1 and M2 are connected to the bit lines. Their substrate terminals are tied to their respective source terminals in order to remove the substrate-bias effect. Notice that each bit line is represented by a large parasitic capacitance. The nMOS transistor M3 is a long-channel device which acts as a current source for both branches, and is controlled by a clock signal. The output inverter is not a part of the differential amplifier, but it is used to drive the output node.

Before the beginning of a "read" operation, the two bit lines (columns) are pulled up for equalization, as discussed earlier. The CLK signal is low during this phase, so that the nMOS transistor M3 remains off. Since both M1 and M2 conduct, the common source node is pulled up, and the output node of the amplifier also goes high. Therefore, the output of the inverter is at a logic-low level initially.

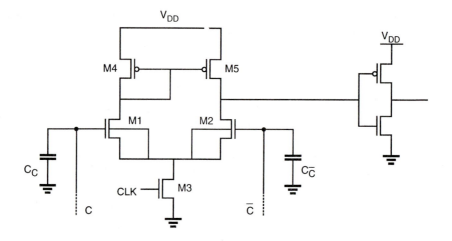

Figure 10.30.   Differential current-mirror amplifier to speed up memory read access time.

Once a memory cell is selected for the "read" operation, the voltage on one of the complementary bit lines will start to drop slightly. At the same time as the row selection signal, the CLK signal driving M3 is also turned on. If the stored data on the selected SRAM cell forces the bit line C to decrease slightly, transistor M1 turns off, and the output voltage of the differential amplifier drops immediately. Consequently, the output voltage of the inverter goes high. Otherwise, if the stored data on the selected memory cell forces the bit line $\overline{C}$ to drop slightly, M2 turns off. Thus, the voltage level at the output node of the differential amplifier remains high in this case, and the inverter also preserves its logic-low output level.

In most high-density SRAM chips, two- or three-stage current-mirror differential sense amplifiers are implemented to further improve the "read" access speed. An example of a two-stage CMOS sense amplifier is shown in Fig. 10.31. Here, the first stage consists of two *complementary* differential sense amplifiers, both of which receive the column (bit line) voltages at their inputs. Since they are operated in an anti-symmetrical configuration, the output voltages of the two sense amplifiers must change into opposite directions. The complementary outputs of the first-stage amplifiers are applied to the input terminals of a second-stage differential amplifier, which generates the output signal.

The dynamic behavior of the one-stage sense amplifier is compared with the dynamic behavior of the two-stage sense amplifier via SPICE simulation, in Fig. 10.32.

Here, one of the column voltages (V_C) starts to drop slightly, while the other column voltage remains constant (not shown in the figure). It can be seen that the output of the one-stage sense amplifier responds to this change with a delay of about 10 ns, whereas the output delay of the two-stage sense amplifier is only about 1 ns.

As mentioned earlier, in the sense amplifier circuit examined here (Fig. 10.30), the substrate terminals of M1 and M2 are connected to the common source node instead of the ground, in order to avoid threshold voltage variations due to the substrate-bias effect. This is possible using twin-tub, silicon-on-sapphire (SOS), or p-well fabrication processes. In an n-well CMOS process, on the other hand, where all nMOS transistors are formed over the common substrate, the circuit cannot be implemented.

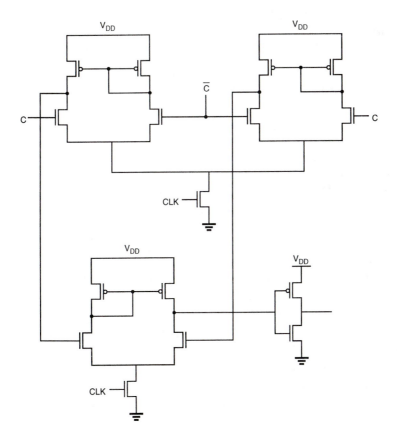

Figure 10.31.   Two-stage differential current-mirror sense amplifier circuit.

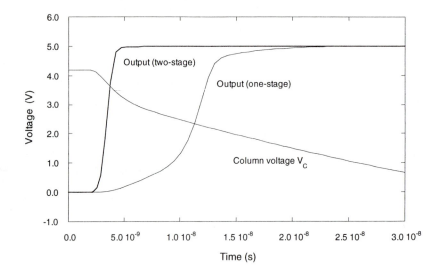

Figure 10.32. Typical dynamic response characteristics of the one-stage sense amplifier and the two-stage sense amplifier circuits.

Another simple circuit topology for sense amplifiers is the cross-coupled latch, shown in Fig. 10.33. Assume that both columns (bit lines) are being pulled up during the precharge cycle and that the voltage on the bit line C starts to drop slightly when the SRAM cell access transistors are activated. Consequently, when the transistor M3 is turned on, the voltage at node $\overline{C}$ is higher than the voltage at node C. Therefore, M1 turns on first and further pulls down the potential at node C. This makes it more difficult for M2 to conduct. Eventually, M2 turns off completely, and M1 keeps conducting, so that the bit line C is discharged through M1 and M3.

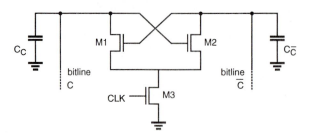

Figure 10.33. Cross-coupled nMOS sense amplifier circuit.

Note that the operation of this circuit is distinctly different from that of the differential amplifier circuit. The cross-coupled sense amplifier does not generate an output voltage level which corresponds to the polarity of the voltage difference between the two bit lines, but it rather *amplifies* the small voltage difference already existing between the bit lines. This voltage difference must still be translated into a logic level, by using a buffer stage. We will see the implementation of a very similar structure also in the dynamic RAM (DRAM) sense amplifiers.

In most SRAM arrays, the cross-coupled sense amplifier circuit is used in conjunction with the differential sense amplifier (Fig. 10.30) examined earlier. In this case, the cross-coupled amplifier serves as a front-end structure to amplify the small voltage difference between the two bit lines, whereas the current-mirror amplifier detects the voltage difference and generates the output level. Figure 10.34 shows the

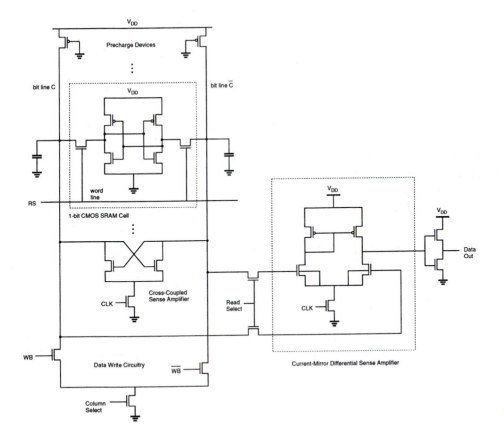

Figure 10.34. Complete circuit diagram of a CMOS static RAM column with data write and data read circuitry.

complete circuit diagram of a CMOS SRAM column with one representative CMOS memory cell, the "data write" circuitry, the cross-coupled sense amplifier, and the differential (main) sense amplifier circuit. Note that the main amplifier is connected to the two complementary bit lines via pass transistors, which are driven by the "read select" signal. This configuration enables the use of one main sense amplifier to read the data out of several columns, one at a time.

Dual-Port Static RAM Arrays

In some cases, the memory array may have to be accessed simultaneously by multiple processors, or by one processor and another peripheral device. This could result in a timing conflict called "contention," which can be resolved only by having one of the processors wait until the SRAM is free. The added wait state, however, significantly reduces the advantages of the high-speed processor. The dual-port RAM architecture is implemented in systems in which a main memory array must serve multiple high-speed processors and peripheral devices with minimum delay.

The ideal dual-port SRAM allows simultaneous access to the same location in the memory array, by using two independent sets of bit lines and associated access switches for each memory cell. The circuit structure of a typical CMOS dual-port SRAM cell is shown in Fig. 10.35. Here, "word line 1" is used to access one set of complementary bit lines (bit line 1), while "word line 2" allows access to the other set of bit lines (bit line 2). The capability of simultaneous access eliminates wait states for the processors during "data read" operations. However, contention may still occur if both external processors accessing the same memory location simultaneously attempt to write data onto the accessed cell, or if one of the processors attempts to read data

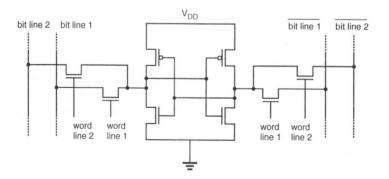

Figure 10.35.   Circuit diagram of the CMOS dual-port SRAM cell.

while the other processor writes data onto the same cell. In most cases, overlapping operations to the same memory location can be eliminated by a contention arbitration logic. It can either allow contention to be ignored and both operations to proceed, or it can arbitrate and delay one port until the operation on the other port is completed.

10.4. Dynamic Read-Write Memory (DRAM) Circuits

All of the static RAM cells examined in the previous section consist of a two-inverter latch circuit, which is accessed for "read" and "write" operations via two pass transistors. Consequently, the SRAM cells require four to six transistors per bit, and four to five lines connecting to each cell, including the power and ground connections. To satisfy these requirements, a substantial silicon area must be reserved for each memory cell. In addition, most SRAM cells have non-negligible standby (static) power dissipation, with the exception of the full CMOS SRAM cell.

As the trend for high-density RAM arrays forces the memory cell size to shrink, alternative data storage concepts must be considered to accommodate these demands. In a *dynamic* RAM cell, binary data is stored simply as charge in a capacitor, where the presence or absence of stored charge determines the value of the stored bit. Note that the data stored as charge in a capacitor cannot be retained indefinitely, because the leakage currents eventually remove or modify the stored charge. Thus, all dynamic memory cells require a periodic refreshing of the stored data, so that unwanted modifications due to leakage are prevented before they occur.

The use of a capacitor as the primary storage device generally enables the DRAM cell to be realized on a much smaller silicon area compared to the typical SRAM cell. Notice that even as the binary data is stored as charge in a capacitor, the DRAM cell must have access devices, or switches, which can be activated externally for "read" and "write" operations. But this requirement does not significantly affect the area advantage over the SRAM cell, since the cell access circuitry is usually very simple. Also, no static power is dissipated for storing charge on the capacitance. Consequently, dynamic RAM arrays can achieve higher integration densities than SRAM arrays. Note that a DRAM array also requires additional peripheral circuitry for scheduling and performing the periodic data refresh operations. The hardware overhead of the refresh circuitry, however, does not overshadow the area advantages gained by the small cell size.

Figure 10.36 shows some of the steps in the historical evolution of the DRAM cell. The four-transistor cell shown in Fig. 10.36(a) is the simplest and one of the earliest dynamic memory cells. This cell is derived from the six-transistor static RAM cell by removing the load devices. The cell has in fact two storage nodes, i.e., the parasitic oxide and diffusion capacitances of the nodes indicated in the circuit diagram. Since no current path is provided to the storage nodes for restoring the charge being lost to leakage, the cell must be refreshed periodically. It is obvious that the four-transistor dynamic RAM cell can have only a marginal area advantage over the six-transistor SRAM cell.

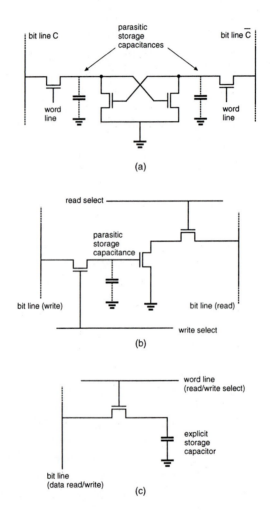

Figure 10.36.   Various configurations of the dynamic RAM cell. (a) Four-transistor DRAM cell with two storage nodes. (b) Three-transistor DRAM cell with two bit lines and two word lines. (c) One-transistor DRAM cell with one bit line and one word line.

The three-transistor DRAM cell shown in Fig. 10.36(b) was the first widely used dynamic memory cell. It utilizes a single transistor as the storage device (where the transistor is turned on or off depending on the charge stored on its gate capacitance), and one transistor each for the "read" and "write" access switches. The cell has two control and two I/O lines. Its separate read and write select lines make it relatively fast, but the four lines with their additional contacts tend to increase the cell area.

The one-transistor DRAM cell shown in Fig. 10.36(c) has become the industry-standard dynamic RAM cell in high-density DRAM arrays. With only one transistor and one capacitor, it has the lowest component count and, hence, the smallest silicon area of all the dynamic memory cells. The cell has one read-write control line (word line) and one I/O line (bit line). We have to emphasize at this point that, unlike in the other dynamic memory cells shown in Figs. 10.36(a) and 10.36(b), the storage capacitance of the one-transistor DRAM cell is *explicit*. This means that a separate capacitor must be manufactured for each storage cell, instead of relying on the parasitic oxide and diffusion capacitances of the transistors for data storage. The word line of the one-transistor DRAM cell is controlled by the row address decoder. Once the selected transistor is turned on, the charge stored in the capacitor can be detected and/or modified through the bit line.

Before we examine the operation of the one-transistor DRAM cell, we will first investigate the three-transistor cell shown in Fig. 10.36(b), which has a very straightforward operation principle. This will help us to illuminate the basic issues involved in the design and operation of dynamic memory cells in general.

Three-Transistor DRAM Cell

The circuit diagram of a typical three-transistor dynamic RAM cell is shown in Fig. 10.37 as well as the column pull-up (precharge) transistors and the column read/write circuitry. Here, the binary information is stored in the form of charge in the parasitic node capacitance C_1. The storage transistor M2 is turned on or off depending on the charge stored in C_1, and the pass transistors M1 and M3 act as access switches for data read and write operations. The cell has two separate bit lines for "data read" and "data write," and two separate word lines to control the access transistors.

The operation of the three-transistor DRAM cell and its peripheral circuitry is based on a two-phase non-overlapping clock scheme. The precharge events are driven by ϕ_1, whereas the "read" and "write" events are driven by ϕ_2. Every "data read" and "data write" operation is preceded by a precharge cycle, which is initiated with the precharge signal PC going high. During the precharge cycle, the column pull-up transistors are activated, and the corresponding column capacitances C_2 and C_3 are charged up to logic-high level. With typical enhancement type nMOS pull-up transistors ($V_{T0} \approx 1.0$ V) and a power supply voltage of 5 V, the voltage level of both columns after the precharge is approximately equal to 3.5 V.

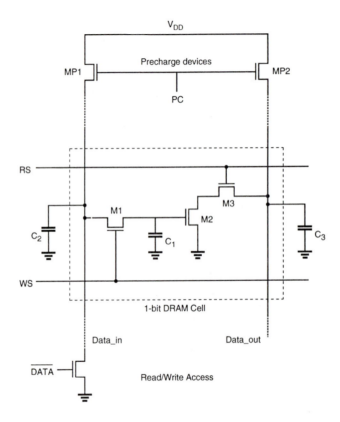

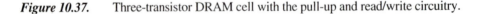

Figure 10.37.   Three-transistor DRAM cell with the pull-up and read/write circuitry.

All "data read" and "data write" operations are performed during the active ϕ_2 phase, i.e., when PC is low. Figure 10.38 depicts the typical voltage waveforms associated with the 3-T DRAM cell during a sequence of four consecutive operations: write "1," read "1," write "0," and read "0." The four precharge cycles shown in Fig. 10.38 are numbered 1, 3, 5, and 7, respectively. Figure 10.39 illustrates the transient currents charging up the two columns (D_{in} and D_{out}) during a precharge cycle. The precharge cycle is effectively completed when both capacitance voltages reach their steady-state values. Note here that the two column capacitances C_2 and C_3 are at least one order of magnitude larger than the internal storage capacitance C_1.

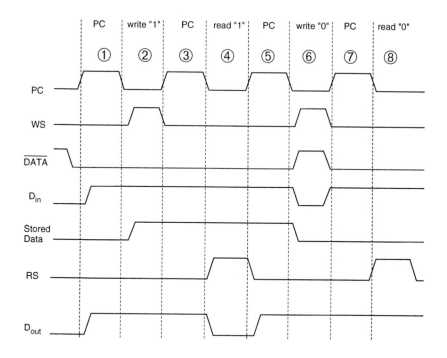

Figure 10.38. Typical voltage waveforms associated with the 3-T DRAM cell during four consecutive operations: write "1," read "1," write "0," and read "0."

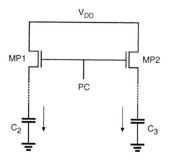

Figure 10.39. Column capacitances C_2 and C_3 are being charged-up through MP1 and MP2 during the precharge cycle.

For the write "1" operation, the *inverse* data input is at the logic-low level, because the data to be written onto the DRAM cell is a logic "1." Consequently, the "data write" transistor MD is turned off, and the voltage level on column D_{in} remains high. Now, the "write select" signal WS is pulled high during the active phase of ϕ_2. As a result, the write access transistor M1 is turned on. With M1 conducting, the charge on C_2 is now shared with C_1 (Fig. 10.40). Since the capacitance C_2 is very large compared to C_1, the storage node capacitance C_1 attains approximately the same logic-high level as the column capacitance C_2 at the end of the charge-sharing process.

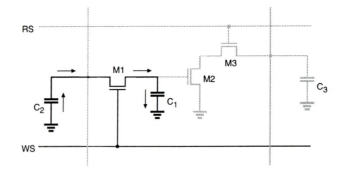

Figure 10.40.   Charge sharing between C_2 and C_1 during the write "1" sequence.

After the write "1" operation is completed, the write access transistor M1 is turned off. With the storage capacitance C_1 charged-up to a logic-high level, transistor M2 is now conducting. In order to read this stored "1," the "read select" signal RS must be pulled high during the active phase of ϕ_2, following a precharge cycle. As the read access transistor M3 turns on, M2 and M3 create a conducting path between the "data read" column capacitance C_3 and the ground. The capacitance C_3 discharges through M2 and M3, and the falling column voltage is interpreted by the "data read" circuitry as a stored logic "1." The active portion of the DRAM cell during the read "1" cycle is shown in Fig. 10.41. Note that the 3-T DRAM cell may be read repeatedly in this fashion without disturbing the charge stored in C_1.

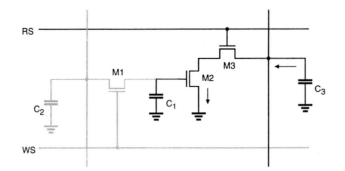

Figure 10.41.   The column capacitance C_3 is discharged through the transistors M2 and M3 during the read "1" operation.

For the write "0" operation, the inverse data input is at the logic-high level, because the data to be written onto the DRAM cell is a logic "0." Consequently, the data write transistor is turned on, and the voltage level on column D_{in} is pulled to logic "0." Now, the "write select" signal WS is pulled high during the active phase of ϕ_2. As a result, the write access transistor M1 is turned on. The voltage level on C_2, as well as that on the storage node C_1, is pulled to logic "0" through M1 and the data write transistor, as shown in Fig. 10.42. Thus, at the end of the write "0" sequence, the storage capacitance C_1 contains a very low charge, and the transistor M2 is turned off since its gate voltage is approximately equal to zero.

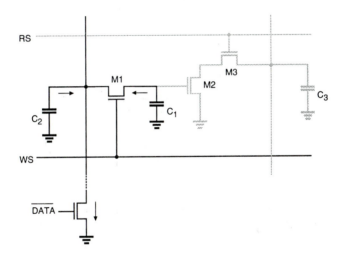

Figure 10.42.   Both C_1 and C_2 are discharged via M1 and the data write transistor during the write "0" sequence.

In order to read this stored "0," the "read select" signal RS must be pulled high during the active phase of ϕ_2, following a precharge cycle. The read access transistor M3 turns on, but since M2 is off, there is no conducting path between the column capacitance C_3 and the ground (Fig. 10.43). Consequently, C_3 does not discharge, and the logic-high level on the D_{out} column is interpreted by the data read circuitry as a stored "0" bit.

As we already pointed out in the beginning of this section, the charge stored in C_1 cannot be held indefinitely, even though the "data read" operations do not significantly disturb the stored charge. The drain junction leakage current of the write access transistor M1 is the main reason for the gradual depletion of the stored charge on C_1. In order to *refresh* the data stored in the DRAM cells before they are altered due to leakage, the data must be periodically read, inverted (since the data output level reflects the inverse of the stored data), and then written back into the same cell location. This refresh operation is performed for all storage cells in the DRAM array every 2 to 4 ms. Note that all bits in one row can be refreshed at once, which significantly simplifies the procedure.

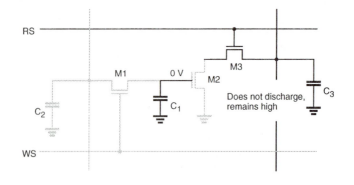

Figure 10.43.   The column capacitance C_3 cannot discharge during the read "0" cycle.

It can be seen that the three-transistor dynamic RAM cell examined here does not dissipate any static power for data storage, since there is no continuous current flow in the circuit. Also, the use of periodic precharge cycles instead of static pull-up further reduces the dynamic power dissipation. The additional peripheral circuitry required for scheduling the non-overlapping control signals and the refresh cycles does not significantly overshadow these advantages of the low-power dynamic memory.

One-Transistor DRAM Cell

The circuit diagram of the one-transistor (1-T) DRAM cell consisting of one explicit storage capacitor and one access transistor is shown in Fig. 10.44. Here, C_1 represents the storage capacitor which typically has a value of 30–100 fF. Similar to the 3-T DRAM cell, binary data are stored as the presence or absence of charge in the storage capacitor. Capacitor C_2 represents the much larger parasitic column capacitance associated with the word line. Charge sharing between this large capacitance and the very small storage capacitance plays a very important role in the operation of the 1-T DRAM cell.

The "data write" operation on the 1-T cell is quite straightforward. For the write "1" operation, the bit line (D) is raised to logic "1" by the write circuitry, while the selected word line is pulled high by the row address decoder. The access transistor M1 turns on, allowing the storage capacitor C_1 to charge up to a logic-high level. For the write "0" operation, the bit line (D) is pulled to logic "0" and the word line is pulled high by the row address decoder. In this case, the storage capacitor C_1 discharges through the access transistor, resulting in a stored "0" bit.

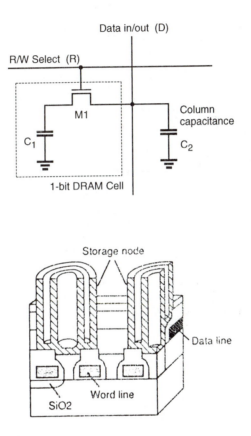

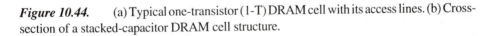

Figure 10.44. (a) Typical one-transistor (1-T) DRAM cell with its access lines. (b) Cross-section of a stacked-capacitor DRAM cell structure.

In order to read stored data out of a 1-T DRAM cell, on the other hand, we have to build a fairly elaborate read-refresh circuit. The reason for this is the fact that the "data read" operation on the one-transistor DRAM cell is by necessity a "destructive readout." This means that the stored data must be destroyed or lost during the read operation. Typically, the read operation starts with precharging the column capacitance C_2. Then, the word line is pulled high in order to activate the access transistor M1.

Charge sharing between C_1 and C_2 occurs and, depending on the amount of stored charge on C_1, the column voltage either increases or decreases slightly. Note that charge sharing inevitably destroys the stored charge on C_1. Hence, we also have to *refresh* data every time we perform a "data read" operation.

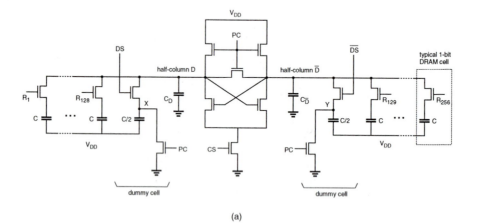

(a)

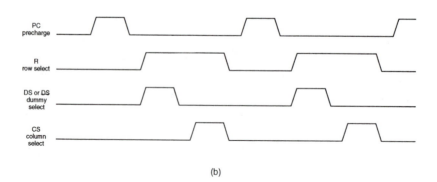

(b)

Figure 10.45.   (a) Data read-restore circuit example for 256 1-T DRAM cells per column. (b) Typical control signal waveforms for two consecutive data read operations.

An example of the 256-cells-per-column DRAM read circuitry is shown in Fig. 10.45, along with typical control signal waveforms. A cross-coupled dynamic latch circuit is used to detect the small voltage differences and to restore the signal levels. The storage array is split in half so that equal capacitances are connected to each side of the cross-coupled latch. This means that half of the cells connected to one bit line (column) are arranged on one side of the latch, and the other half of the cells connected to the same column are arranged on the other side. As shown in Fig. 10.45, each half-column in the array also has a dummy cell which contains a capacitance half of the storage capacitance value. The capacitors C_D and $C_{\bar{D}}$ in Fig. 10.45 represent the relatively large parasitic column capacitances associated with the half-columns.

The "read-refresh" operation occurs in three stages. First, the precharge devices are turned on during the active phase of PC. Both column capacitances C_D and $C_{\bar{D}}$ are charged-up to the same logic-high level, whereas the dummy nodes X and Y are pulled to logic-low level. The devices involved in the precharge operation are highlighted in Fig. 10.46. Note that during this phase, all other signals are inactive.

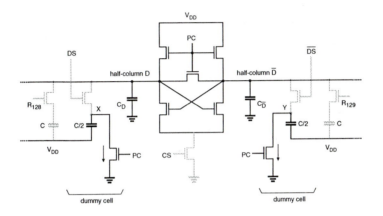

Figure 10.46.   The half-columns are being charged-up during the precharge phase.

Next, one of the 256 word lines is raised to logic "1" during the row selection phase. At the same time, the dummy cell on the other side is also selected by raising either DS or $\overline{DS}$. This situation is depicted in Fig. 10.47, where only the selected

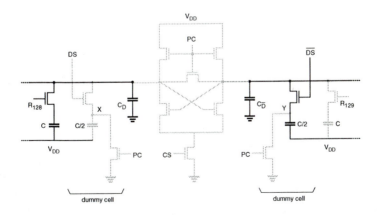

Figure 10.47.   Two complementary column voltages are determined through charge sharing, on the one side between the selected storage cell and the half-column capacitance, on the other side between the dummy cell and the other half-column capacitance.

DRAM cell (left) and the corresponding dummy cell (right) are highlighted. If a logic "1" is stored in the selected cell, the voltage on the half-column D will rise slightly, while the voltage on half-column $\overline{D}$ drops, because the dummy cell is being charged up. If a logic "0" is stored in the selected cell, however, the voltage on the half-column D will also drop, and the drop in V_D will be larger than the drop in $V_{\overline{D}}$. Consequently, there will be a detectable difference between a stored "0" and a stored "1."

The final stage of the "read-refresh" operation is performed during the active phase of CS, the column-select signal. As soon as the cross-coupled latch is activated, the slight voltage difference between the two half-columns is amplified, and the latch forces the two half-columns into opposite states (Fig. 10.48). Thus, the stored data on the selected DRAM cell is refreshed while it is being read by the "read-refresh" circuitry.

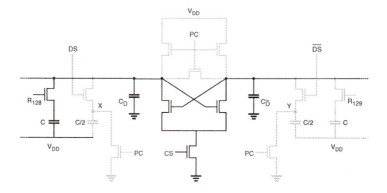

Figure 10.48.   The cross-coupled latch circuit is used for detection of the voltage difference between the half-columns and for restoring the voltage level on the accessed cell.

Example 10.2.

The data read operation of the typical one-transistor DRAM cell is simulated with SPICE in the following. The sense-refresh circuit used in this simulation is shown in Fig. 10.45. Two different cases are being considered: a read "0" operation and a read "1" operation. Depending on the stored data bit, the cross-coupled sense amplifier circuit detects the small voltage difference between the two half-columns, and pulls down one of the column voltages to zero.

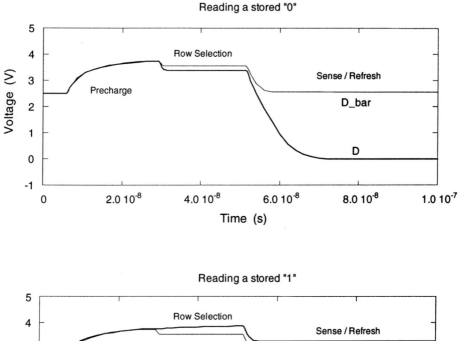

Reading a stored "0"

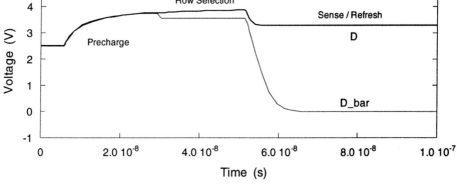

Reading a stored "1"

References

1. G. Luecke, J. P. Mize, and W. N. Carr, *Semiconductor Memory Design and Application*, New York, NY: McGraw-Hill Book Company, 1973.

2. P. R. Gray and R. G. Meyer, *Analysis and Design of Analog Integrated Circuits*, third edition, New York, NY: John Wiley & Sons, Inc., 1993.

3. B. Prince, *Semiconductor Memories*, second edition, New York, NY: John Wiley & Sons, Inc., 1993.

Exercise Problems

10.1   Consider the DRAM circuit shown in Figure 10.45 in the text. The
threshold voltage of the two precharge transistors is 2 V. Calculate the
steady-state voltages of V_D in Region I and Region II of Fig. P10.1 by
assuming the following:

i) $C = 50$ fF
ii) $C_D = 400$ fF
iii) $V(C) = V(C/2) = V_Y = 0$ V in region I
iv) While PC is high, no other transistor connected to D or $\bar{D}$ is on.

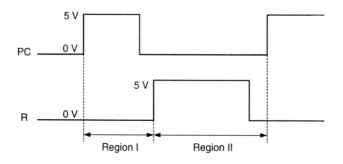

Figure P10.1

10.2   A single-transistor DRAM cell is represented by the following circuit
diagram. The bit line can be precharged to $V_{DD}/2$ by using a clocked precharge
circuit. Also the WRITE circuit is assumed here to bring the potential of the
bit line to V_{DD} or 0 V during the WRITE operation with word line at V_{DD}.
Using the parameters given:

- $V_{T0} = 1.0$ V
- $\gamma = 0.3$ V$^{1/2}$
- $|2\phi_F| = 0.6$ V

(a)   Find the maximum voltage across the storage capacitor C_S after
WRITE-1 operation, i.e., when the bit line is driven to $V_{DD} = 5$ V.

(b)   Assuming zero leakage current in the circuit, find the voltage at the
bit line during READ-1 operation after the bit line is first precharged
to $V_{DD}/2$.

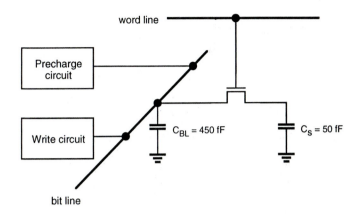

Figure P10.2

10.3 A dynamic CMOS Read Only Memory (ROM) has been designed with a core array consisting of 64 rows with a pitch of 12 μm and 64 columns with a pitch of 10 μm, as shown on the following page. Each column is precharged to 5 V by a pMOS transistor during the interval of zero clock phase and then is pulled down after the clock signal switches to 5 V by one or more nMOS transistors (i.e., NOR implementation) with high gate inputs on the appropriate rows. All of the nMOS transistors have channel widths $W = 4$ μm and source/drain lengths $Y = 5$ μm. As a designer, you are to determine the propagation delay time from a particular input (on row 64) going high to a particular bit-line output (on column 64) going low t_{PHL} between 50% points. Assume that the input signal to row 64 becomes valid-high only after the precharge operation is finished, as shown in the timing diagram. Also, assume that row 64 is running over 30 nMOS transistors and column 64 has 20 nMOS transistors connected to it. For delay calculation, assume that only one nMOS transistor is pulling down. Also assume that pMOS is strong enough to fully charge the precharging node during the precharge phase of the clock signal and to neglect its drain parasitic capacitance. Device parameters are given:

- $C_{jsw} = 250$ pF/m
- $C_{j0} = 80$ μF/m²
- $C_{ox} = 350$ μF/m²
- $L_D = 0.5$ μm
- $K_{eq} = 1.0$ for worst-case capacitance
- $C_{metal} = 2.0$ pF/cm and $R_{metal} = 0.03$ Ω/sq.
- $C_{poly} = 2.2$ pF/cm and $R_{poly} = 25$ Ω/sq.

- Polysilicon line width = 2 μm
- Metal line width = 2 μm
- $k'_n = 20\ \mu A/V^2$
- $k'_p = 10\ \mu A/V^2$
- $V_{T,n} = -V_{T,p} = 1.0\ V$

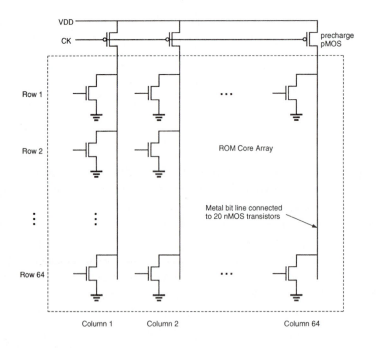

Figure P10.3

10.4 Consider the CMOS SRAM cell shown in Fig. P10.4. Transistors M1 and M2 have (*W/L*) values of 4/4. Transistors M3 and M4 have (*W/L*) values of 2/4. M5 and M6 are to be sized such that the state of the cell can be changed for $V_C \le 0.5\ V$. Assuming that M5 and M6 are the same size, calculate the required (*W/L*). Use the following parameters:

- $V_{T0,n} = 0.7\ V$
- $V_{T0,p} = -0.7\ V$
- $k'_n = 20\ \mu A/V^2$
- $k'_p = 10\ \mu A/V^2$
- $\gamma = 0.4\ V^{1/2}$
- $|2\phi_F| = 0.6\ V$

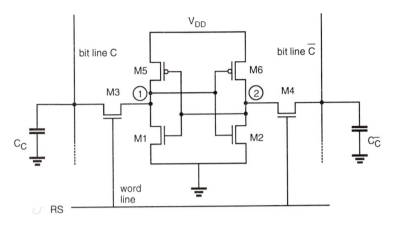

Figure P10.4

10.5 Draw circuit diagrams of the row decoder and the column decoder for an EPROM with 4 rows and 2 columns. Use nMOS technology. Develop formulas for the row and column delays in the EPROM. Define any terms in your formulas which are not obvious.

10.6 Consider an 8K x 8K SRAM. An 8K x 8K SRAM has 64K (= 65536) memory cells and 8 output lines. In the particular SRAM under discussion, 7 address bits go to the row decoder and 6 address bits go to the column decoder. Bit lines are precharged to $V_{DD} = 5$ V before each read operation. A read operation is complete when the bit line has discharged by 0.5 V. A memory cell can provide 1.0 mA of pull-down current to discharge the bit line.

(a) Word line resistance is 390 Ω per memory cell. What formula was used to calculate this resistance?

(b) Word line capacitance is 22 fF per memory cell. What formula was used to calculate this capacitance?

(c) Bit line capacitance is 6 fF per memory cell. What formula was used to calculate this capacitance?

(d) Calculate the access time (row delay + column delay) for this SRAM.

(e) Describe the operation and design of the word line decoder and the bit line decoder.

CHAPTER 11

BiCMOS
LOGIC CIRCUITS

11.1. Introduction

The signal propagation delay due to large interconnect capacitances is a major factor which limits the performance of CMOS digital integrated circuits. The system speed is ultimately restricted by the current-driving capability of CMOS gates that drive large capacitive loads, such as the word lines in memory arrays, or data bus lines between large logic blocks. The problem of driving large off-chip as well as on-chip loads is traditionally solved by using specific CMOS buffer circuits with enhanced current driving capabilities, which will be examined in further detail in Chapter 12. Most of these buffer configurations require a significant amount of silicon area for improvement in the signal propagation delay.

Consider a large capacitive load being driven by a CMOS inverter circuit, which in turn is driven by another CMOS inverter, or any CMOS logic gate in general, as shown in Fig. 11.1. Assume that the propagation delay of the output buffer is represented by τ_{out}, and the propagation delay of the entire two-stage structure is represented by τ_{total}. Simple dynamic analysis suggests that, to drive a large capacitive load with a specified propagation delay, we have to design the output buffer with sufficiently large transistor channel widths, W_n and W_p. However, simply increasing the transistor dimensions will not necessarily help to reduce the propagation delay time.

As we make the channel widths of the nMOS and the pMOS transistors larger, the proportionally increasing drain parasitics will eventually annihilate the speed improvement to be gained by larger transistor sizing. Thus, the output stage propagation delay may not be reducible beyond a certain limit due to the increase in drain parasitics, as illustrated in Fig. 11.2. In fact, the overall delay of the two-stage structure shown in Fig. 11.1 will even *increase* with larger W_n and W_p, because of the large gate capacitance being imposed on the first stage.

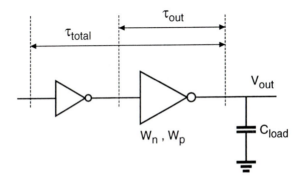

Figure 11.1. A two-stage CMOS buffer structure used for driving a large capacitive load.

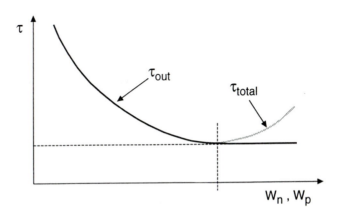

Figure 11.2. Propagation delay of the CMOS buffer vs. transistor size in the output stage.

The only feasible solution using conventional CMOS structures is to build a tapered or scaled buffer chain, consisting of several stages which gradually increase in size from the input stage toward the output. The detailed design issues associated with scaled buffer chains will be investigated in Chapter 12. Yet scaled or tapered buffer structures typically require a large silicon area to implement, which increases the overall cost, especially if many on-chip and off-chip loads must be considered.

In comparison, bipolar junction transistors (BJTs) have more current driving capability, and hence, can overcome such speed bottlenecks using less silicon area. However, the power dissipation of bipolar logic gates is typically one or two orders of magnitude larger than that of comparable CMOS gates. Therefore, such all-bipolar high-speed VLSI circuits are difficult to realize and require very elaborate heat-sink arrangements.

An alternative solution to the problem of driving large capacitive loads can be provided by merging CMOS and bipolar devices (BiCMOS) on chip. Taking advantage of the low static power consumption of CMOS and the high current driving capability of the bipolar transistor during transients, the BiCMOS configuration can combine the "best of both worlds" (Fig. 11.3). In view of the limited driving capabilities

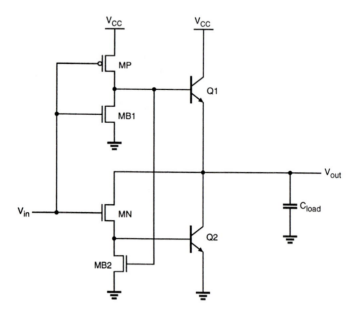

Figure 11.3.   A typical BiCMOS inverter circuit, with four MOSFETs and two BJTs.

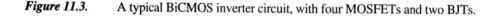

of MOS transistors in general, the BiCMOS combination has significant advantages to offer, such as improved switching speed and less sensitivity with respect to the load capacitance. In general, BiCMOS logic circuits are not bipolar-intensive, i.e. most logic operations are performed by conventional CMOS subcircuits, while the bipolar transistors are used only when high on-chip or off-chip drive capability is required.

The most significant drawback of the BiCMOS circuits lies in the increased fabrication process complexity. The fabrication of the bipolar transistors involves more process steps beyond the CMOS process. But many of the conventional CMOS process steps can be utilized to concurrently fabricate bipolar transistor structures along with the MOS transistors, as shown in the simplified cross-section in Fig. 11.4. For example, the process used to create the n-well (n-tub) on a p-type substrate can also be used to create the n-type collector region of the npn bipolar transistor. The source and drain diffusion steps in the CMOS process can be used to form the emitter region and the base contact region of the bipolar transistor. The silicided polysilicon gate of the MOS transistor forms the emitter contact. In fact, the only bipolar fabrication step which cannot be adopted from the CMOS process is the creation of the base region. The BiCMOS fabrication process typically requires only 3–4 masks in addition to the well-established CMOS process.

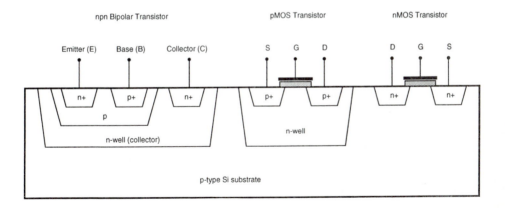

Figure 11.4.   Simplified cross-section showing an npn bipolar transistor, an n-channel MOS transistor, and a p-channel MOS transistor fabricated on the same p-type silicon substrate. Notice that many standard CMOS process steps can be used to create bipolar and MOS transistors side-by-side on the chip.

In the following sections, we will present the basic building blocks of BiCMOS logic circuits, discuss the static and dynamic behavior of BiCMOS circuits, and examine some applications. To provide the necessary analytical basis for the upcoming discussions, we start by investigating the structure and the operation of the bipolar junction transistor (BJT).

11.2. Bipolar Junction Transistor (BJT): Structure and Operation

The simplified cross-section of an npn bipolar junction transistor (BJT) is shown in Fig. 11.5. The structure consists of several regions and layers of doped silicon, which essentially form the three terminals of the device: emitter (E), base (B), and collector (C). The npn-type bipolar transistor shown in Fig. 11.5 is fabricated on a p-type Si substrate. In full-bipolar ICs, a lightly doped n-type epitaxial layer created on top of the p-type substrate serves as the collector (note: in BiCMOS ICs, the n-well is used as the collector region). The collector contact region is doped with a higher concentration of n-type impurities in order to reduce the contact resistance. The base region is created by forming a p-type diffusion in the n-type epitaxial layer, and the heavily doped n-type emitter is formed within this p-type base diffusion region.

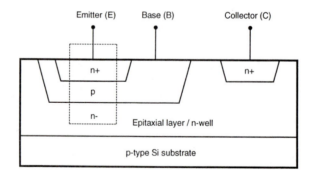

Figure 11.5.   Cross-section of an npn-type bipolar junction transistor (BJT).

The rectangular slice highlighted in Fig. 11.5 represents the basic functional BJT device, which essentially consists of two back-to-back connected pn junctions. This basic structure is shown separately in Fig. 11.6 with applied bias voltages, V_{BE} and V_{CB}. Notice that with the voltage source polarities shown here, the base-emitter junction is forward-biased, whereas the base-collector junction is reverse-biased. We will call this particular operating mode the *forward active mode*. The circuit symbol of the npn transistor is also shown in Fig. 11.6.

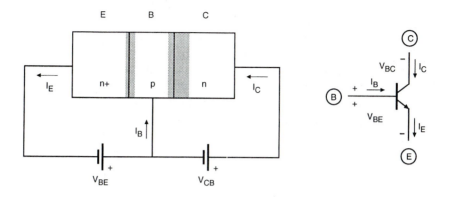

Figure 11.6. The simplified view of the bipolar junction transistor (BJT) biased in the forward active mode with V_{BE} and V_{CB}, and the corresponding circuit symbol. Note that the depletion regions at both pn junctions are also highlighted.

BJT Operation: A Qualitative View

Under the bias conditions shown in Fig. 11.6, the positive base-emitter voltage V_{BE} causes the base-emitter pn junction to become forward-biased and to conduct a current which corresponds to the emitter terminal current I_E. The emitter current consists of two components: electrons injected from the emitter into the base region, and holes injected from the base into the emitter. Since the doping concentration of the emitter region is much higher than that of the base region, the magnitude of the injected electron current component is larger than the injected hole current component.

The injected electrons which enter the very thin base region are the *minority carriers* in the p-type base. Their concentration is highest at the emitter side of the base and lowest (zero) at the collector side of the base region. The reason for the zero concentration at the collector side is that the positive collector voltage V_{CB} causes the electrons at the collector-base junction to be swept across the depletion region, and into the collector. Thus, electrons injected from the emitter into the base region will *diffuse* toward the collector, since the local electron concentration profile *decreases* in that direction. The diffusion current made up of electrons in the base is directly proportional to the minority carrier concentration difference between the emitter side and the collector side of the base. Electrons that diffuse through the thin base region and reach the collector junction will be swept into the collector, making up the collector current I_C. Therefore, we can see that the amount of the collector current depends on the amount of electrons which successfully diffuse through the base region.

Ideally, it is preferred that almost all of the electrons injected into the base at the emitter junction diffuse through the base region, reach the collector junction and are swept into the collector region, thus making the collector current I_C approximately equal to the emitter current I_E. In order to meet this requirement, the thickness W of the base region must be much smaller than the diffusion length L_D of electrons in the base, i.e., $W \ll L_D$. Also, the doping concentration of the emitter region must be much larger than that of the base region, so that the emitter terminal current is dominated by the electron injection current component. If these conditions are satisfied, the terminal behavior of the BJT will be a good approximation of a controlled current source. The emitter current and therefore the collector current flow can be controlled by modifying the bias at the base-emitter junction.

Note that some of the electrons that are diffusing through the base region will recombine with holes, i.e., with the majority carriers in the base. Also, a small number of holes will be injected from the base into the emitter through the forward-biased base-emitter junction. The injected holes constitute the base current I_B. Although the magnitude of the base current is very small in comparison to the collector current, it nevertheless plays an important role in the current-controlled operation. It can be shown that the collector current is in fact proportional to the base current, with a very large proportionality factor, also called the *current gain*. Thus, by simply modifying the amount of the small base current, one can control the collector current of a bipolar transistor operating in the forward active mode.

This brief discussion is not intended to provide a complete picture of the operation of BJTs, but rather to illustrate qualitatively the pertinent characteristics of current-controlled operation in bipolar devices. Our objective is to provide a sufficient basis for understanding the analytical models to be presented next.

BJT Current-Voltage Models

The operational characteristics of the npn bipolar transistor under various conditions will be analyzed in the following by using the well-known Ebers-Moll model. We will not attempt to derive the model equations here; they are simply presented as a complete set of equations describing the current-voltage behavior of the bipolar transistor under certain bias conditions.

The collector current I_C and the emitter current I_E are expressed as

$$I_C = \alpha_F I_F - I_R \tag{11.1}$$

$$I_E = I_F - \alpha_R I_R \tag{11.2}$$

where α_F and α_R are two dimensionless coefficients, and the two currents I_F and I_R are given by the following diode current equations.

$$I_F = I_{ES}\left(e^{\frac{qV_{BE}}{kT}} - 1\right) \tag{11.3}$$

$$I_R = I_{CS}\left(e^{\frac{qV_{BC}}{kT}} - 1\right) \tag{11.4}$$

Here, I_{ES} and I_{CS} represent the reverse saturation currents of the base-emitter and the base-collector junctions, respectively. The equivalent circuit diagram corresponding to the Ebers-Moll model equations is shown in Fig. 11.7.

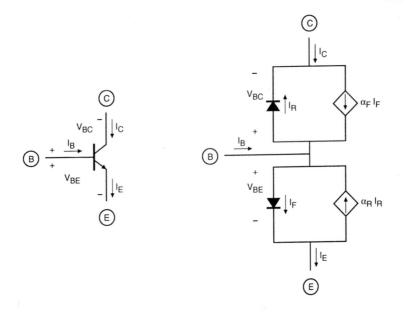

Figure 11.7.   The Ebers-Moll equivalent circuit diagram of the npn BJT.

Note that the BJT terminal current directions used in this chapter do *not* match the current directions used in SPICE BJT models, where all terminal currents are directed into the device. By combining (11.1) through (11.4), we obtain the following form of the Ebers-Moll current-voltage model equations for the bipolar transistor.

$$I_C = \alpha_F I_{ES} \left(e^{\frac{V_{BE}}{V_T}} - 1 \right) - I_{CS} \left(e^{\frac{V_{BC}}{V_T}} - 1 \right) \tag{11.5}$$

$$I_E = I_{ES} \left(e^{\frac{V_{BE}}{V_T}} - 1 \right) - \alpha_R I_{CS} \left(e^{\frac{V_{BC}}{V_T}} - 1 \right) \tag{11.6}$$

where V_T represents the thermal voltage,

$$V_T = \frac{kT}{q} \tag{11.7}$$

It is also obvious that the sum of the three terminal currents must always be equal to zero, therefore,

$$I_E = I_C + I_B \tag{11.8}$$

Now we can examine the external current-voltage behavior of the npn bipolar junction transistor under different terminal bias conditions, also called *operating modes*. The four possible modes of operation for the BJT are listed below as functions of the bias directions (polarities) applied to the two junctions.

BE junction	BC junction	Operating Mode
forward	reverse	forward active
reverse	forward	reverse active
forward	forward	saturation
reverse	reverse	cut-off

In analog circuit applications where linear current and/or voltage amplification is the primary concern, the bipolar transistor is usually operated only in the forward active mode. In digital circuit applications where the BJT is used primarily as a switching device, on the other hand, all four operating modes may be involved.

Forward Active Mode

In the forward active mode, the base-emitter junction is forward-biased, with $V_{BE} > 0$, and the base-collector junction is reverse-biased, with $V_{BC} < 0$. Under these conditions, the current across the reverse-biased base-collector junction is limited to the reverse saturation current I_{CS}, which is relatively small in magnitude (on the order of approx. 10^{-15} A). The Ebers-Moll equivalent circuit diagram for the forward active mode is shown in Fig. 11.8.

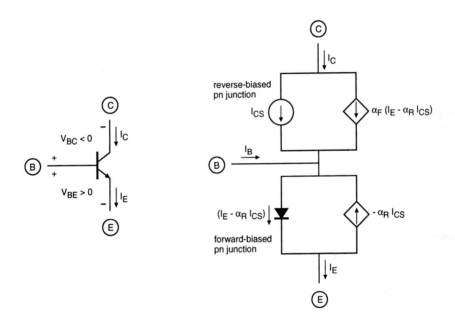

Figure 11.8. The Ebers-Moll equivalent circuit diagram of the npn BJT operating in the forward active mode.

As seen from Fig. 11.8, the two diode current components I_F and I_R used in the original model can now be expressed as

$$I_F = I_E - \alpha_R I_{CS}$$

$$I_R = -I_{CS}$$

(11.9)

The collector current is then found as follows.

$$I_C = \alpha_F \left(I_E - \alpha_R I_{CS} \right) + I_{CS} \tag{11.10}$$

$$I_C = \alpha_F I_E + I_{CO} \quad \text{where} \quad I_{CO} = I_{CS} \left(1 - \alpha_F \alpha_R \right) \tag{11.11}$$

Using (11.8), the collector current can also be written in terms of the base current.

$$I_C = \alpha_F \left(I_C + I_B \right) + I_{CO} \tag{11.12}$$

$$I_C = \frac{\alpha_F}{1 - \alpha_F} I_B + \frac{I_{CO}}{1 - \alpha_F} \tag{11.13}$$

At this point, we will define a new current coefficient, β_F. Since the magnitude of α_F is usually less than but very close to unity, the magnitude of β_F is on the order of 100 –1000 for a typical npn BJT.

$$\beta_F = \frac{\alpha_F}{1 - \alpha_F} \tag{11.14}$$

Finally, the collector current of the npn bipolar transistor operating in the forward active mode is found as

$$I_C = \beta_F I_B + \left(1 + \beta_F \right) I_{CO} \approx \beta_F I_B \tag{11.15}$$

This current equation shows that the collector current in the forward active mode is proportional to the much smaller base current, with a large proportionality (current gain) factor of β_F.

Reverse Active Mode

In the reverse active mode, the base-emitter junction is reverse-biased, with $V_{BE} < 0$, and the base-collector junction is forward-biased, with $V_{BC} > 0$. Note that this bias condition can be viewed as the complementary mode to the forward active mode examined above. Under these conditions, the current across the reverse-biased base-emitter junction is limited to the small reverse saturation current I_{ES}. The Ebers-Moll equivalent circuit diagram of the BJT operating in reverse active mode is shown in Fig. 11.9.

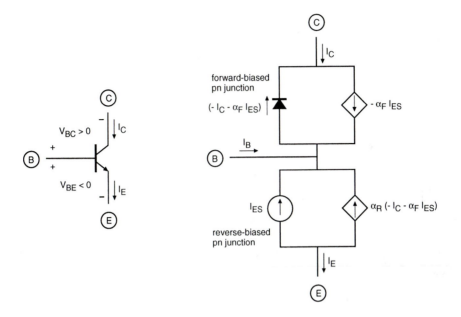

Figure 11.9. The Ebers-Moll equivalent circuit diagram of the npn BJT operating in reverse active mode.

It is seen from Fig. 11.9 that the two diode current components I_F and I_R can now be expressed as follows.

$$I_F = -I_{ES}$$

$$I_R = -I_C + \alpha_F I_{ES} \tag{11.16}$$

The emitter current is then found from (11.2):

$$I_E = -\alpha_R \left(-I_C - \alpha_F I_{ES} \right) - I_{ES} \tag{11.17}$$

$$I_E = \alpha_R I_C - I_{EO} \qquad \text{where} \qquad I_{EO} = I_{ES} \left(1 - \alpha_R \alpha_F \right) \tag{11.18}$$

Using (11.8), the emitter current I_E can also be written in terms of the base current I_B as follows.

$$I_E = \alpha_R \left(I_E - I_B \right) - I_{EO} \tag{11.19}$$

$$I_E = \frac{-\alpha_R}{1 - \alpha_R} I_B - \frac{I_{EO}}{1 - \alpha_R} \tag{11.20}$$

At this point, let us define a new current coefficient, β_R, with

$$\beta_R = \frac{\alpha_R}{1 - \alpha_R} \tag{11.21}$$

Typical values for α_R are between 0.4 and 0.9, for β_R between 0.6 and 10. The emitter current of the npn bipolar transistor operating in the reverse active mode is found as

$$I_E = -\beta_R I_B - \left(1 + \beta_R \right) I_{EO} \approx -\beta_R I_B \tag{11.22}$$

This equation shows us that the emitter current in the reverse active mode is proportional to the much smaller base current, with a proportionality factor of β_R.

It can be observed that there is a close analogy between the current equations derived for the forward active mode and those for the reverse active mode. From a practical point of view, the emitter and collector terminals of the bipolar transistor appear to be swapped, and the current flow directions for the emitter and the collector currents are reversed accordingly. Note, however, that unlike the MOS transistor, the physical structure of the BJT is not symmetrical. Since the doping concentrations in the collector and the emitter regions are quite different, the reverse active mode characteristics of the BJT will not be exact replicas of the forward active mode characteristics. In fact, the reverse active mode current gain is significantly smaller than the forward active mode current gain.

Cut-off Mode

In this case, both the base-emitter and the base-collector junctions are reverse-biased, i.e., $V_{BE} < 0$ and $V_{BC} < 0$. Consequently, the pn junction currents I_F and I_R are both reduced to the corresponding reverse saturation current values.

$$I_F = -I_{ES}$$

$$\tag{11.23}$$

$$I_R = -I_{CS}$$

The Ebers-Moll equivalent circuit diagram of the BJT operating in the cut-off mode is shown in Fig. 11.10.

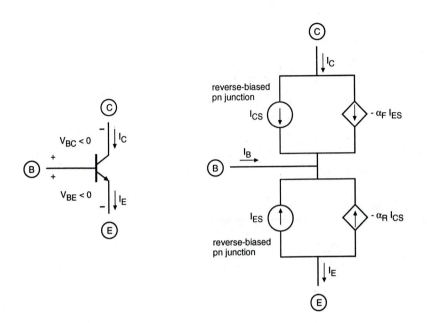

Figure 11.10.   The Ebers-Moll equivalent circuit diagram of the npn BJT operating in cut-off mode.

The emitter and the collector currents of the bipolar transistor in cut-off mode are expressed as follows.

$$I_E = -I_{ES} + \alpha_R I_{CS}$$

$$I_C = -\alpha_F I_{ES} + I_{CS}$$

(11.24)

Using the following equality

$$\alpha_F I_{ES} = \alpha_R I_{CS}$$

(11.25)

the terminal currents can be found as

$$I_E = -\left(1-\alpha_F\right)I_{ES}$$

(11.26)

$$I_C = \left(1-\alpha_R\right)I_{CS}$$

Both the emitter current and the collector current of the bipolar transistor in cut-off mode are even smaller than the junction reverse saturation currents, I_{CS} and I_{ES}. Although the non-ideal effects of leakage and thermal generation in the depletion regions are neglected in this derivation, the actual values of the emitter and collector currents of the BJT in cut-off mode are nevertheless very small. It can thus be assumed that there is an open circuit between each of the transistor terminals.

Saturation Mode

In the saturation mode, both the base-emitter junction and the base-collector junction are forward-biased, with $V_{BE} > 0$ and $V_{BC} > 0$. The Ebers-Moll equivalent circuit diagram of the BJT operating in the saturation mode is shown in Fig. 11.11.

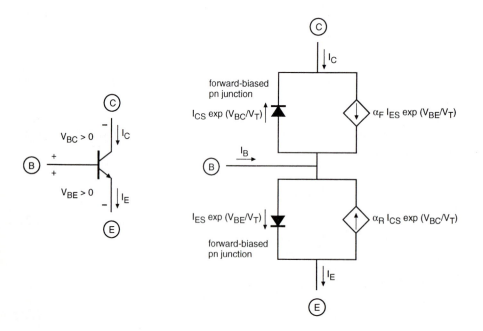

Figure 11.11.   The Ebers-Moll equivaient circuit diagram of the npn BJT operating in saturation mode.

Since both of the junctions are conducting in the forward mode, the currents I_F and I_R can be expressed as

$$I_F = I_{ES}\, e^{\frac{V_{BE,sat}}{V_T}}$$

$$I_R = I_{CS}\, e^{\frac{V_{BC,sat}}{V_T}}$$

(11.27)

In this case, the emitter and the collector currents are found as follows.

$$I_{E,sat} = I_F - \alpha_R I_R$$

$$I_{C,sat} = \alpha_F I_F - I_R$$

(11.28)

$$I_{E,sat} = I_{ES}\, e^{\frac{V_{BE,sat}}{V_T}} - \alpha_R I_{CS}\, e^{\frac{V_{BC,sat}}{V_T}}$$

$$I_{C,sat} = \alpha_F I_{ES}\, e^{\frac{V_{BE,sat}}{V_T}} - I_{CS}\, e^{\frac{V_{BC,sat}}{V_T}}$$

(11.29)

For a simple analysis of the bipolar transistor operating in saturation, we can assume the voltages across both forward-biased junctions to be constant. Since the collector-emitter voltage V_{CE} of the bipolar transistor appears as the output voltage in most digital circuit configurations, we also have to determine the exact value of V_{CE} in the saturation mode. Writing the simple voltage loop equation around the three terminals of the bipolar transistor, we obtain

$$V_{CE,sat} = V_{BE,sat} - V_{BC,sat}$$

(11.30)

Combining the current equations (11.28) and (11.29) yields

$$I_{E,sat} - \alpha_R I_{C,sat} = I_{ES}\, e^{\frac{V_{BE,sat}}{V_T}} \left(1 - \alpha_F \alpha_R\right)$$

(11.31)

which allows us to express the base-emitter voltage in the saturation mode in terms of the collector and emitter currents, as follows.

$$
\begin{aligned}
V_{BE,sat} &= V_T \ln\left(\frac{I_{E,sat} - \alpha_R I_{C,sat}}{I_{ES}(1 - \alpha_F \alpha_R)}\right) \\
&= V_T \ln\left(\frac{I_{B,sat} + I_{C,sat}(1 - \alpha_R)}{I_{EO}}\right)
\end{aligned}
\tag{11.32}
$$

Similarly, the following current equation can be used to express the base-collector voltage in the saturation mode in terms of the collector and emitter currents.

$$
I_{C,sat} - \alpha_F I_{E,sat} = -I_{CS}\, e^{\frac{V_{BC,sat}}{V_T}}(1 - \alpha_F \alpha_R)
\tag{11.33}
$$

$$
\begin{aligned}
V_{BC,sat} &= V_T \ln\left(\frac{I_{C,sat} - \alpha_F I_{E,sat}}{-I_{CS}(1 - \alpha_F \alpha_R)}\right) \\
&= V_T \ln\left(\frac{\alpha_F I_{B,sat} - I_{C,sat}(1 - \alpha_F)}{I_{CO}}\right)
\end{aligned}
\tag{11.34}
$$

Finally, the collector-emitter voltage of the BJT in saturation can be found using (11.30), (11.32), and (11.34).

$$
V_{CE,sat} = V_T \ln\left(\frac{\dfrac{1}{\alpha_R} + \left(\dfrac{I_{C,sat}}{I_{B,sat}}\right)\dfrac{1 - \alpha_R}{\alpha_R}}{1 - \left(\dfrac{I_{C,sat}}{I_{B,sat}}\right)\dfrac{1 - \alpha_F}{\alpha_F}}\right)
\tag{11.35}
$$

Here, the ratio of the saturation collector current $I_{C,sat}$ and the saturation base current $I_{B,sat}$ is called the *saturation β*, or *forced β*.

$$
\left(\frac{I_{C,sat}}{I_{B,sat}}\right) \equiv \beta_{sat} < \beta_F
\tag{11.36}
$$

Note that the value of β_{sat} is not a constant for a transistor operating in saturation, and that it changes with the particular operating point.

With $\alpha_F = 0.99$, $\alpha_R = 0.66$, and $(I_{C,sat}/I_{B,sat}) = 10$, the value of the collector-emitter voltage in saturation can be found by using (11.35) as $V_{CE,sat} = 55\,mV$. We must emphasize that the $V_{CE,sat}$ value calculated with (11.35) is the *intrinsic saturation voltage*, which does not take into account the additional voltage drops across the emitter and collector bulk resistances. The actual value of $V_{CE,sat}$ measured between the collector and emitter terminals is therefore larger, typically around 200 mV.

BJT Inverter Circuit: Static Characteristics

Now that we have introduced the basic current-voltage model for the bipolar junction transistor, we can embark on analyzing the simplest bipolar digital circuit, the resistive-load inverter. The circuit diagram of the resistive-load BJT inverter circuit is shown in Fig. 11.12. The circuit consists of one npn bipolar transistor, which has a resistor R_C connected between its collector terminal and the power supply voltage source V_{CC}, and another resistor R_B between its base terminal and the input node.

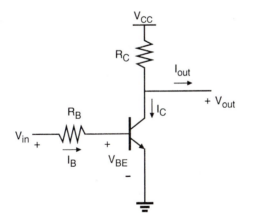

Figure 11.12.   The resistive-load BJT inverter circuit.

Assuming that the inverter is not *loaded* by another similar BJT inverter, i.e., assuming that the output current I_{out} is equal to zero, we can calculate the output voltage V_{out} as a function of the collector current I_C.

$$V_{out} = V_{CC} - R_C\,I_C \tag{11.37}$$

In the following, we will examine the voltage transfer characteristic of the BJT inverter circuit, and highlight its main features.

If the input voltage V_{in} is smaller than the base-emitter junction turn-on voltage ($V_{BE,turn-on}$), the transistor is in the cut-off mode. Hence, the collector current will be approximately equal to zero, and the logic-high output voltage V_{OH} will be equal to the power supply voltage V_{CC}, according to (11.37). Note that the logic-high output voltage level will change if there is a non-zero output current flowing through the output node, i.e., if the inverter is loaded by another stage in cascade.

Once the applied input voltage is increased above $V_{BE,turn-on}$, the transistor enters the forward active mode and starts conducting a positive collector current. The output voltage, therefore, starts to decrease after this point. Thus, for a first-order approximation, we can assume that the voltage V_{IL} is defined as the input voltage at which the transistor enters the forward active operation mode.

As the input voltage is further increased, the output voltage continues to decrease, until the transistor enters the saturation region. At this point, the collector-emitter voltage of the transistor, which is also equal to the output voltage of the inverter, assumes a relatively small and constant value, corresponding to the logic-low output voltage level of V_{OL}. The typical static voltage transfer characteristic of the BJT inverter circuit is illustrated in Fig. 11.13, showing the significant voltage points and the corresponding operating modes of the transistor.

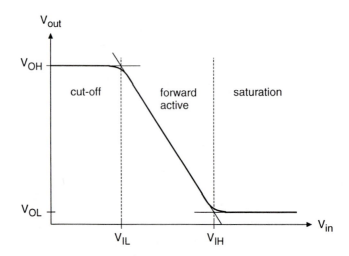

Figure 11.13. Typical input-output voltage transfer characteristic of the resistive-load BJT inverter circuit.

The values of the three important voltage points on the VTC are summarized below, based on the simple discussion outlined in the preceding paragraphs.

$$V_{OH} = V_{CC} \ (no \ output \ load)$$

$$V_{OL} = V_{CE,sat}$$

$$V_{IL} = V_{BE,turn\text{-}on}$$

(11.38)

To calculate the value of the fourth voltage point (V_{IH}) on the VTC, we have to determine the input voltage value at which the transistor enters the saturation mode. Notice that the input voltage V_{in} can be expressed as

$$V_{in} = V_{BE} + R_B \, I_B$$

(11.39)

The collector current of the transistor at the edge of saturation mode is found as follows.

$$I_C \ (edge \ of \ saturation) = \frac{V_{CC} - V_{CE,sat}}{R_C}$$

(11.40)

Note that, since at this point the transistor is also operating at the edge of the forward active mode, by definition, the collector current is still proportional to the base current.

$$I_C \ (edge \ of \ saturation) = \beta_F \, I_B \ (edge \ of \ saturation)$$

(11.41)

Using (11.39), the voltage V_{IH} can be expressed as follows.

$$V_{IH} = V_{BE,sat} + R_B \, I_B \ (edge \ of \ saturation)$$

(11.42)

$$V_{IH} = V_{BE,sat} + \frac{R_B}{R_C} \frac{1}{\beta_F} \left(V_{CC} - V_{CE,sat} \right)$$

(11.43)

The simple VTC analysis above has been carried out by assuming that the inverter under consideration has no output load; therefore, $I_{out} = 0$. If we assume more realistic conditions where one inverter is driving one or more inverters in cascade, however, the logic-high output voltage level will decrease to a lower value, since subsequent logic stages draw a non-zero output current from the inverter.

11.3. Dynamic Behavior of BJTs

To examine the dynamic behavior of bipolar junction transistors under transient terminal voltage conditions, we have to use a simple, universal model that describes the current-voltage relationships of the transistor regardless of the particular operating mode. For this reason, the charge-control model of the BJT is introduced in the following.

Charge-Control Model

The minority carrier concentrations in the emitter, base, and collector regions of an npn-type BJT operating in forward active mode are shown in Fig. 11.14. Here, $p_{eo}, n_{bo},$ and p_{co} represent the minority carrier concentrations at equilibrium, and $p_e(x)$, $n_b(x)$, and $p_c(x)$ represent the actual concentration profiles with applied bias. Since the operation of the BJT primarily depends on minority carrier transport through the base region, we will consider this region first.

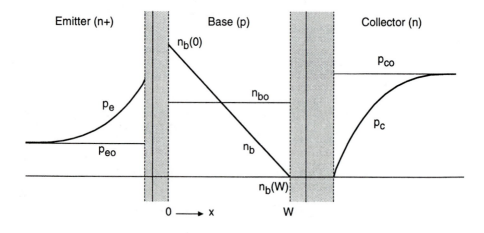

Figure 11.14.   Minority carrier concentrations in the emitter, base, and collector of an npn-type BJT operating in forward active mode.

The equilibrium electron concentration in the base is given as

$$n_{bo} = \frac{n_i^2}{N_A(base)}$$

(11.44)

With applied bias, the *excess minority carrier concentration* in the base region is described by

$$n_b' = n_b(x) - n_{bo} \qquad (11.45)$$

Note that the boundary condition for the excess electron concentration at the base-emitter junction is

$$n_b'(x=0) = n_{bo}\left(e^{\frac{V_{BE}}{V_T}} - 1\right) \qquad (11.46)$$

In forward active operation mode, the collector current I_C is related to the excess minority carrier concentration as follows.

$$I_C = q\,A\,D_b\,\frac{dn_b'(x)}{dx} \approx q\,A\,D_b\,\frac{n_b'(x=0)}{W} \qquad (11.47)$$

$$I_C = q\,A\,D_b\,\frac{n_{bo}}{W}\left(e^{\frac{V_{BE}}{V_T}} - 1\right) \qquad (11.48)$$

where A represents the cross-sectional area of the junction, and D_b is the base diffusion constant. Let us now define the excess minority carrier charge in the base region, Q_F.

$$Q_F = \frac{q\,A\,W\,n_b'(x=0)}{2} = \frac{q\,A\,W\,n_{bo}\left(e^{\frac{V_{BE}}{V_T}} - 1\right)}{2} \qquad (11.49)$$

Comparing (11.49) with (11.47), we see that the excess minority carrier charge can also be written in terms of the collector current.

$$Q_F = \tau_F\,I_C \quad \text{where} \quad \tau_F = \frac{W^2}{2\,D_b} \qquad (11.50)$$

Here, the parameter τ_F is called the *mean forward transit time* of electrons in the base. Thus, we obtain a very simple equation for the collector current, as follows.

$$I_C = \frac{Q_F}{\tau_F} \tag{11.51}$$

Similarly, the base current in forward active mode can also be expressed in terms of the excess minority carrier charge Q_F.

$$I_B = \frac{I_C}{\beta_F} = \frac{Q_F}{\tau_F \beta_F} = \frac{Q_F}{\tau_{BF}} \qquad \text{where} \qquad \tau_{BF} = \tau_F \beta_F \tag{11.52}$$

Here, τ_{BF} represents the time constant which combines the effects of base recombination and carrier injection into the emitter. Thus, Q_F may be viewed as an independent variable which controls the terminal currents of the BJT.

For transient analysis, we have to express the instantaneous current values in terms of the excess minority carrier charge. Note that any time-dependent variation of the base minority carrier charge must be attributed to the instantaneous base current, which actually supplies the carriers responsible for charge variation. For this reason, an additional derivative term is used in the base current expression. The emitter current is simply obtained from i_C and i_B, by using Kirchhoff's current law.

$$i_C(t) = \frac{Q_F(t)}{\tau_F}$$

$$i_B(t) = \frac{Q_F(t)}{\tau_{BF}} + \frac{dQ_F(t)}{dt} \tag{11.53}$$

$$i_E(t) = \frac{Q_F(t)}{\tau_F} + \frac{Q_F(t)}{\tau_{BF}} + \frac{dQ_F(t)}{dt}$$

The analysis until this point considered only the forward active mode operation. Using simple analogy, we can define the excess minority carrier charge in the base for reverse active mode as

$$Q_R = \frac{q A W n_b{}'(x=W)}{2} = \frac{q A W n_{bo} \left(e^{\frac{V_{BC}}{V_T}} - 1 \right)}{2} \tag{11.54}$$

Also, we define the reverse transit time for minority carriers in the base as follows.

$$\tau_{BR} = \beta_R \, \tau_R \quad \text{where} \quad \beta_R = \frac{\alpha_R}{1 - \alpha_R} \tag{11.55}$$

Now, we can devise a set of current equations which describe the terminal currents of the BJT operating in the reverse active mode.

$$i_{E,reverse}(t) = -\frac{Q_R(t)}{\tau_R}$$

$$i_{B,reverse}(t) = \frac{Q_R(t)}{\tau_{BR}} + \frac{dQ_R(t)}{dt} \tag{11.56}$$

$$i_{C,reverse}(t) = -\frac{Q_R(t)}{\tau_R} - \frac{Q_R(t)}{\tau_{BR}} - \frac{dQ_R(t)}{dt}$$

Figure 11.15 shows the minority carrier charge distribution in the base region for forward active and for reverse active modes. In general, the forward active mode base charge Q_F is much larger than the reverse active mode charge Q_R. Also note that the reverse-mode excess minority carrier charge reaches its maximum value at the base-collector junction, as expected.

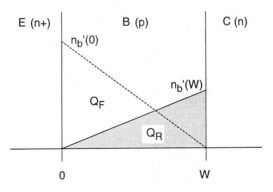

Figure 11.15. Minority carrier charge distribution in the base region for forward active and for reverse active modes.

The charge-controlled current equations derived for the forward active mode and for the reverse active mode can now be combined into one unified set of model equations that is applicable to all possible cases. Note that the time-dependent variation of the depletion region charges (Q_{jC} and Q_{jE}) at both junctions are also included in the equations below, to complete the picture.

$$i_C(t) = \underbrace{\frac{Q_F(t)}{\tau_F}}_{forward} - \underbrace{Q_R(t)\left(\frac{1}{\tau_R}+\frac{1}{\tau_{BR}}\right) - \frac{dQ_R(t)}{dt}}_{reverse} - \underbrace{\frac{dQ_{jC}(t)}{dt}}_{depletion\ charge}$$

$$i_B(t) = \underbrace{\frac{Q_F(t)}{\tau_{BF}}+\frac{dQ_F(t)}{dt}}_{forward} + \underbrace{\frac{Q_R(t)}{\tau_{BR}}+\frac{dQ_R(t)}{dt}}_{reverse} + \underbrace{\frac{dQ_{jC}(t)}{dt}+\frac{dQ_{jE}(t)}{dt}}_{depletion\ charge} \qquad (11.57)$$

$$i_E(t) = \underbrace{Q_F(t)\left(\frac{1}{\tau_F}+\frac{1}{\tau_{BF}}\right)+\frac{dQ_F(t)}{dt}}_{forward} - \underbrace{\frac{Q_R(t)}{\tau_R}}_{reverse} + \underbrace{\frac{dQ_{jE}(t)}{dt}}_{depletion\ charge}$$

If the transistor is operating in the saturation mode, the total excess minority carrier charge in the base region is the superposition of the forward and reverse mode base charges Q_F and Q_R, since both junctions are forward-biased and inject carriers into the base region. Because the voltages across the two junctions are nearly constant in saturation, the depletion charge variation terms are usually negligible. Thus, the collector current and the base current of a BJT operating in saturation mode can be expressed as

$$i_C = \frac{Q_F}{\tau_F} - Q_R\left(\frac{1}{\tau_R}+\frac{1}{\tau_{BR}}\right) - \frac{dQ_R}{dt}$$

$$\qquad (11.58)$$

$$i_B = \frac{Q_F}{\tau_{BF}} + \frac{Q_R}{\tau_{BR}} + \frac{d}{dt}(Q_F+Q_R)$$

The total base region excess minority charge Q_{base} is equal to the sum of the forward mode and the reverse mode excess minority charges. This sum can also be expressed as the sum of two charges, Q_A and Q_S, as follows (Fig. 11.16).

$$Q_{base} = Q_F + Q_R = Q_A + Q_S \tag{11.59}$$

where the new charge components are defined as

$$Q_A = \tau_F \, I_{C,edge\,of\,saturation} \tag{11.60}$$

and

$$Q_S = \tau_S \, I_{B,overdrive} \tag{11.61}$$

Note that Q_A represents the amount of base charge which is required to bring the transistor to the edge of saturation, according to (11.60). The charge component Q_S represents the *overdrive* base charge that drives the BJT into saturation, with

$$I_{B,overdrive} = I_{B,saturation} - I_{B,edge\,of\,saturation} \tag{11.62}$$

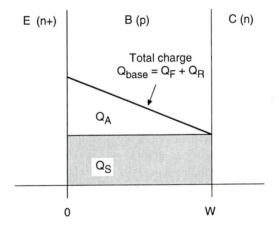

Figure 11.16.   Total excess minority carrier charge in the base region of a BJT operating in saturation.

Now, the base current in the saturation mode can be expressed using only one time constant, i.e., the saturation time constant τ_S

$$i_B(t) = \frac{I_{C,edge\,of\,saturation}}{\beta_F} + \frac{Q_S}{\tau_S} + \frac{dQ_S}{dt} \tag{11.63}$$

where τ_S is expressed by

$$\tau_S = \frac{\alpha_F\left(\tau_F + \alpha_R\,\tau_R\right)}{1 - \alpha_F\,\alpha_R} \qquad (11.64)$$

The saturation time constant τ_S, which characterizes the recombination process for the overdrive base charge, is typically much larger than the forward transit time τ_F. This causes a significant switching delay for the bipolar transistor if it is already operating in saturation mode, i.e., if there is a non-zero overdrive base charge present in the base region. The charge-controlled current equations derived in this section can now be applied to transient analysis of bipolar logic circuits.

BJT Inverter Delay Times

Consider the resistive-load BJT inverter circuit with no output load shown in Fig. 11.17, which is driven with an ideal pulse voltage waveform applied to the input terminal. The output voltage waveform is expected to follow the low-to-high and the high-to-low transitions of the input voltage, with certain time delay. Typical input and output voltage waveforms associated with the inverter are plotted in Fig. 11.18, and the relevant delays are identified on the time axis.

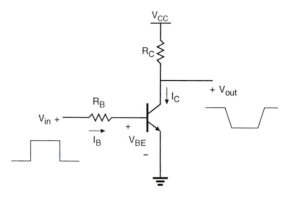

Figure 11.17.   BJT inverter with ideal pulse voltage waveform applied to the input.

When the input voltage switches from low to high, the transistor is initially in cut-off. After the switching of the input voltage, the transistor remains in cut-off mode, and the output voltage remains at its logic-high level for a time period of τ_1, during which the base-emitter and the base-collector junction capacitances are charged up by the base current, so that the BJT can subsequently operate in forward active mode.

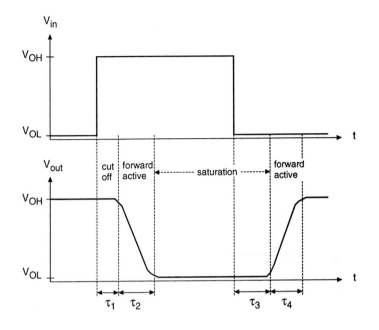

Figure 11.18. Typical input and output voltage waveforms of the BJT inverter during transient operation.

Once the transistor enters forward active operation mode, a positive collector current starts to flow through the collector resistor, and the output voltage falls accordingly. The time delay τ_2 is primarily due to junction capacitance effects, and also due to time required for minority carrier charge buildup in the base region. By the time the excess minority carrier charge in the base becomes sufficiently large, the BJT goes into saturation, and the output voltage stabilizes at the logic-low value of $V_{OL} = V_{CE,sat}$.

The BJT will remain in saturation as long as the input voltage is equal to V_{OH}. When the input voltage falls from high to low, the transistor cannot leave the saturation mode immediately. A certain saturation recovery time delay (τ_3) is required for the removal of the overdrive base charge, through the base current. Only after the saturation charge (Q_S) is removed from the base region can the transistor enter the forward active mode again. Once in forward active mode, the collector current of the BJT starts to decrease, causing the output voltage to rise from logic-low to logic-high voltage level. After a time delay of τ_4, the output voltage stabilizes again at $V_{out} = V_{OH}$.

Note that the delay times examined here are primarily due to finite carrier transit times and internal junction capacitances of the BJT. These delay times can therefore be called *intrinsic delays* of the transistor. If the BJT inverter is driving a capacitive load, on the other hand, the time required to charge and discharge the external load capacitance must also be taken into consideration. Assuming that the output load capacitance is connected to the collector node, it will appear in parallel with the collector junction capacitance in (11.57), thereby increasing all delay times except the saturation recovery delay time τ_3.

11.4. Basic BiCMOS Circuits: Static Behavior

The obvious advantages of using npn-type bipolar transistors jointly with MOS devices in various digital circuit structures were briefly outlined in Section 11.1. Having examined the operation of the bipolar transistor in detail in the preceding sections, we can now consider static and dynamic operating characteristics of the basic BiCMOS logic circuits.

Consider the simple BiCMOS inverter circuit shown in Fig. 11.19, which consists of two MOS transistors and two npn-type bipolar transistors which drive a large output capacitance C_{load}. The operation concept of this circuit can be very briefly summarized as follows.

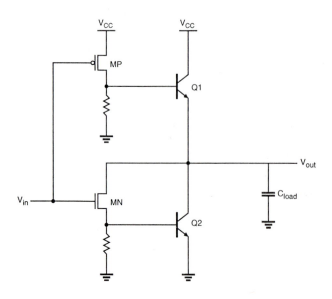

Figure 11.19.   Simple BiCMOS inverter circuit with resistive base pull-down.

The complementary pMOS and nMOS transistors MP and MN supply base currents to the bipolar transistors and thus act as "trigger" devices for the bipolar output stage. The bipolar transistor Q1 can effectively pull up the output voltage in the presence of a large output capacitance, whereas Q2 pulls down the output voltage, similar to the well-known *totem pole* configuration. Depending on the logic level of the input voltage, either MN or MP can be turned on in steady state, therefore assuring a fully complementary push-pull operation mode for the two bipolar transistors. In this very simplistic configuration, two resistors are used to remove the base charge of the bipolar transistors when they are in cut-off mode.

In general, the superiority of the BiCMOS gate lies in the high current drive capability of the bipolar output transistors, the zero static power dissipation, and the high input impedance provided by the MOSFET configuration. To reduce the turn-off times of the bipolar transistors during switching, two minimum-size nMOS transistors (MB1 and MB2) are usually added to provide the necessary base discharge path, instead of the two resistors. The resulting six-transistor inverter circuit, shown in Fig. 11.20, is the most widely used conventional BiCMOS inverter configuration.

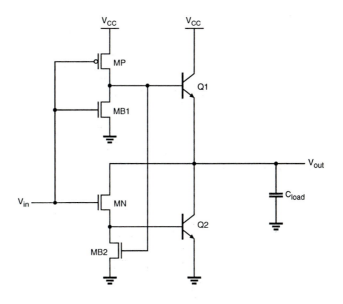

Figure 11.20.   Conventional BiCMOS inverter circuit with active base pull-down.

The typical input-output voltage transfer characteristic of a BiCMOS inverter circuit is shown in Fig. 11.21. When the input voltage is very close to zero, the nMOS transistors MN and MB1 are off, whereas the pMOS transistor MP and thus the nMOS transistor MB2 are on. Since the base voltage of Q2 is equal to zero, the bipolar output pull-down transistor is in cut-off mode. Consequently, both of the bipolar transistors Q1 and Q2 are not able to conduct any current at this point. Notice that the pMOS transistor MP is operating in the linear region, but its drain-to-source voltage is equal to zero because no base current is flowing through Q1.

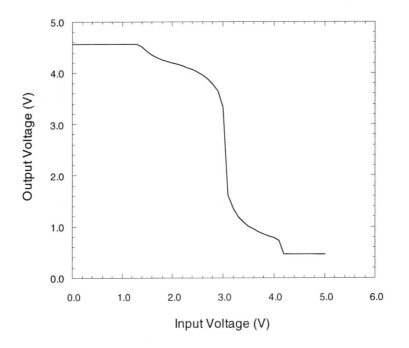

Figure 11.21.   Typical voltage transfer characteristic of the BiCMOS inverter circuit.

When the input voltage is increased beyond $(V_{T,n} + V_{BE,2})$, the nMOS transistor MN starts to conduct a non-zero base current for Q2. Thus, the bipolar transistors Q1 and Q2 enter the forward active region. The base-emitter junction voltages of both transistors rise very sharply and cause the observed *step-down* in the DC voltage transfer characteristic. The second step-down in the VTC occurs when the pMOS transistor is turned off, and both bipolar transistors are abruptly driven into cut-off mode. For higher input voltage levels, the drain-to-source voltage of MN is equal to zero and no base current can be supplied to Q2.

The logic threshold voltage of the BiCMOS inverter circuit can be adjusted by modifying the aspect ratio between the pMOS and nMOS transistors, MN and MP.

11.5. Switching Delay in BiCMOS Logic Circuits

As we already mentioned in the preceding sections, the large current-driving capability of the BiCMOS inverter is one of its most significant advantages over the conventional CMOS buffer circuits for driving large capacitive loads. Figure 11.22 shows the simulated transient output voltage waveforms of a conventional CMOS buffer circuit and a BiCMOS buffer circuit, both of which occupy the same amount of silicon area, and drive the same capacitive load of 5 pF. It can be seen that the BiCMOS buffer can pull up the output voltage in about one-fourth of the time required for the CMOS inverter to pull up the output.

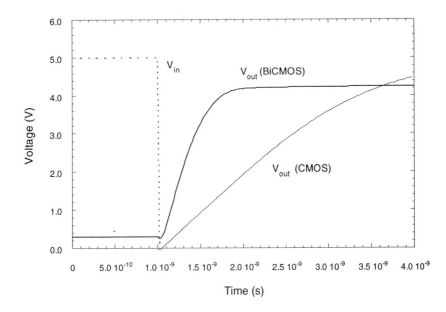

Figure 11.22. Simulated output voltage waveforms of a conventional CMOS inverter and a BiCMOS inverter, both driving a capacitive load of 5 pF. Note that both circuits also occupy approximately the same amount of silicon area.

It should be emphasized that the speed advantage of BiCMOS circuits becomes more pronounced for larger capacitive loads. In fact, for driving small load capacitances, the conventional CMOS inverter still provides a better and more viable option than the BiCMOS buffer (Fig. 11.23). The reason for this is that the parasitic junction capacitances associated with the bipolar transistors tend to cancel out any speed improvement to be gained by using a BiCMOS buffer when driving a small capacitive load. For larger capacitive loads, the influence of internal device parasitics becomes negligible, and the current-driving advantage of BiCMOS starts to dominate.

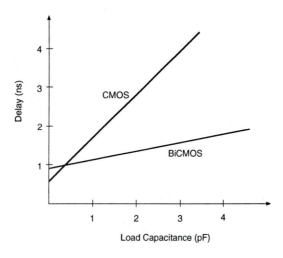

Figure 11.23.   Delay vs. load capacitance for conventional CMOS and BiCMOS buffers.

Since the typical BiCMOS logic gate does not dissipate any significant amount of static power during steady-state operation, it also enjoys a substantial advantage in terms of power consumption over the conventional bipolar logic circuit

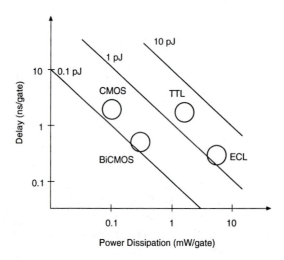

Figure 11.24.   Power-delay products of BiCMOS, CMOS, and conventional bipolar logic families, using 2-μm technology.

families, e.g. TTL or ECL. Figure 11.24 shows that the BiCMOS logic family has about the same power-delay product as the conventional CMOS, but the gate delay is much smaller. Compared with fully bipolar alternatives, the power dissipation of the BiCMOS is at least one order of magnitude smaller for the same time delay.

In the following, we will examine in detail the transient behavior of the BiCMOS inverter circuit. Consider first the output pull-up transient response, which starts with the input voltage abruptly falling from V_{OH} to V_{OL} at $t = 0$. The initial condition of the output node voltage is assumed to be $V_{out} = V_{OH}$. The inverter circuit during this switching event is depicted in Fig. 11.25, where the active (conducting) devices are highlighted.

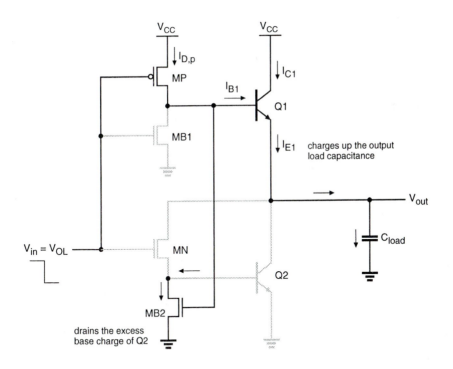

Figure 11.25.   BiCMOS inverter during transient output pull-up event. The active devices in the circuit are highlighted (darker).

As the input voltage drops, the pMOS transistor MP is turned on and starts operating in the saturation region. The nMOS transistors MN and MB1 are turned off; thus, the lower "pull-down" part of the inverter circuit can be ignored except for the corresponding parasitic capacitances of the nMOS transistors and the bipolar transistor Q2. The base pull-down transistor MB2 is turned on, which effectively drains the excess base minority carrier charge of Q2 and assures that Q2 remains in cut-off mode. At the same time, MP is supplying the base current of Q1, which starts to charge up C'_{load} with its emitter current. Here, C'_{load} is the combined output capacitance consisting of the external load and the parasitic capacitances of Q1. The pull-up circuit, which is responsible for charging up the output load capacitance during the pull-up transient event, can be represented by the equivalent circuit shown in Fig. 11.26.

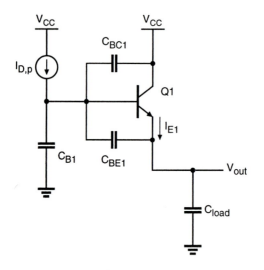

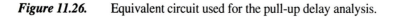

Figure 11.26.   Equivalent circuit used for the pull-up delay analysis.

It is obvious that both the pMOS transistor MP and the bipolar transistor Q1 will switch operating modes more than once during the pull-up transient event. To simplify the delay analysis, the output pull-up time period can be partitioned into three time segments depending on the operating modes of the transistors. Figure 11.27 shows the typical output voltage waveform during the pull-up event, and the three time segments $(\tau_1, \tau_2, \text{and } \tau_3)$.

The three time segments are defined according to the operating modes of the transistors, as follows:

- During τ_1, the base-emitter voltage of Q1 increases from its initial value to $V_{BE,turn-on}$. The bipolar transistor is in cut-off during this time segment, and the pMOS transistor operates in the saturation region.

- During τ_2, the bipolar transistor starts conducting a non-zero current in the forward active mode, and the pMOS transistor MP operates in saturation.

- During τ_3, the pMOS operates in the linear region, and the bipolar transistor continues to operate in the forward active mode.

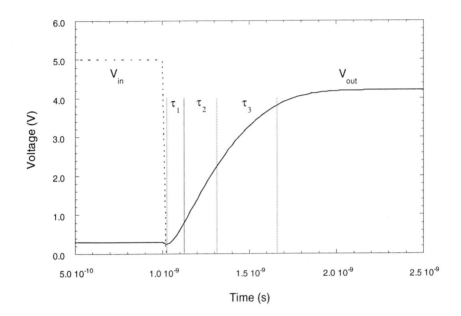

Figure 11.27. Output voltage waveform during the pull-up transient event, and the partitioning of the pull-up delay time into three segments.

The differential equation describing the base-emitter voltage variation of the bipolar transistor during τ_1 is as follows.

$$\frac{dV_{BE}}{dt} = I_{D,p} \cdot \frac{C'_{load}}{C'_{load} \cdot C_{BE1} + (C_{B1} + C_{BC1})(C'_{load} + C_{BE1})} \qquad (11.65)$$

At the beginning of the time segment τ_2, the bipolar transistor Q1 enters the forward active mode while the pMOS is operating in saturation. The combined output capacitance C'_{load} begins to charge up due to the emitter current. The differential equations describing the time dependence of the output and the base voltages are given in the following.

$$\frac{dV_{out}}{dt} = \frac{i_C + i_B}{C'_{load}} \tag{11.66}$$

$$\frac{dV_B}{dt} = \frac{I_{D,p} - i_B}{C_{B1} + C_{BC1}} \tag{11.67}$$

Note that high-level injection phenomena were found to be dominant in bipolar transistors during the time segments τ_2 and τ_3. The dependence of the current gain factor upon the collector current under high-level injection conditions is described by the Gummel-Poon model as

$$\beta_F = \frac{\beta_{FO}}{1 + \frac{I_C}{I_k}} \tag{11.68}$$

where I_k is called the *knee current*. Hence, the relationship between the forward transit time and the collector current can be described by

$$\tau_F = \tau_{FO}\left(1 + \frac{I_C}{I_k}\right) \tag{11.69}$$

The time-dependent variation of the output voltage during τ_2 can thus be found accurately by combining (11.66) and (11.67) with the modified charge-control model equations given in (11.57). Finally, the time segment τ_3 is calculated by replacing the saturation current with the linear region current expression for the pMOS transistor in (11.67).

Now consider the output pull-down transient response, which starts with the input voltage abruptly rising from V_{OL} to V_{OH} at $t = 0$. The initial condition of the output node voltage is assumed to be $V_{out} = V_{OL}$. The inverter circuit during this switching event is depicted in Fig. 11.28, where the active (conducting) devices are highlighted.

As the input voltage rises, the pMOS transistor MP is turned off and the nMOS transistors MN and MB1 are turned on. The bipolar pull-up transistor Q1 immediately ceases to conduct because its base current drops to zero, and MB1 starts to remove the excess minority carrier base charge of Q1. The nMOS transistor MN operates initially in the saturation region and supplies the base current of the bipolar pull-down transistor

Q2. Note that the base pull-down transistor MB2 will continue to conduct during the initial phase of this transition, because MB1 cannot pull down the base node of Q1 immediately after the input pulse arrives.

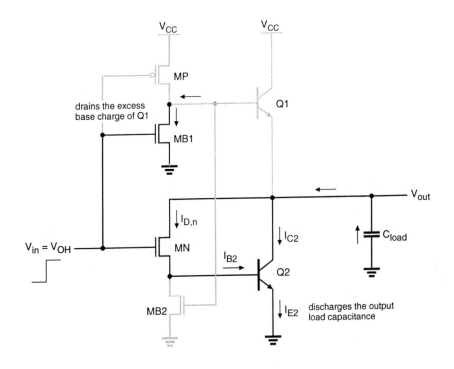

Figure 11.28. BiCMOS inverter during transient output pull-down event. The active devices in the circuit are highlighted (darker).

The pull-down circuit which is responsible for discharging the output capacitance during the pull-down transient event can be represented by the equivalent circuit shown in Fig. 11.29. The calculation of the pull-down delay time is analogous to the calculation of the pull-up delay, which is outlined above.

11.6. BiCMOS Applications

The basic BiCMOS inverter structure, which is studied extensively in Sections 11.4 and 11.5, can be used as the starting point to construct complex logic gates, simply by replacing the pMOS and nMOS base driver transistors MP and MN of the basic inverter circuit with complementary pMOS and nMOS logic blocks.

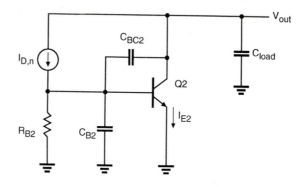
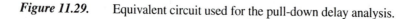

Figure 11.29.   Equivalent circuit used for the pull-down delay analysis.

Figure 11.30 shows the circuit diagram of a BiCMOS NOR2 gate. Here, the base of the bipolar pull-up transistor Q1 is being driven by two series-connected pMOS transistors. Therefore, the pull-up device can be turned on only if both of the inputs are logic-low.

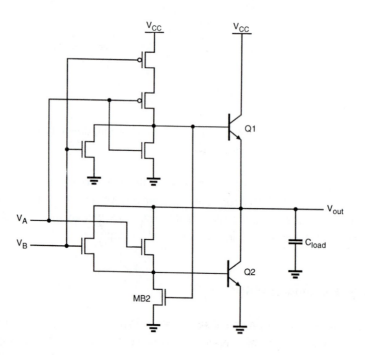

Figure 11.30.   Circuit diagram of BiCMOS NOR2 gate.

The base of the bipolar pull-down transistor Q2 is driven by two parallel-connected nMOS transistors. Therefore, the pull-down device can be turned on if either one or both of the inputs are logic-high. Also, the base charge of the pull-up device is removed by two minimum-size nMOS transistors connected in parallel between the base node and the ground. Notice that only one nMOS transistor, MB2, is being used for removing the base charge of Q2, when both inputs are logic-low.

Figure 11.31 shows the circuit diagram of a BiCMOS NAND2 gate. In this case, the base of the bipolar pull-up transistor Q1 is being driven by two parallel-connected pMOS transistors. Hence, the pull-up device is turned on when either one or both of the inputs are logic-low. The bipolar pull-down transistor Q2, on the other hand, is driven by two series-connected nMOS transistors between the output node and the base. Therefore, the pull-down device can be turned on only if both of the inputs are logic-high. For the removal of the base charges of Q1 during turn-off, two series-connected nMOS transistors are used, whereas only one nMOS transistor is utilized for removing the base charge of Q2.

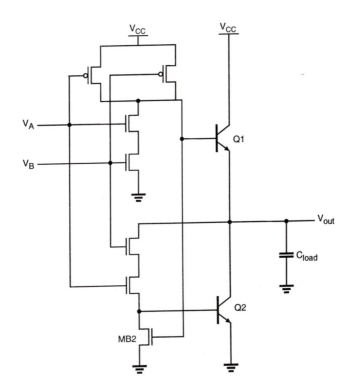

Figure 11.31.   Circuit diagram of BiCMOS NAND2 gate.

The complex logic gate concept developed in Chapter 7 for nMOS and CMOS circuits can also be implemented in BiCMOS circuits with minimal hardware overhead, as illustrated in Fig. 11.32. Note that in this example, the base pull-down of the bipolar transistor Q2 is provided by a simple resistor, instead of an active base pull-down device used in the previous examples.

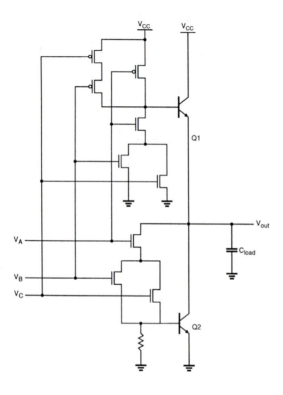

Figure 11.32.   Circuit diagram of a BiCMOS complex logic gate.

Finally, consider the following BiCMOS implementation in a high-performance bit-line read circuitry for memory arrays. It was shown in Chapter 10 that high-speed column sense-amplifier circuits are crucial for reducing the data access times in large memory arrays. Since the basic operations required from the column sense amplifier are to detect a small voltage difference between the two column (bit-line) voltages *and* to drive the data output during the read cycle, the well-known bipolar differential-pair (common emitter amplifier) circuit appears to be a good alternative.

The circuit diagram of a typical column pair in a memory array with the attached BiCMOS sense amplifier is shown in Fig. 11.33. During the data read cycle, the bit-line differential voltage appears between the respective base terminals of the bipolar differential pair. Collector current flows in the bipolar device that has the higher base current. Each bipolar device in the differential pair arrangement has its collector terminal connected to one of the local data lines.

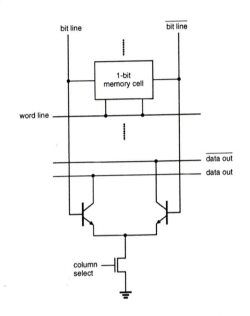

Figure 11.33.   BiCMOS column sense amplifier for high-speed memory array.

Note that although each column pair in the memory array has its own differential-pair amplifier, current can flow only in the sense amplifier of the selected column. The current source of the bipolar differential pair is provided by an nMOS device, which is controlled by the column select signal. Also, the column pull-up operation is usually provided by bipolar transistors in this BiCMOS memory array.

The application areas for BiCMOS circuit techniques have grown steadily. However, the industry has also experienced that careful process technology development is required to achieve high-performance bipolar transistors. In this section, we provided some circuit examples in order to illustrate the circuit design possibilities and the performance advantages of the BiCMOS. Interested readers are referred to the growing literature on BiCMOS applications for a more comprehensive treatment of this subject.

References

1. H.C. Lin, J.C. Ho, R.R. Iyer, and K. Kwong, "Complementary MOS-bipolar transistor structure," *IEEE Transactions on Electron Devices*, vol. ED-16, pp. 945-951, November 1969.

2. H. Momose, H. Shibata, S. Saitoh, J. Miyamoto, K. Kanzaki, and S. Kohyama, "1.0-μm n-well CMOS/bipolar technology," *IEEE Transactions on Electron Devices*, vol. ED-32, pp. 217-223, February 1985.

3. T. Ikeda, A. Watanabe, Y. Nishio, I. Masuda, N. Tamba, M. Odaka, and K. Ogiue, "High-speed BiCMOS technology with a buried twin well structure," *IEEE Transactions on Electron Devices*, vol. ED-34, pp. 1304-1309, June 1987.

4. H.J. De Los Santos and B. Hoefflinger, "Optimization and scaling of CMOS bipolar drivers for VLSI interconnects," *IEEE Transactions on Electron Devices*, vol. ED-33, pp. 1722-1729, November 1986.

5. G.P. Rosseel and R.W. Dutton, "Influence of device parameters on the switching speed of BiCMOS buffers," *IEEE Journal of Solid-State Circuits*, vol. 24, pp. 90-99, February 1989.

6. C.H. Diaz, S.M. Kang, and Y. Leblebici, "An accurate analytical delay model for BiCMOS driver circuits," *IEEE Transactions on Computer-Aided Design*, vol. 10, pp. 577-588, May 1991.

7. S.H.K. Embabi, A. Bellaouar, and M.I. Elmasry, *Digital BiCMOS Integrated Circuit Design*, Boston, MA: Kluwer Academic Publishers, 1993.

8. A.R. Alvarez, ed., *BiCMOS Technology and Applications*, second edition, Boston, MA: Kluwer Academic Publishers, 1994.

Exercise Problems

11.1   A BJT transistor has $\alpha_F = 0.99$ and $\alpha_R = 0.2$. Calculate the collector junction reverse saturation current when its emitter junction reverse saturation current is 10^{-14} A. Determine the area of the collector junction compared to the area of the emitter junction.

11.2   A circuit for generating a reference voltage V_R is shown below. Find the value of V_R by assuming that the voltage drop across each diode and base-emitter junction in Q1 is 0.7 V.

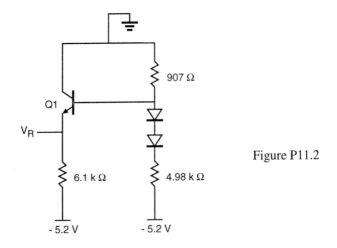

Figure P11.2

11.3   Derive the expression for τ_S in equation (11.64) in the text.

11.4   For the BJT inverter circuit in Figure 11.12 with the following parameters:

- $V_{CC} = 5$ V
- $R_B = 10$ kΩ
- $R_C = 1$ kΩ
- $\beta_F = 100$
- $V_{BE\,(on)} = 0.7$ V
- $V_{BE\,(sat)} = 0.8$ V
- $V_{CE\,(sat)} = 0.1$ V

Calculate V_{IL}, V_{IH}, and noise margins NM_L, NM_H.

11.5   Derive equation (11.65) for the equivalent circuit in Figure 11.26.

11.6   Consider a logic gate for the Boolean function $Z = \overline{ABCDE + FGH}$.

(a)   Design a BiCMOS circuit to implement the Boolean function Z.
(b)   Design a domino CMOS circuit to implement the Boolean function $\overline{Z}$.
(c)   Compare the pros and cons of the BiCMOS over the domino CMOS implementation.

11.7   For the bipolar transistor shown in Fig. P11.7, use the following parameters and neglect any junction capacitances.

- $I_{ES} = 10^{-16}$ A
- $I_{CS} = 2 \times 10^{-16}$ A
- $\alpha_F = 0.98$
- $\alpha_R = 0.49$
- $\tau_F = 0.2$ ns
- $\tau_R = 20$ ns

(a)   Use the Ebers-Moll equations to find I_B, I_E, and I_C before and (long) after the transition.

(b)   Use the simplified charge-control equations to sketch V_B as a function of time. Include values for all important voltages and time intervals on the graph.

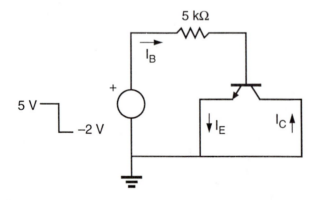

Figure P11.7

11.8   Compare the basic BiCMOS inverter and the basic BiNMOS inverter. Specifically, address voltage swing and propagation delay.

11.9   Use the following parameters for your calculations, assuming the emitter junction is an abrupt junction and the collector junction is a gradual junction:

- $V_{BE\,(on)} = 0.7$ V
- $V_{BE\,(sat)} = 0.8$ V
- $V_{CE\,(sat)} = 0.1$ V
- $\tau_F = 0.2$ ns
- $\tau_{BF} = 15$ ns
- $\tau_S = 20$ ns
- $C_{je0} = 0.5$ pF   (emitter junction capacitance at zero bias)
- $C_{jc0} = 0.25$ pF (collector junction capacitance at zero bias)
- $\phi_e = 0.9$ V
- $\phi_c = 0.7$ V

(a)   For the bipolar inverter shown in Fig. P11.9a, calculate the DC transfer characteristics (i.e., $V_{OH}, V_{OL}, V_{IH}, V_{IL}, NM_H, NM_L$). Repeat for the circuit shown in Fig. P11.9b.

(b)   For the circuit shown in Fig. P11.9a, use hand calculations to solve for t_{PLH} and t_{PHL}.

Repeat for the circuit shown in Fig. P11.9b. Describe how the pull-down resistor at the base affects propagation delay.

(c)   Compare your calculations for both circuits with SPICE results.

Note: Run SPICE to simulate both DC and transient responses.

(d)   For the circuit shown in Fig. P11.9a, let $R_C = 0$. Calculate the size of the speed-up capacitor to be placed in parallel with R_B to minimize the propagation delay (neglect charge due to junction capacitance). Repeat for the circuit shown in Fig. P11.9b.

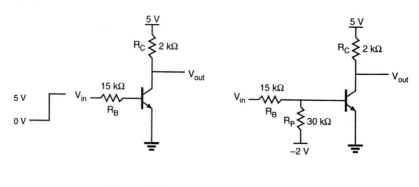

Figure P11.9a   Figure P11.9b

11.10   Calculate the t_{PLH} and t_{PHL} values for the BiCMOS gate shown in Figure 11.20. Assume that charging of the MOSFET's parasitic capacitances can be neglected. Use the following parameters for MOSFETs and BJTs. Ignore the bias dependencies of the capacitances. Explain any simplifying assumptions that you make.

- $V_{CC} = 5$ V
- $V_{BE(on)} = 0.7$ V
- $V_{T0} = 0.8$ V
- $k'_n = 200$ µA/V^2
- $k'_p = 100$ µA/V^2
- $\beta_F = 100$
- $C_{je} = 20$ fF
- $C_{jc} = 22$ fF
- $C_L = 0.5$ pF
- $(W/L)_p = 6$
- $(W/L)_n = 3$
- $r_c = 75$ Ω

CHAPTER 12

CHIP INPUT AND OUTPUT (I/O) CIRCUITS

12.1. Introduction

The input/output (I/O) circuits, clock generation, and distribution circuits are essential to VLSI chip design. The design quality of these circuits is a critical factor that determines the reliability, signal integrity, and interchip communication speed of the chip in a systems environment. If the package is considered a protection layer of the silicon chip, then the I/O frame containing input and output circuits and clock circuits can be considered a second protection layer. Any external hazards such as electrostatic discharge (ESD) and noises should be filtered out before propagating to the internal circuits for their protection. Also, some chips have to communicate with Transistor-Transistor Logic (TTL) or Emitter-Coupled Logic (ECL) bipolar chips, and in such cases, the input or output circuit must provide proper level shifting so that the transmitted signal contents can be correctly received or sent by the CMOS chip.

Most VLSI chips receive clock signals from a common clock source and then in turn must generate internal clock signals. Although the ideal location of such a clock module is the center of the chip, in most cases the clock module is placed in the I/O frame due to wire-bonding constraints. However, some recent chips which use bare flip-chip bonding onto printed circuit boards (PCBs) or multichip modules (MCMs) have their clock circuits placed in the center for on-chip distribution of clock signals to various locations with less skew. In this chapter, we discuss the design of electrostatic discharge damage-protection circuits, input circuits, on-chip clock generation and distribution, output circuits, on-chip noise due to parasitic inductance in bonding wires of output pads, super buffer circuit design, and the latch-up phenomenon in the I/O frame due to parasitic bipolar transistors in CMOS chips and its prevention method.

12.2. ESD Protection

Electrostatic discharge is one of the most prevalent causes for chip failures in both chip manufacturing and field operation. ESD can occur when the charges stored in machines or the human body are discharged to the chip on contact or by static induction. Figure 12.1 shows different models for ESD testing, namely the human body model (HBM), the machine model (MM), and the charged device model (CDM).

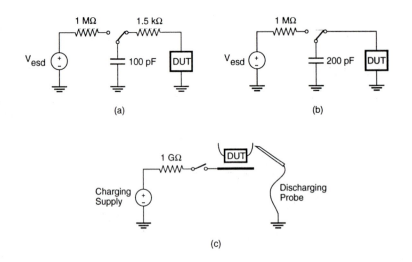

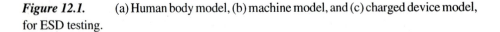

Figure 12.1.   (a) Human body model, (b) machine model, and (c) charged device model, for ESD testing.

A human walking across synthetic carpet in 80% relative humidity can potentially induce 1.5 kV of static voltage stress. In the HBM (MIL-STD-883C, Method 3015, 1988) shown in Fig. 12.1(a), a touch of a charged person's finger is simulated by discharging a 100-pF capacitor through a 1.5-kΩ resistor. It is important that some protection network be designed into the I/O circuits of the chip so that the

ESD effect can be filtered out before its propagation to the internal logic circuit. Effective protection networks can withstand as high as 8-kV HBM ESD stress.

In addition to human handling, contact with other machines can also cause ESD stress. Since body resistance is absent, the stress can be more severe with higher current levels. The schematic diagram of the machine model is shown in Fig. 12.1(b).

The third model is the charged device model shown in Fig. 12.1(c). It is intended to model the discharge of the packaged integrated circuits. The charge can be accumulated either during the chip assembly process or in the shipping tubes. The CDM ESD testers electrically charge the device under test (DUT) and then discharge it to ground, thus probing the high short-duration current pulse to DUT.

Simplified lumped-circuit element models of both HBM and MM ESD testers are shown in Fig. 12.2 along with corresponding parameter values.

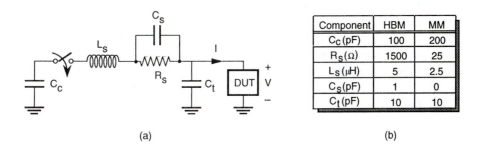

Component	HBM	MM
C_C (pF)	100	200
R_S (Ω)	1500	25
L_S (μH)	5	2.5
C_S (pF)	1	0
C_t (pF)	10	10

(a)   (b)

Figure 12.2.   (a) Simplified lumped-element model of HBM-ESD and MM-ESD testers. (b) Model parameter values.

The protection network (PN) usually consists of a *diffused* resistor-diode structure as shown in Fig. 12.3 along with its equivalent circuit model. The input resistance is normally between 1 and 3 kΩ. This resistance in conjunction with the capacitances in diffusion, diodes, and gate capacitance in input transistors integrates and clamps the voltage to a safe level. The RC time constant, however, should be small enough not to increase the circuit delay significantly.

In essence, the diodes clamp the signal level within a certain voltage range, in order to minimize the impact of ESD.

$$-0.7 \text{ V} < V_A < V_{DD} + 0.7 \text{ V} \qquad\qquad (12.1)$$

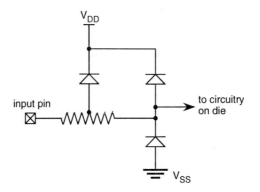

Figure 12.3.   ESD protection network example.

This practice is adopted to meet the industry standard (JEDEC Standard No. 7) which is intended to avoid user-related damage to the chips. In order not to permanently damage the diode structure, the current through the diode should be limited to less than several tens of milliamperes. Past attempts to use polysilicon series resistors failed due to dielectric breakdown under high electric fields. The use of additional thick-oxide nMOS transistors as shown in Fig. 12.4 has proven to be very effective and yielded protection in excess of 3 kV in the HBM-ESD test. In this circuit, M1 is a thick-oxide punch-through device, M2 is a thick-oxide nMOS transistor, and M3 is a thin-oxide nMOS transistor operating in saturation mode. For positive input transients, M1 and M2 have threshold values of 20 to 30 V.

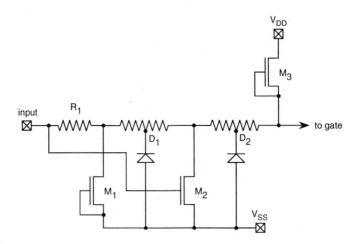

Figure 12.4.   Protection network with thick-oxide transistor.

Figure 12.5 shows typical ESD failure modes caused by ESD-induced heat dissipation in an nMOS transistor along with a scanning electron microscopy (SEM) photograph of a failed nMOS transistor. Similar protection circuits can be used for the output circuit, although large driver transistors have intrinsic protection capability through diffusion and substrate or tub structures.

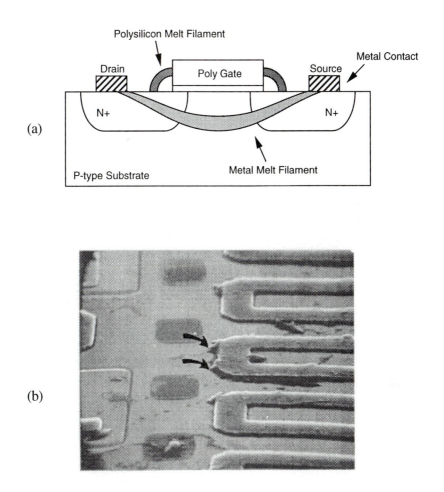

Figure 12.5.   (a) Typical ESD failure modes. (b) SEM photograph of a failed nMOS transistor (from Reference 7).

12.3. Input Circuits

A simple input circuit consisting of a transmission gate activated by an enable (E) signal and its complement is shown in Fig. 12.6.

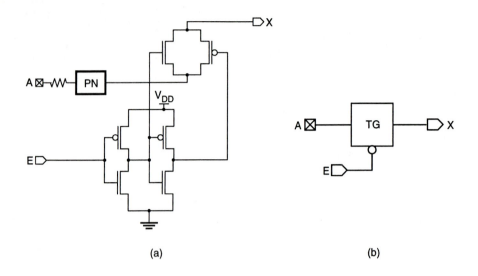

(a)   (b)

Figure 12.6.   (a) Input series transmission gate circuit and (b) its symbolic representation.

The incoming signal A is fed into the transmission gate through the protection network (PN) from the bonding pad of the chip. The enable signal is generated on-chip and controls the gating of the input signal as

$\quad\quad X = A$, when $E = 0$ and
$\quad\quad X = $ high-impedance state, otherwise.

Any unused chip input terminals should be tied to V_{DD} or V_{SS} using pull-up or pull-down resistors externally. Some input pad circuit modules have a built-in internal pull-up or pull-down resistor or active load (normally-on transistor) with a resistance of 200 kΩ to 1 MΩ.

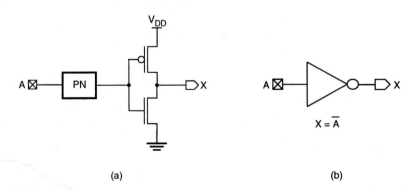

(a)   (b)

Figure 12.7.   Inverting input circuit with (a) protection network, and (b) symbolic view.

Figure 12.7 shows an inverting input circuit consisting of the protection network and a CMOS inverter. Typical values for V_{IL} and V_{IH} are $0.3 V_{DD}$ and $0.7 V_{DD}$, respectively for about 30% noise margins.

This basic input circuit can be designed to receive TTL signals for CMOS logic circuits by adjusting the ratio of the channel widths in pMOS and nMOS transistors in the inverter. Figure 12.8 shows the principle of level shifting from TTL to CMOS logic. In TTL, the worst-case output signal levels are

- $V_{OL} = 0.8$ V
- $V_{OH} = 2.0$ V

Therefore, input voltages less than or equal to 0.8 V should be interpreted low and input voltages greater than or equal to 2.0 V should be interpreted high.

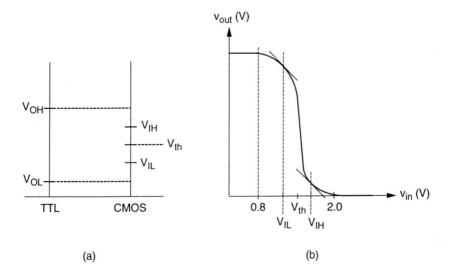

(a)   (b)

Figure 12.8.   (a) TTL to CMOS level shifting and (b) the corresponding voltage transfer characteristic curve.

After the input protection circuit, the incoming signals have to be level-shifted to a desirable level, depending on their voltage levels. For instance, if the incoming signal is from a TTL driver, then its low voltage can be as high as 0.8 V and its high output voltage can be as low as 2.0 V. Therefore a careful level shifting has to be done to translate such logic levels to corresponding MOS gate voltage levels as shown in Fig.12.8.

The level shifting between a TTL driver and a CMOS gate can be achieved by properly designing the ratio between pMOS and nMOS transistors of the receiving CMOS inverter gate. A practical method is to adjust the transistor ratio in the inverter gate such that the saturation voltage at which both transistors operate in saturation region is set at the midpoint between 0.8 V and 2.0 V. By using first-level models of MOS transistors, it can be shown that the saturation voltage of the inverter gate can be expressed by

$$V_{th} = \frac{V_{DD} + V_{Tp} + rV_{Tn}}{1+r} \tag{12.2}$$

$$r = \sqrt{\frac{\mu_n C_{ox} W_n / L_n}{\mu_p C_{ox} W_p / L_p}} \tag{12.3}$$

From these two equations, we find that

$$\frac{W_n / L_n}{W_p / L_p} = \frac{\mu_p}{\mu_n} \left[\frac{V_{DD} + V_{Tp} - V_{sat}}{V_{sat} - V_{Tn}} \right]^2 \tag{12.4}$$

For example, if $\mu_n = 3\mu_p$ and $V_{Tn} = -V_{Tp} = 1.0$ V and $V_{DD} = 5$ V, then in order to achieve

$$V_{sat} = \frac{0.8 + 2.0}{2} = 1.4 \text{ V}$$

the nMOS-to-pMOS ratio must be

$$\frac{W_n / L_n}{W_p / L_p} = \frac{1}{3} \left[\frac{5 - 1 - 1.4}{1.4 - 1} \right]^2 = \frac{169}{12}$$

From the above calculation, we determine that $r = 6.5$ and

$$V_{IL} = \frac{2V_{out} - V_{DD} + r^2 V_{Tn} + V_{Tp}}{r^2 + 1} = \frac{2V_{out} + 36.25}{43.25}$$

where V_{out} satisfies the following current equation:

$$\frac{r^2}{2}(V_{IL} - V_{Tn})^2 = (V_{DD} - V_{IL} + V_{Tp})(V_{DD} - V_{out}) - \frac{1}{2}(V_{DD} - V_{out})^2$$

or

$$21.125(V_{IL} - 1)^2 = (4 - V_{IL})(5 - V_{out}) - \frac{1}{2}(5 - V_{out})^2.$$

Combining these two equations, we obtain

$$21.125\left[\frac{2V_{out} - 7}{43.25}\right]^2 = \left[\frac{136.75 - 2V_{out}}{43.25}\right](5 - V_{out}) - \frac{1}{2}(5 - V_{out})^2$$

$$V_{out} = 4.97 \text{ V}$$

and, hence

$$V_{IL} = \frac{2 \times 4.97 + 36.25}{43.25} = 1.07 \text{ V}$$

Likewise,

$$V_{IH} = \frac{r^2(2V_{out} + V_{Tn}) + V_{DD} + V_{Tp}}{r^2 + 1} = \frac{84.5V_{out} + 47.25}{43.25}$$

where V_{out} satisfies the following current equation:

$$\frac{1}{2}(V_{DD} - V_{IH} + V_{Tp})^2 = r^2\left[(V_{IH} - V_{Tn})V_{out} - \frac{1}{2}V_{out}^2\right]$$

or

$$\frac{1}{2}(4 - V_{IH})^2 = 6.5^2\left[(V_{IH} - 1)V_{out} - \frac{1}{2}V_{out}^2\right]$$

Combining these two equations, we obtain

$$\frac{1}{2}\left(4 - \frac{84.5V_{out} + 47.25}{43.25}\right)^2 = 42.25\left[\left(\frac{84.5V_{out} + 4}{43.25}\right)V_{out} - \frac{1}{2}V_{out}^2\right]$$

Solving for V_{out} and V_{IH} yields :

$$V_{out} = 0.206 \text{ V} \quad \text{and} \qquad V_{IH} = 1.47 \text{ V}$$

This design appears to meet the design objective of a level-shifting CMOS inverter, providing logic 1 output level for TTL input voltages of up to 0.8 V (less than $V_{IL} = 1.07$ V) and logic 0 output level for TTL input voltages not less than 2.0 V. The output voltage of 0.206V at $V_{in} = 1.47$ V is much less than the n-channel threshold voltages of the next stage. However, to assure that the circuit would function properly under all circumstances, careful circuit simulation should be performed by considering the variations in process conditions, device temperature, and power supply voltage level. Note that due to process variations, some chips can have strong pMOS (PH)-weak nMOS (NL), or weak pMOS (PL)-strong nMOS (NH) combinations for which the level-shift circuit performance would be somewhat different. This variation is illustrated in Fig. 12.9.

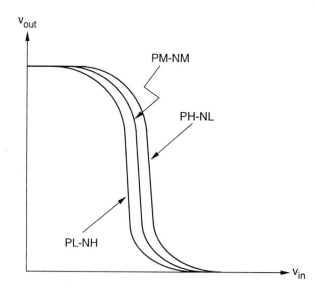

Figure 12.9.   Variation of the level-shifter VTC due to process variations. Statistical analysis and design methods to overcome such variations will be discussed in Chapter 14.

Figure 12.10 shows another non-inverting TTL level-shifting circuit. In this circuit, the level shifting is accomplished in the first stage, which is followed by the second-stage inverter.

Figure 12.11 shows an input pad circuit with a Schmitt trigger circuit and a 70-kΩ pull-down resistor. This circuit provides a negative-going logic threshold voltage of 1 V and a positive-going logic threshold voltage of 4 V, for a 5-V power supply.

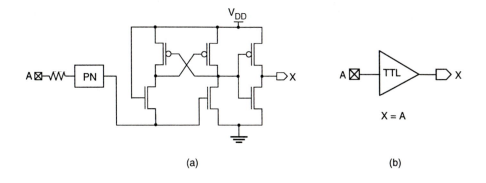

(a)

(b)

Figure 12.10.   (a) Non-inverting TTL level-shifting circuit and (b) its symbolic view.

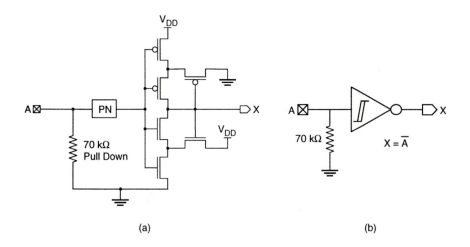

(a)

(b)

Figure 12.11.   (a) Input pad circuit with Schmitt trigger and (b) its symbolic view.

12.4. Output Circuits and *L(di/dt)* Noise

The output circuits of VLSI chips are designed to be tristable as shown in Fig. 12.12. The circuit implementation (b) requires more transistors (12 transistors) than the circuit implementation (c), in which only four transistors are required if polarity is ignored. In terms of silicon area, however, the implementation in (b) may require less than the circuit in (c) since the last-stage transistors have to be sized large to provide sufficient current sinking and sourcing capability and also to reduce delay times. Unfortunately, such a requirement demands a high rate of change in the current di/dt and can cause significant on-chip noise problems due to the $L(di/dt)$ drop across the bonding wire connecting the output pad to the package.

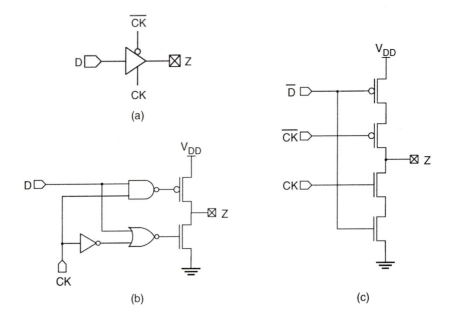

Figure 12.12.   (a) Symbolic view of a tristable output circuit and (b)–(c) two different circuit implementations.

For illustration, consider a case in which the capacitor load is initially charged to $V_{DD} = 5\ V$ and the clock signal is set to turn on the nMOS transistor to sink the current into ground. Figure 12.13 shows the current waveform during the switching period. The solid line represents a realistic current waveform whereas the dotted triangular line represents a simple approximation of the current waveform. By approximation as used in Reference 1

$$I_{max}\frac{t_s}{2} = C_{load}V_{DD} \tag{12.5}$$

Also,

$$\left[\frac{di}{dt}\right]_{max} \geq \frac{I_{max}}{t_s/2} = \frac{2I_{max}}{t_s} \tag{12.6}$$

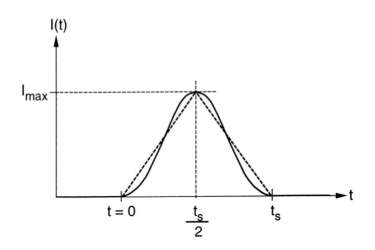

Figure 12.13.   Typical output circuit current waveform during switching.

Therefore, the following inequality holds.

$$\left[\frac{di}{dt}\right]_{max} \geq \frac{4C_{load}V_{DD}}{t_s^2} \tag{12.7}$$

For example, if $C_{load} = 100$ pF and $t_s = 5$ ns, then

$$\left[\frac{di}{dt}\right]_{max} \geq \frac{4 \times 100 \times 10^{-12} \times 5}{\left(5 \times 10^{-9}\right)^2} = 80 \ \frac{mA}{ns}$$

and for a bonding wire with $L = 2$ nH, the $L(di/dt)$ drop can be as high as

$$L\left[\frac{di}{dt}\right]_{max} \geq 160 \ mV$$

It should be noted that this voltage drop would be quadrupled if t_s were reduced by a factor of two. This shows a serious trade-off problem between the delay time and the noise. It has been observed in a 1.2-μm CMOS process chip that a current surge can be as high as 1100 mA/ns at power and ground terminals.

In high-end microprocessor chips with 32 bits or higher number of data bus lines, the noise problem can be significantly escalated if all output drivers are driven simultaneously. In such cases, it is desirable to stagger the switching times with built-in delays in the clock distribution network, which amounts to reducing the noise at the expense of chip speed.

An interesting circuit technique for reducing di/dt is shown in Fig. 12.14. This circuit requires an additional strobe signal and hence, complicates the timing design, but reduces the magnitude of di/dt significantly.

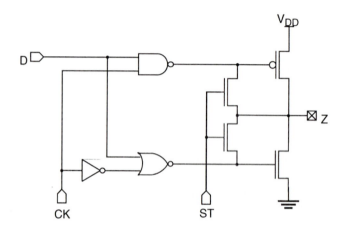

Figure 12.14.   Circuit structure for reducing (di/dt) noise.

The role of two nMOS transistors controlled by the strobe signal (ST) is to precharge the gate potentials of the last-stage driver transistors at an approximate midpoint between the initial and final potentials of the load capacitor. For instance, if $r = 1$ for the pMOS and nMOS driver pair, then when ST is high, the gate voltages can be precharged to $V_{DD}/2$ before CK goes to high.

Another technique for resolving the output driver problem is to adopt a basic driver circuit that sends out only changes in the data pattern, as shown in Fig. 12.15. With a delay element, the circuit produces pulses at nodes B and C only when the polarity of the input signal changes. As a result the driver transmits only differential signals rather than full digital waveforms. As shown in Fig. 12.15(b), the reference output voltage level is maintained at $V_{DD}/2$ during the quiescent periods, which are equivalent to tristate periods. The output driver uses a phase splitter to generate differential pairs. The corresponding receiver circuit has to sense, latch, and level-shift the differential data. The circuit shown in Fig. 12.16 performs these functions.

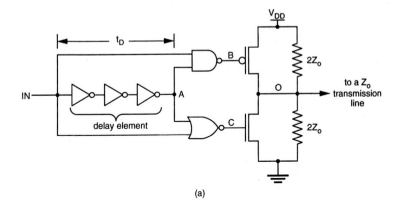

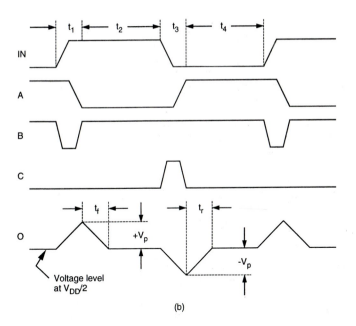

Figure 12.15.   (a) Basic driver circuit that transmits only differential signals. (b) Timing diagram showing the voltage waveforms associated with the driver circuit.

A pair of input and output circuits can be combined into a single bidirectional I/O pad circuit as shown in Fig. 12.17. The layout of a sample bidirectional I/O pad is presented in Fig. 12.18, showing the bonding pad, protection diodes, diffusion resistor, and input and output circuits.

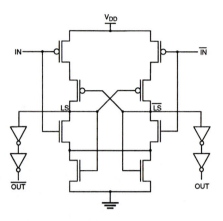

Figure 12.16.   Receiver circuit designed to sense, latch, and level-shift differential data.

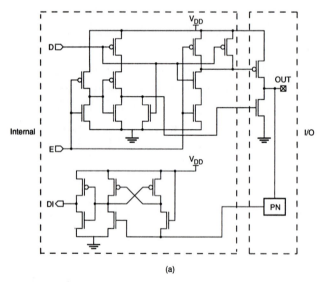

(a)

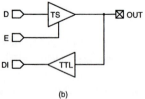

(b)

Figure 12.17.   Bidirectional buffer circuit with TTL input capability.

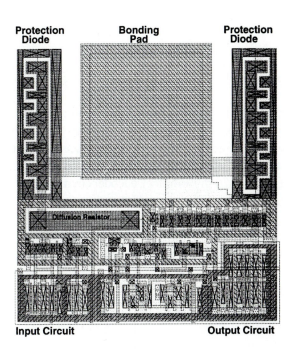

Figure 12.18.   Layout of a bidirectional I/O pad circuit (courtesy of MOSIS).

12.5. On-Chip Clock Generation and Distribution

Clock signals are the heartbeats of digital systems. Hence, the stability of clock signals is highly important. Ideally, clock signals should have minimum rise/fall times, specified duty cycles, and zero skew. In reality, clock signals have nonzero skews and noticeable rise/fall times; duty cycles can also vary. In fact, as much as 10% of a machine cycle time is expended to allow realistic clock skews in large computer systems. The problem is no less serious in VLSI chip design. A simple technique for on-chip generation of a primary clock signal would be to use a ring oscillator as shown in Fig. 12.19. Such a clock circuit has been used in low-end microprocessor chips.

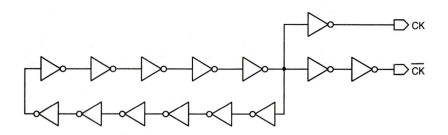

Figure 12.19.   Simple on-chip clock generation circuit using a ring oscillator.

However, the generated clock signal can be quite process-dependent and unstable. As a result, separate clock chips which use crystal oscillators have been used for high-performance VLSI chip families. Usually a VLSI chip receives one or more primary clock signals from an external clock chip and, in turn, generates necessary derivatives for its internal use. Figure 12.20 shows a clock decoder circuit that takes in the primary clock signals and generates four phase signals.

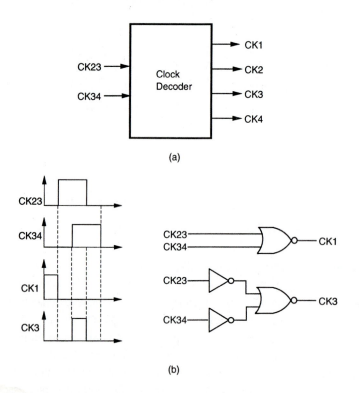

Figure 12.20.   Clock decoder circuit: (a) symbolic representation and (b) sample wave-forms and gate-level implementation.

Since clock signals are required almost uniformly over the chip area, it is desirable that all clock signals are distributed with a uniform delay. An ideal distribution network would be the H-tree structure shown in Fig. 12.21. In such a structure, the distances from the center to all branch points are the same and hence, the signal delays would be the same. However, this structure is difficult to implement in practice due to routing constraints and different fanout requirements. A more practical approach for clock-signal distribution is to route main clock signals to macroblocks and use local clock decoders to carefully balance the delays under different loading conditions.

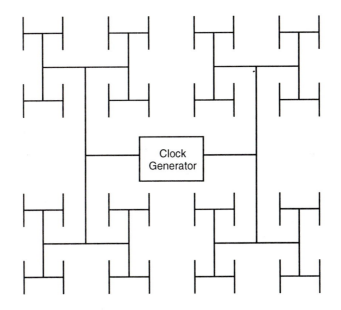

Figure 12.21.   General layout of an H-tree clock distribution network.

12.6. Super Buffer Design

The term *super buffer* has been used to describe a chain of inverters designed to drive a large capacitive load with minimal signal propagation delay time. To reduce delay time, it is necessary for the buffer circuit to provide quickly a large amount of pull-up or pull-down current to charge or discharge the load capacitor. One seemingly obvious method would be to use large pMOS and nMOS transistors in the inverter driving the load capacitor. However, such a large buffer has a large input capacitance, which in turn creates a large load for the previous stage. Then an alert designer would suggest increasing the transistor sizes in the previous stage. If so, then what about the sizing of the transistors in the stage prior to the previous stage? Thus the effect of the large load can be propagated to many gates preceding the last-stage driver, and indeed such fine tuning of transistors is practiced in custom design. An alternative method of handling a large capacitive load is to use a super buffer between a logic gate facing the large load and the load itself as shown in Fig. 12.22.

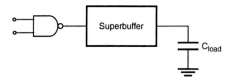

Figure 12.22.   Using a super buffer circuit to drive a large capacitive load.

Now a major objective of super buffer design becomes:

Given the load capacitance faced by a logic gate, design a scaled chain of N inverters such that the delay time between the logic gate and the load capacitance node is minimized.

To solve this problem, let us first introduce an equivalent inverter for the logic gate (NAND2 in this case). For simplicity, it is assumed that the pull-up and pull-down delays of the first-stage inverter driving an identical inverter are the same, say τ_0. The next design task is to determine the following:

- the number of stages, N
- the optimal scale factor, α

To determine these quantities, the following observations can be made under uniform α-scaling of inverters from one stage to the next in the super buffer shown in Fig. 12.23.

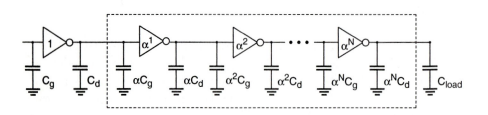

Figure 12.23.   Scaled super buffer circuit consisting of N inverter stages.

For the super buffer, the following observations can be made:

- C_g denotes the input capacitance of the first stage inverter
- C_d denotes the drain capacitance of the first stage inverter
- the inverters in the chain are scaled up by a factor of α per stage

- $C_{load} = \alpha^{N+1} C_g$ $\qquad\qquad\qquad\qquad\qquad\qquad\qquad\qquad$ (12.8)

- all inverters have identical delay of $\tau_0 \left(C_d + \alpha C_g \right) / \left(C_d + C_g \right)$ $\qquad$ (12.9)

where τ_0 represents the per-gate delay in the ring oscillator circuit with load capacitance $(C_d + C_g)$. Thus the total delay time from the input terminal to the load capacitance node becomes

$$\tau_{total} = (N+1)\tau_0 \left(\frac{C_d + \alpha C_g}{C_d + C_g} \right) \qquad\qquad (12.10)$$

There are two unknowns in this equation. To solve for these unknowns, consider the relationship between α and N in (12.8), i.e.,

$$(N+1) = \frac{\ln\left(\dfrac{C_{load}}{C_g} \right)}{\ln \alpha} \qquad\qquad (12.11)$$

Combining (12.10) and (12.11), the following delay relationship can be derived.

$$\tau_{total} = \frac{\ln\left(\dfrac{C_{load}}{C_g} \right)}{\ln \alpha} \tau_0 \left(\frac{C_d + \alpha C_g}{C_d + C_g} \right) \qquad\qquad (12.12)$$

To minimize the delay, we set the derivative of (12.12) with respect to α equal to zero and solve for α.

$$\frac{\partial \tau_{total}}{\partial \alpha} = \tau_0 \ln\left(\frac{C_{load}}{C_g} \right) \left[-\frac{\dfrac{1}{\alpha}}{(\ln \alpha)^2} \left(\frac{C_d + \alpha C_g}{C_d + C_g} \right) + \frac{1}{\ln \alpha} \left(\frac{C_g}{C_d + C_g} \right) \right] = 0 \qquad (12.13)$$

Solving for α in (12.13) we obtain the following condition for the optimal scale factor.

$$\alpha(\ln\,\alpha-1)=\frac{C_d}{C_g} \qquad\qquad (12.14)$$

A special case of the above equation occurs when the drain capacitance is neglected, i.e., $C_d=0$. In that case, the optimal scale factor becomes the natural number $e=2.718$. However, in reality the drain parasitics cannot be ignored and hence, (12.13) should be considered instead.

Example 12.1

For $C_d=5$ fF, $C_g=10$ fF, determine α and N for $C_{load}=50$ pF.

From (12.14),

$$\alpha\,(\ln\,\alpha-1)=0.5$$

hence, $\alpha=3.18$ and $N=\ln\,(50000/5)/\ln\,3.18-1\doteq6.96$.

Thus a chain of seven inverters with a scale factor of 3.18 has to be cascaded for the inverter to drive the capacitive load of 50 pF in this problem.

It should be noted that when the chip area minimization is also considered along with the delay minimization, rather than using a uniform scaling factor, each transistor size can be treated as a variable and solved. For instance, for the above problem, a total of ten transistor sizes can be tuned to minimize both the delay and the chip area. Since this topic is beyond the scope of this book, the detailed discussion will be omitted here.

12.7. Latch-Up and Its Prevention

Latch-up is defined as the generation of a low-impedance path in CMOS chips between the power supply rail and the ground rail due to interaction of parasitic pnp and npn bipolar transistors. These BJTs form a silicon-controlled rectifier (SCR) with positive feedback and virtually short circuit the power rail to ground, thus causing excessive current flows and even permanent device damage. It is perhaps one of the most negative attributes of CMOS technology. Although the use of an epitaxial layer and other process improvements have lessened the severity of latch-up problems, the reliability concerns persist on latch-up, especially in I/O circuits, since its packing density also increases with decreasing feature sizes and spacings. The latch-up susceptibility is inversely proportional to the product of the substrate doping level and the square of the spacing. In other words, if the spacing is reduced by half and the substrate doping is increased by two times, then the latch-up susceptibility would increase by a factor of two. Latch-up can also occur in internal circuits, although its frequency would be much lower than that occurring in I/O circuits. Figure 12.24 shows a cross-sectional view of a CMOS inverter circuit with identification of parasitic npn and pnp bipolar transistors.

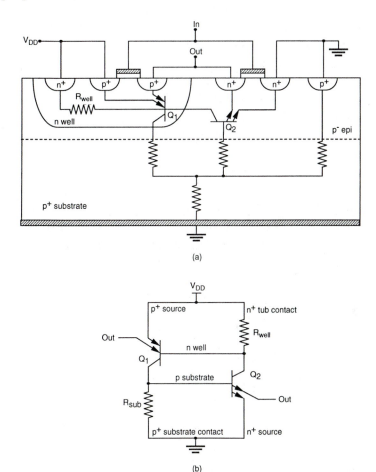

Figure 12.24.   (a) Cross-sectional view of a CMOS inverter with parasitic bipolar transistors. (b) Circuit model for SCR formed of parasitic BJTs.

In the equivalent circuit, Q1 is a vertical double-emitter pnp transistor whose base is formed by the n-well with its base-to-collector current gain (β_1) as high as several hundreds. Q2 is a lateral double-emitter npn transistor with its base formed by the p-type substrate. The base-to-collector current gain (β_2) of this lateral transistor may range from a few tenths to tens. R_{well} represents the parasitic resistance in the n-well structure with its value ranging from 1 kΩ to 20 kΩ. The substrate resistance R_{sub} strongly depends on the substrate structure, whether it is a simple p$^-$ or p$^-$ epitaxial layer grown on top of the p$^+$ substrate which acts as a ground plane. In the former case R_{sub} can be as high as several hundred ohms, whereas in the latter case the resistance can be as low as a few ohms.

To examine the latch-up event, first assume that the parasitic resistances R_{well} and R_{sub} are sufficiently large so that they can be neglected (open circuit). Unless the SCR is triggered by an external disturbance, the collector currents of both transistors consist of the reverse leakage currents of the collector-base junctions and therefore, their current gains are very low. If the collector current of one of the transistors is temporarily increased by an external disturbance, however, the resulting feedback loop causes this current perturbation to be multiplied by ($\beta_1 \cdot \beta_2$). This event is called the *triggering* of the SCR. Once triggered, each transistor drives the other transistor with positive feedback, eventually creating and sustaining a low-impedance path between the power and the ground rails, resulting in latch-up. It can be seen that if the condition

$$\beta_1 \cdot \beta_2 \geq 1$$

is satisfied, both transistors will continue to conduct a high (saturation) current, even after the triggering perturbation is no longer available. This latch-up condition can also be written in terms of the collector-emitter current gains as follows.

$$\frac{\alpha_1}{1-\alpha_1} \cdot \frac{\alpha_2}{1-\alpha_2} \geq 1 \quad \Rightarrow \quad \alpha_1 + \alpha_2 \geq 1 \tag{12.15}$$

Figure 12.25 shows the *I-V* characteristics of a typical SCR. At the onset of latch-up, the voltage drop across the SCR becomes

$$V_H = V_{BE1,sat} + V_{CE2,sat} = V_{BE2,sat} + V_{CE1,sat}$$

where V_H is called the *holding voltage*. The low impedance state is sustained as long as the current through SCR is greater than I_H, the *holding current* value which is determined by the device structure. Also note that the slope of the *I-V* curve is determined by the total parasitic resistance in the current path, R_T.

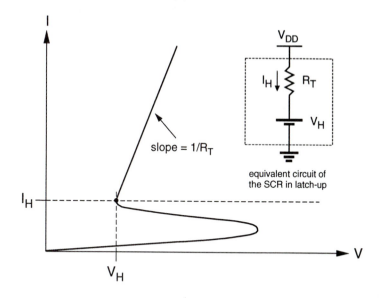

Figure 12.25.   Current-voltage characteristics of a typical SCR.

Some of the causes for latch-up are:

- Slewing of V_{DD} during initial start-up can cause enough displacement currents due to the well junction capacitance in the substrate and the well. If the slew rate is large enough, latch-up can be induced. But, the SCR can have a *dynamic recovery* before latch-up when the slew rate is not very high.

- Large currents in the parasitic SCR in CMOS chips can occur when the input or output signal swings either far beyond the V_{DD} level or far below the V_{SS} (ground) level, thus injecting a triggering current. Such disturbance can happen due to impedance mismatches in transmission lines of high-speed circuits.

- ESD stress can also cause latch-up by the injection of minority carriers from the clamping device in the protection circuit into either the substrate or the well.

- Sudden transients in power or ground buses due to simultaneous switching of many drivers may turn on a BJT in SCR.

- Leakage currents in well junctions can cause large enough lateral currents.

- Radiation due to X-rays, cosmic rays, or alpha particles may generate enough electron-hole pairs in both the substrate and well regions and thus trigger the SCR.

Below, we derive an expression for the SCR holding current I_H in terms of current gains in parasitic transistors, Q1 and Q2. For simple illustration, the circuit in Fig. 12.24(b) is redrawn in Fig. 12.26, with important circuit parameters. It can be observed that

$$I = I_{E1} + I_{RW} \tag{12.16}$$

$$I = I_{E2} + I_{RS} \tag{12.17}$$

From the collector-emitter current gain (α) relations of Q1 and Q2,

$$I_{C1} = \alpha_1 I_{E1} = \alpha_1^0 I \tag{12.18}$$

$$I_{C2} = \alpha_2 I_{E2} = \alpha_2^0 I \tag{12.19}$$

where α_1^0 and α_2^0 denote the equivalent collector-to-emitter current gains absorbing the effects of parasitic resistances into transistors. Thus, when α_1^0 and α_2^0 are used, the resistors in Fig. 12.26 are effectively open-circuited. The SCR current, I, can be exressed by

$$I = I_{C1} + I_{C2} + (I_{CBO1} + I_{CBO2}) \tag{12.20}$$

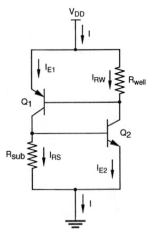

Figure 12.26. Equivalent circuit model for the SCR.

where I_{CBO1} and I_{CBO2} represent the collector-base junction leakage currents, which can be combined into a single term I_{CBO}. Combining (12.16) through (12.20), we obtain the following relationship.

$$I = \frac{I_{CBO} - (I_{RS}\alpha_1 + I_{RW}\alpha_2)}{1 - (\alpha_1 + \alpha_2)} \qquad (12.21)$$

The holding current, I_H, is defined as the current with zero I_{CBO}, i.e.,

$$I_H = \frac{I_{RS}\alpha_1 + I_{RW}\alpha_2}{\alpha_1 + \alpha_2 - 1} \qquad (12.22)$$

Now it is clear that when the sum of the collector-to-emitter gains of Q1 and Q2 is close to unity, the value of the holding current will be large. Thus it is important to keep the gains of parasitic BJTs low. The parasitic resistances R_{sub} and R_{well} also play an important role in latch-up since their currents actually *reduce* the base currents of the parasitic transistors and thus, *weaken* the feedback loop which leads to latch-up. Therefore, reducing these resistances may prevent latch-up. Consider the SCR current at the onset of latch-up

$$I \geq I_H = (V_{DD} - V_H)/R_T \qquad (12.23)$$

where both transistors are at the saturation boundary and hence, the holding voltage is $V_H = 2V_{BE}$, with $V_{BE1} = V_{BE2} = V_{BE}$. Here, the SCR is modeled by a DC voltage source of magnitude V_H in series with a resistor R_T. Combining the SCR current expression (12.22) with (12.23), and using $I_{RW} = V_{BE}/R_{well}$ and $I_{RS} = V_{BE}/R_{sub}$ in (12.22), we obtain

$$\alpha_1 + \alpha_2 \geq 1 + \left(\frac{\dfrac{R_T}{R_{well}}\alpha_1 + \dfrac{R_T}{R_{sub}}\alpha_2}{\left(\dfrac{V_{DD}}{V_{BE}} - 2\right)} \right) \qquad (12.24)$$

as the condition for the occurrence of latch-up in the presence of parasitic resistances. Compare this equation with the simple latch-up condition given in (12.15). The extra term on the right-hand side of (12.24) dictates that the sum of the two current gains must be *larger* than unity by that amount, in order to satisfy the latch-up condition and to trigger the SCR. Therefore, to avoid latch-up, this extra term should be made as large as possible, i.e., the resistances R_{sub} and R_{well} should be reduced as much as possible.

The following simulation example (Fig. 12.28) illustrates latch-up in the CMOS inverter structure shown in Fig. 12.27, which is triggered by a pulse at the output node of the circuit.

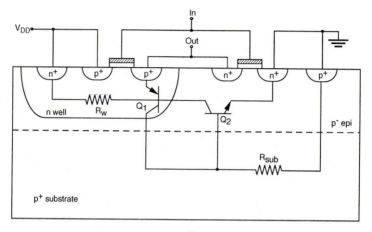

(a)

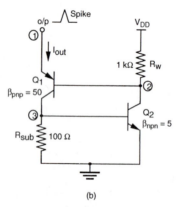

(b)

Figure 12.27.   CMOS inverter circuit used in the latch-up simulation example.

Guidelines for Avoiding Latch-Up

- Reduce the gains of BJTs by lowering the minority carrier lifetime through gold doping of the substrate (but without causing excessive leakage currents) or reducing the minority carrier injection efficiency of BJT emitters by using Schottky source/drain contacts.

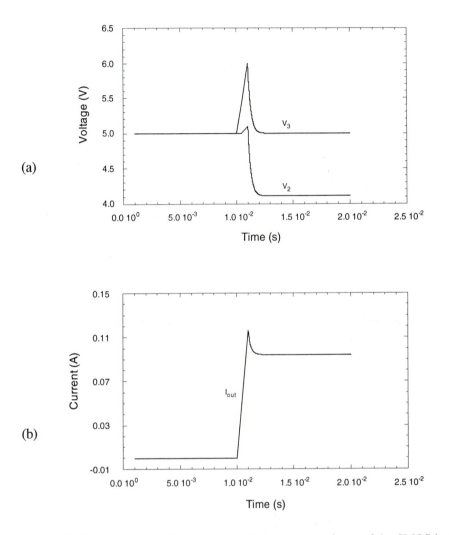

Figure 12.28. Simulated (a) voltage and (b) current waveforms of the CMOS inverter circuit during latch-up triggered through the output node.

- Use p^+ guardband rings connected to ground around nMOS transistors and n^+ guard rings connected to V_{DD} around pMOS transistors to reduce R_w and R_{sub} and to capture injected minority carriers before they reach the base of the parasitic BJTs.

- Place substrate and well contacts as close as possible to the source connections of MOS transistors to reduce the values of R_w and R_{sub}.

- Use minimum area p-wells (in case of twin-tub technology or n-type substrate) so that the p-well photocurrent can be minimized during transient pulses.

- Source diffusion regions of pMOS transistors should be placed so that they lie along equipotential lines when currents flow between V_{DD} and p-wells. In some n-well I/O circuits, wells are eliminated by using only nMOS transistors.

- Avoid the forward biasing of source/drain junctions so as not to inject high currents; the use of a lightly doped epitaxial layer on top of a heavily doped substrate has the effect of shunting lateral currents from the vertical transistor through the low-resistance substrate.

- Lay out n- and p-channel transistors such that all nMOS transistors are placed close to V_{SS} and pMOS transistors are placed close to V_{DD} rails. Also maintain sufficient spacings between pMOS and nMOS transistors.

An I/O cell layout which is designed using some of the latch-up guidelines listed here is shown in Fig. 12.29, where the same types of transistors are placed side-by-side.

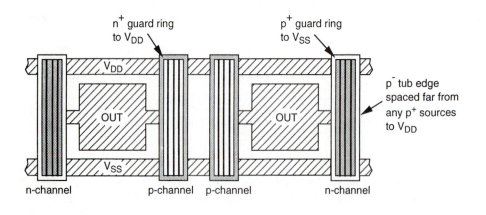

Figure 12.29.   I/O cell layout with latch-up guidelines.

For prevention of latch-up, chip manufacturers typically specify limiting operating conditions. As an example, Mitel's octal interface device, MD74SC540AC, requires 1.9 V over V_{DD} and 200 mA, and 1.0 V below V_{SS} and 90 mA to trigger output latch-up.

References

1.   M. Shoji, *CMOS Digital Circuit Technology*, Englewood Cliffs, NJ: Prentice-Hall, 1988.

2.   Harris Semiconductor, *SC3000 1.5-Micron CMOS Standard Cells*, 1989.

3.   N. Hedenstierna and K. O. Jeppson, "Comments on the optimum CMOS tapered buffer problem," *IEEE Journal of Solid-State Circuits*, vol. 29, no. 2, pp. 155-158, February 1994.

4.   S. S. Sapatnekar and S. M. Kang, *Design Automation for Timing-Driven Layout Synthesis*, Norwell, MA: Kluwer Academic Publishers, 1993

5.   Digital/Analog Communications Handbook, Issue 9, *Mitel Semiconductor*, 1993.

6.   W. D. Greason, *Electrostatic Damage in Electronics: Devices and Systems*, Somerset, England: Research Studies Press, Ltd.,1987.

7.   C. H. Diaz, S. M. Kang, and C. Duvvury, *Modeling of Electrical Overstress in Integrated Circuits*, Norwell, MA: Kluwer Academic Publishers, 1994.

8.   L. A. Glasser and D. W. Dobberpuhl, *The Design and Analysis of VLSI Circuits*, Reading, MA: Addison-Wesley Publishing Co., 1985.

9.   N. H. E. Weste and K. Eshraghian, *Principles of CMOS VLSI Design*, second edition, Reading, MA: Addison-Wesley Publishing Co., 1993.

10.   M. Shoji, *Theory of CMOS Digital Integrated Circuits and Circuit Failures*, Princeton, NJ: Princeton University Press, 1992.

Exercise Problems

12.1   For low-power design, multiple power supply voltages may be used on a chip by using on-chip voltage converters. A chip may take in a 5-V power supply and then in turn generate and use 3.3-V power rails besides 5-V power rails. Design a level shifter which can interface 3.3-V logic with a 5-V logic circuit. Use $|V_{T0}| = 1.0$ V, $\mu_n/\mu_p = 3$ in your calculation.

12.2   The distribution of clock signals without clock skew is usually desirable in order to lessen the design complexity. However, in some cases, clock skews can be utilized to resolve very tight timing budget problems. Find an example wherein clock skews can be utilized.

12.3   Design a clock decoder circuit which generates four clock phases from two primary clock signals.

12.4   Since the fanout count of a typical clock signal is very high, it is important to size the interconnection wire dimensions properly. The parasitic interconnection resistance and capacitance are discussed in Chapter 6 using formulas (6.34), (6.36), and (6.37). The parasitic resistance in metallic wire is assumed to be 0.03 Ω/square.

   (a)   For $t = 0.4\,\mu m$, $h = 1\,\mu m$, l (length) $= 1000\,\mu m$, and w (width) $= 2\,\mu m$, calculate the interconnection delay for a fanout capacitance load of 5 pF by using the Elmore delay formula. For consideration of distributed parasitic effect, the total length can be divided into 10 segments of 100-μm length.

   (b)   Verify the answer in part (a) using SPICE simulation.

12.5   The bonding pads in I/O circuits are implemented in the topmost metal layer with a dimension of 75 μm x 75 μm. If the separation of the topmost metal layer with SiO_2 from the common substrate layer (ground plane) is 1 μm,

   (a)   What is the parasitic capacitance of the bonding pad?

   (b)   What is the total parasitic capacitance of the bonding pad node if it is connected to a CMOS inverter gate ($W_p = 10\,\mu m$, $W_n = 5\,\mu m$, $L_M = 1\,\mu m$) and also to the output of a tristatable buffer ($W_p = 1000\,\mu m$, $W_n = 500\,\mu m$, $L_M = 1\,\mu m$).

   The other dimension of the drain regions is 3 μm and the parasitic capacitance in the drain is $C_{j0} = 0.3\ fF/\mu m^2$, $C_{jsw} = 0.5\ fF/\mu m$.

12.6   Verify the correctness of the TTL to CMOS level shifter by SPICE simulation.

12.7   The transistors in the final stage of the chip output buffer are usually chosen to be very large in order to provide sufficient current-driving capability. Discuss the layout strategy for implementing such huge transistors in the bonding pad area.

12.8   The switching noise in the power and ground rails of the chip output driver circuit can be very large to the extent that it may upset the logic level of nearby internal circuits due to coupled noise. Discuss whether this problem can be avoided with the use of separate power and ground rails for I/O circuits (noisy) and internal circuits (quiet).

12.9   The bonding wire inductance is 2 nH, load capacitance is 100 pF, and the 50% switching delay time is 5 ns.

    (a)   Estimate the maximum $L(di/dt)$ noise.

    (b)   Explain how this noise would change at a lower operating temperature and higher power supply voltage.

    (c)   Calculate the total noise voltage peaks when 32 such output pads are switching simultaneously and when 32 output pads are switching with 3.2 ns skew from the first bit to the last bit.

    (d)   Verify your results using proper models and SPICE simulation.

12.10   Discuss how the sensitivity of the level-shifting I/O circuit to process variations, especially the variation in the channel length, can be reduced by the mask design of particular (W/L) values. Would you choose the minimum L allowed?

12.11   Discuss the pros and cons of having pull-up and pull-down resistors connected to I/O pads in view of impedance matching in high-speed circuits.

CHAPTER 13

VLSI DESIGN METHODOLOGIES

The rate of progress in VLSI development has been phenomenal. Practically hundreds of team members are involved in the development of VLSI technology, computer-aided design (CAD) tools, chip design, fabrication, packaging and testing, and reliability qualification. In this chapter, we will examine the overall flow of the design activities, various design styles, quality of design, and the CAD technology.

13.1. Design Complexity vs. Design Cycle Time

The number of recent applications of integrated circuits in high-performance computing, communications, and consumer electronics has been rising at a fast pace. The level of integration as measured by the number of logic gates in a monolithic chip has also been steadily increasing for almost three decades, mainly due to the rapid progress in processing technology and interconnect technology. The monolithic integration provides:

- less area and thus compactness
- less power consumption
- less testing
- higher reliability, mainly due to improved on-chip interconnects
- higher speed, due to significantly reduced interconnection length
- significant cost savings

Figure 13.1 shows the level of integration vs. time for memory chips and logic chips. It can be observed that in terms of transistor count, logic chips contain significantly fewer transistors in any given year, mainly due to large consumption of chip area for complex interconnects. Memory circuits are highly regular and thus more cells can be integrated with much less area for interconnects.

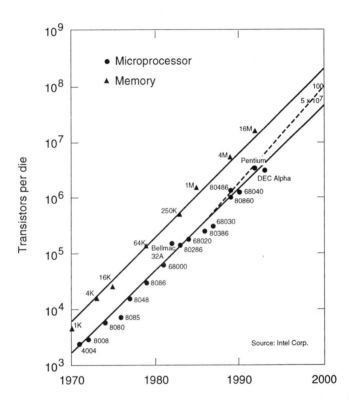

Figure 13.1.   Level of integration vs. time, for memory chips and logic chips.

Generally speaking, logic chips such as microprocessor chips and digital signal processing (DSP) chips contain not only large arrays of memory (SRAM) cells, but also many different functional units. As a result, their design complexity is considered much higher than that of memory chips, although advanced memory chips contain some sophisticated logic functions. The design complexity of logic chips increases almost exponentially with the number of transistors to be integrated. This is translated into an increase in the design cycle time, which is the time period from the start of chip development until the mask-tape delivery time. However, in order to make the best use of the current technology, the chip development time has to be short enough to allow the maturing of chip manufacturing and timely delivery to customers. As a result, the level of actual logic integration tends to fall short of the integration level achievable with the current processing technology. As will be discussed in Section 13.6, more sophisticated computer-aided design (CAD) tools and methodologies have to be developed and applied in order to manage the rapidly increasing design complexity.

13.2. Design Flow

The design process, at various levels, is usually evolutionary in nature. It starts with a given set of requirements. Initial design is developed and tested against the requirements. When requirements are not met, the design has to be improved. If such improvement is either not possible or too costly, then the revision of requirements and its impact analysis must be considered. The Y-chart (first introduced by D. Gajski) shown in Fig. 13.2 illustrates a design flow for most logic chips, using design activities on three different axes (domains) which resemble the letter Y.

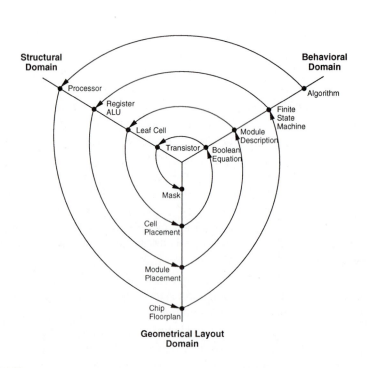

Figure 13.2. Typical VLSI design flow in three domains (Y-chart representation).

The Y-chart consists of three major domains, namely

- behavioral domain,
- structural domain,
- geometrical layout domain.

The design flow starts from the *algorithm* that describes the behavior of the target chip. The corresponding *architecture of the processor* is first defined. It is mapped onto the chip surface by *floorplanning*. The next design stage in the behavioral domain defines *finite state machines* (FSMs) which are structurally implemented with *functional modules* such as registers and arithmetic logic units (ALUs). These modules are then geometrically placed onto the chip surface using CAD tools for automatic *module placement* followed by routing, with a goal of minimizing the interconnect area and signal delays. The third stage starts with a behavioral *module description*. Individual modules are then implemented with *leaf cells*. At this stage the chip is described in terms of logic gates (leaf cells), which can be placed and interconnected by using a *cell placement & routing* program. The last stage involves a detailed *Boolean description* of leaf cells followed by a transistor-level implementation of leaf cells and *mask generation*. In standard-cell–based design, leaf cells are already pre-designed and stored in a library for logic design use.

Although the design process has been described in linear fashion for simplicity, in reality there are many iterations back and forth, especially between any two neighboring steps, and occasionally even remotely separated pairs. Although top-down design flow provides excellent design process control, in reality, there is no truly unidirectional top-down design flow. Both top-down and bottom-up approaches have to be combined. For instance, if a chip architect defined an architecture without close estimation of the corresponding chip area, then it is very likely that the resulting chip layout would exceed the area limit of the available technology. In such a case, in order to fit the architecture into the allowable chip area, some functions may have to be removed and the design process must be repeated. Such changes may require significant modification of the original requirements. Thus, it is very important to feed forward low-level information to higher levels (bottom up) as early as possible.

13.3. Design Styles

Several design styles can be considered for chip implementation of specified algorithms or logic functions. Each design style has its own merits and shortcomings, and thus a proper choice has to be made by designers in order to provide the functionality at low cost.

Field Programmable Gate Array (FPGA)

The fabricated chip containing thousands of logic gates or even more, with programmable interconnects, is available to users for their custom hardware programming to realize desired functionality. This design style provides a means for fast prototyping and also for cost-effective chip design, especially for low-volume applications. A typical field programmable gate array (FPGA) chip consists of I/O buffers, an array of configurable logic blocks (CLBs), and programmable interconnect structures. The programming of the interconnects is implemented by programming of RAM cells whose output terminals are connected to the gates of MOS pass transistors. A general architecture of FPGA from Xilinx is shown in Fig. 13.3.

A simple CLB (model XC2000 from Xilinx) is shown in Fig. 13.4. It consists of four signal input terminals (A, B, C, D), a clock signal terminal, user-programmable multiplexors, an SR-latch, and a look-up table (LUT). The LUT is a digital memory that stores the truth table of the Boolean function. Thus, it can generate any function of up to four variables or any two functions of three variables. The control terminals of multiplexors are not shown explicitly in Fig. 13.4.

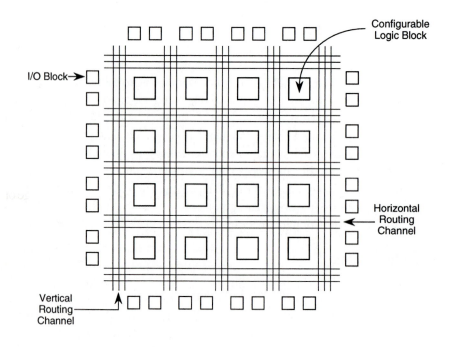

Figure 13.3.   General architecture of Xilinx FPGAs.

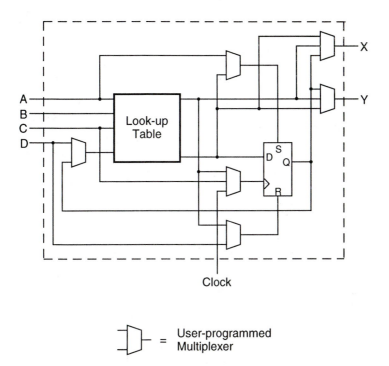

Clock

= User-programmed Multiplexer

Figure 13.4.   XC2000 CLB of the Xilinx FPGA.

The CLB is configured such that many different logic functions can be realized by programming its array. More sophisticated CLBs have been introduced recently. The design flow of a typical FPGA chip is shown in Fig. 13.5.

The chip design is inputted either by using the logic schematic capture or, alternatively, by using Boolean expressions or finite state machine (FSM) language or by mixing these two approaches. Such design information is translated and merged into a single netlist format. This netlist is then technology-mapped (or partitioned) into circuit or logic cells. At this stage, the chip design is completely described in terms of available logic cells. Next, the place-and-route step assigns individual logic cells to FPGA chip sites and determines the routing patterns among the cells in accordance with the netlist. After routing is completed, the chip performance of the design can be simulated and verified before downloading the design for programming of the FPGA chip. The programming of the chip will remain valid as long as the chip is powered or until new programming is done. In most cases, the full utilization of FPGA chip area is rarely achieved and many cell sites may remain unused.

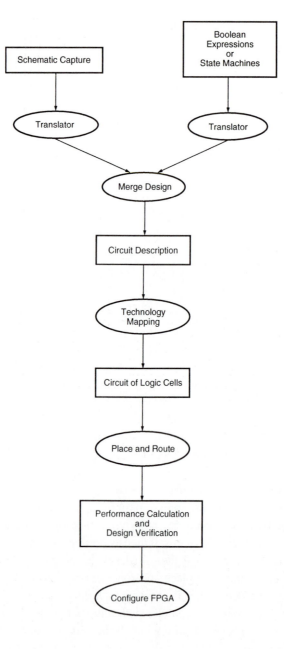

Figure 13.5.   Design flow of a typical FPGA chip.

Gate Array

In view of the fast prototyping capability, the gate array (GA) comes after the FPGA. While the design implementation of the FPGA chip is done with user programming, that of the gate array is done with metal mask design and processing. Since the patterning of metallic interconnects is done at the end of the chip fabrication, the turnaround time can still be short, a few days to a few weeks. Figure 13.6 shows a corner of a gate array chip which contains bonding pads on its left and bottom edges, diodes for I/O protection, nMOS transistors and pMOS transistors for chip output driver circuits in the neighboring areas of bonding pads, arrays of nMOS transistors and pMOS transistors, underpass wire segments, and power and ground buses along with contact windows.

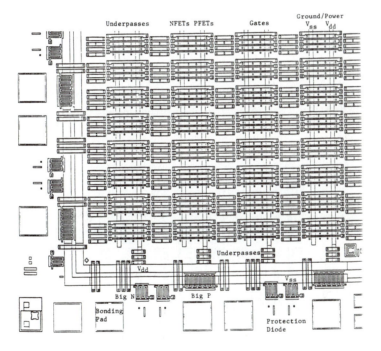

Figure 13.6.   A corner of a typical gate array chip (Copyright © 1987 Prentice Hall, Inc.).

Figure 13.7 shows a magnified portion of the internal array with metal mask design (metal lines highlighted in dark) to realize a double buffer consisting of two parallel inverters. Typical gate array platforms allow dedicated areas, called channels, for intercell routing as shown in Fig. 13.6 between rows or columns of MOS transistors. However, with the use of multiple interconnect layers, the routing can be achieved over the active cell areas; thus, the channels can be removed as in Sea-of-Gates (SOG) chips. The platform of a SOG chip is shown in Fig. 13.8.

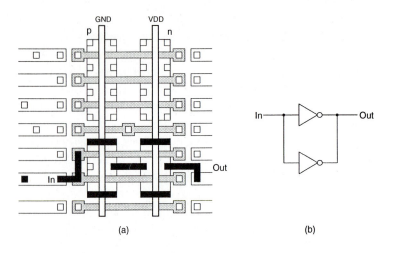

Figure 13.7.   (a) A metal mask design for a double buffer on gate array. Note that the power and ground buses running vertically are on the same metal layer as the interconnects highlighted in dark. (b) Gate-level diagram of the double buffer.

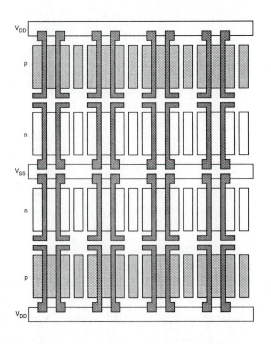

Figure 13.8.   The platform of a Sea-of-Gates (SOG) chip.

In general, the GA chip utilization factor, as measured by the used chip area divided by the total chip area, is higher than that of the FPGA and so is the chip speed, since more customized design can be achieved with metal mask designs. The current gate array chips can implement as many as hundreds of thousands of logic gates.

Standard-Cells–Based Design

The standard-cells–based design is one of the most prevalent full-custom design styles that require development of a full-custom mask set. The standard cell is also called the polycell. In this design style, all of the commonly used logic cells are developed, characterized, and stored in a standard cell library. A typical library may contain a few hundred cells including inverters, NAND gates, NOR gates, complex AOI, OAI gates, D-latches, and flip-flops. Each gate type can have multiple implementations to provide adequate driving capability for different fanouts. For instance, the inverter gate can have standard size, double size, and quadruple size so that the chip designer can choose the proper size to achieve high circuit speed and layout density. The characterization of each cell is done for several different categories. It consists of

- delay time vs. load capacitance
- circuit simulation model
- timing simulation model
- fault simulation model
- cell data for place-and-route
- mask data

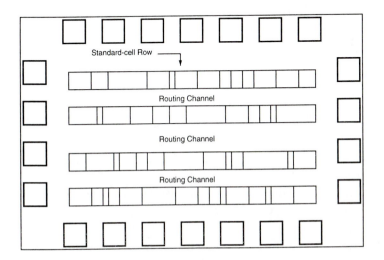

Figure 13.9.   A floorplan of standard-cells–based design.

Figure 13. 9 shows a floorplan for standard-cell–based design. Inside the I/O frame, which is reserved for I/O cells, the chip area contains rows or columns of standard cells. Between cell rows are channels for dedicated intercell routing. As in the case of Sea-of-Gates, with over-the-cell routing, the channel areas can be reduced or even removed provided that the cell rows offer sufficient routing space. The physical design and layout of logic cells ensure that when cells are placed into rows, their heights are matched and neighboring cells can be abutted side-by-side, which provides natural connections for power and ground lines in each row. The signal delay, noise margins, and power consumption of each cell should also be optimized with proper sizing of transistors using circuit simulation. Figure 13.10 shows a layout of a standard cell OAI23 along with important layout parameters.

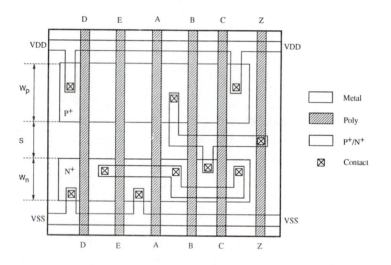

Figure 13.10.   A simplified standard-cell layout example.

After chip logic design is done using standard cells in the library, the most challenging task is to place individual cells into rows and interconnect them in a way that meets stringent design goals for circuit speed, chip area, and power consumption. Many advanced CAD tools for place-and-route have been developed and used to achieve such goals. Also from the chip layout, circuit models which include intercon-nect parasitics can be extracted and used for timing simulation and analysis to identify timing critical paths. For timing critical paths, proper gate sizing is often practiced to meet the timing requirements. In many VLSI chips, such as microprocessors and digital signal processing chips, standard-cells based design is used to implement complex control logic modules. Some full custom chips can also be implemented exclusively with standard cells.

Full Custom Design

Although the standard-cells–based design is often called full-custom design, in a strict sense, it is somewhat less than fully custom since the cells are predesigned for general use and the same cells are utilized in many different chip designs. In a *fuller* custom design, the entire mask design is done anew without use of any library. However, the development cost of such a design style is becoming prohibitively high. Thus, the concept of *design reuse* is becoming popular in order to reduce design cycle time and development cost. The most rigorous full-custom design can be the design of a memory cell, be it static or dynamic. Since the same layout design is replicated, there would not be any alternative to high-density memory chip design. For logic chip design, a good compromise can be achieved by using the symbolic layout method in which the symbol can take the form of stick or character. In the following, we will consider a particular symbolic layout method, called *gate matrix,* which was first developed in the late 1970s for AT&T's 32-bit CMOS microprocessor chip design.

Gate Matrix Layout Style

Given a logic or circuit diagram, the most common practice would be to lay out transistors for logic gates and then try to interconnect them. In such an approach, interconnection wires and transistors are laid out separately, which turns out to be costly in view of the chip area. In fact, as Alex Lopez tried to do the same for the layout of the 32-bit ALU unit, it was not possible to implement the ALU in a specified chip area. After trying various alternatives, he realized that a more compact layout can be achieved by using a matrix structure of polysilicon columns running vertically to form both transistor gates and their interconnects and horizontal rows of diffusion runners under polysilicon columns to form transistors, using horizontal metal runners for interconnects. To ease the use of such a layout method, Lopez and Law introduced (P) and (N) symbols for pMOS and nMOS transistors, (*) symbol for contacts and vias, (|) symbol for vertical column lines, and (—) symbol for horizontal metal runners, etc. Figure 13.11 shows a representative symbolic layout and the corresponding mask data.

The mask data can be generated automatically by running the symbolic layout file through a conversion program that generates the layout according to the particular technology's layout design rules. Once the symbolic layout file is documented, then the technology update of the layout is easily done by rerunning the conversion program with new layout design rules. This design style has been successfully used for design of both data-path and control logic blocks and for I/O circuits.

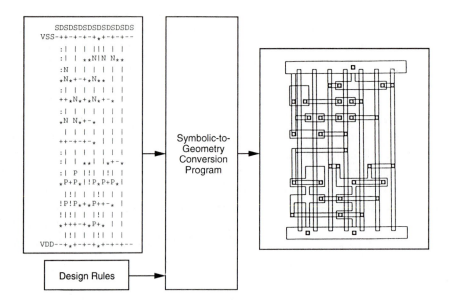

Figure 13.11.   Gate matrix symbolic layout style.

13.4. Design Quality

It is desirable to measure the quality of design in order to improve the chip design. Although no universally accepted metric exists to measure the design quality, the following criteria are considered to be important:

- testability
- yield and manufacturability
- reliability
- technology updateability

Testability

Developed chips are eventually inserted into printed circuit boards or multichip modules for system applications. The correct functionality of the system hinges upon the correct functionality of the chips used. Therefore, fabricated chips should be fully testable to ensure that all the chips passing the specified chip test can be inserted into the system, either in packaged or in bare die form, without causing failures. Such a goal requires

• generation of good test vectors
• availability of reliable test fixture at speed
• design of testable chip

In fact, some chip projects have had to be abandoned after chip fabrication due to inadequate testability of the design. As the complexity of the chips increases with the increasing level of monolithic integration, additional circuitry has to be included to ensure that the fabricated chips can be fully tested. This translates into the increased chip area and some speed penalty, but such a trade-off will become unavoidable in VLSI design. The design-for-testability issues will be discussed in Chapter 15.

Yield and Manufacturability

If we assume that the test procedure is flawless, the chip yield can be calculated by dividing the number of good tested chips by the total number of tested chips. However, this calculation may not correctly reflect the quality of the design or the processing. The most strict definition of the yield can be the number of good tested chips divided by the total number of chip sites available at the start of wafer processing. However, since some wafers may be scrapped in the process line due to mishandling or for other reasons, such a metric may not reflect the design quality. Also, poor design of the wafer array for chips may cause some chips to fail routinely due to uncontrollable process variations and handling problems. On the other hand, poor chip design can cause processing problems and, therefore, dropouts during the processing. In such a case, the first yield metric will overestimate the design quality. The chip yield can be further divided into the following subcategories:

• functional yield
• parametric yield

The functional yield is obtained by testing the functionality of the chip at a speed usually lower than the required chip speed. The functional test weeds out the problems of shorts, opens, and leakage current, and can detect logic and circuit design faults.

The parametric test is usually performed at the required speed for the chips that passed the functional test. All the delay testing is performed at this stage. Poor design that failed to consider uncontrollable process variations that cause significant variations in chip performance may cause poor parametric yield, thus, significant manufacturing problems. In order to achieve high chip yield, chip designers should consider manufacturability of the chip by considering realistic fluctuations in device parameters that cause performance fluctuations.

Reliability

The reliability of the chip depends on the design and process conditions. The major causes for chip reliability problems can be categorized into the following:

- electrostatic discharge (ESD) and electrical overstress (EOS)
- electromigration
- latch-up in CMOS I/O and internal circuits
- hot-carrier induced aging
- oxide breakdown
- single event upset
- power and ground bouncing
- on-chip noise and crosstalk

Usually the wafer lots with poor yields also cause reliability problems. For example, when the processing of a particular wafer is poorly controlled, thus causing aluminum overetch, many chips on the wafer may suffer from open-circuited metallic interconnects. Some chips with severely overetched but not fully open-circuited interconnects may pass the test. But, under current stress, such interconnects can be open-circuited due to electromigration problems, causing chip and system failures in the field. Any good manufacturing practice should weed out such potential failures during the accelerated reliability test.

Nevertheless, for any specified process, chip design can be improved to overcome such process-dependent reliability problems. For example, knowing such potential aluminum overetch problems, alert designers may choose to widen the metal width beyond the minimum width allowed. Similarly, to avoid the transistor aging problem due to hot-carrier aging, designers can improve the circuit reliability with proper sizing of transistors or by reducing the rise time of signals feeding into the nMOS transistor gate. The protection of I/O circuits against electrostatic damages (ESD) and latch-up is another example. Some of these specific points have already been discussed in Chapter 12 for design of reliable I/O circuits.

Technology Updateability

Despite rapid progress in process technology development, the lifespan of a given technology generation has remained almost constant even for submicron technologies. Yet, the time pressure to develop increasingly more complex chips in a shorter time is constantly increasing. Under such circumstances, the chip products often have to be technology-updated to new design rules. Even without any change in the chip's functionality, the task of updating the mask to new design rules is very formidable. The so-called "dumb shrink" by uniform scaling of mask dimensions is rarely practiced due to nonideal scaling of device feature sizes and technology parameters. Thus, the design style should be chosen such that the technology update of the chip or functional modules for design reuse can be achieved quickly with minimal cost. One way of achieving such a goal is through the use of a symbolic layout system as explained in Section 13.3. Alternatively, designers can develop and use advanced CAD tools that can automatically generate the physical layout, the so-called *silicon compilation*, which meets the timing requirements with proper gate sizing or transistor sizing.

13.5. Packaging Technology

Novice designers often fail to give enough consideration to the packaging technology, especially in early stages of the chip development. However, many high-performance VLSI chips can fail stringent test specifications after packaging if chip designers have not included various effects of packaging constraints and parasitics in their design.

The numbers of ground planes and power planes and the bonding pads greatly affect the behaviors of the on-chip power and ground buses. Also, the length of the bonding wire between the chip and the package and the lead length in the package determine the inductive voltage drop in the output circuit. An equally important consideration is thermal problems. Good packages should provide low thermal resistance and, hence, low temperature rise due to unit power wattage beyond the ambient temperature.

Since the choice of proper packaging technology is critical to the success of the chip development, chip designers should work closely with package designers from the start of the project, especially for full custom designs. Also, since the final cost of the packaged chip depends largely on the package cost itself, for low-cost chip development, designers must ensure enough design margins that the chips can function properly in low-cost packages with more parasitic effects and less thermal conductivity. Some of the important packaging concerns are

- hermeticity to prevent the penetration of moisture
- thermal conductivity
- thermal expansion coefficient
- pin density
- parasitic inductance and capacitance
- α-particle protection

Various types of packages are available for integrated circuit chips:

- dual in-line package (DIP) in plastic or ceramic material
- leaded plastic chip carrier
- leadless ceramic chip carrier
- pin grid array (PGA)
- tape automated bonding (TAB)

The following table provides some data on the characteristics of packages.

Parameter	Package type				
	DIP (cer.)	DIP (plstc.)	PGA	Leadless chip carrier	Leaded chip carrier
Max. lead R (Ω)	1.1	0.1	0.2	0.2	0.1
Max. lead C (pF)	7	4	2	2	2
Max. lead L (nH)	22	36	7	7	7
Thermal R (°C/W)	32	35	20	13	28
PCB area (cm²)	18.7	18.7	6.45	6.45	6.45

Table 13.1 Characteristics of 64-68 pin packages

13.6. Computer-Aided Design Technology

Computer-aided design (CAD) tools are essential for timely development of integrated circuits. Although CAD tools cannot replace the creative and inventive parts of the design activities, the majority of time-consuming and computation-intensive mechanistic parts of the design can be executed by using CAD tools. The CAD technology for VLSI chip design can be categorized into the following areas:

- high-level synthesis
- logic synthesis

- circuit optimization
- layout
- simulation
- design rules checking

The high-level synthesis tools using hardware description languages (HDLs) address the automation of the design phase in the top level of the design hierarchy. With accurate estimation of lower-level realization, such as chip area and signal delay, it can be very effective in determining the types and quantities of modules to be included in the chip development from HDLs. Many tools have also been developed for logic synthesis and optimization, customized to particular design needs, especially for area minimization, low power, high speed, or their weighted combination.

The tools for circuit optimization are concerned with transistor sizing for minimization of delays and sensitivities to process variations, noises, and reliability hazards. The layout CAD includes floorplanning, place-and-route, and module generation. Sophisticated layout tools are goal driven and include some degree of optimization functions. For example, timing-driven layout tools are intended to produce layouts which meet timing specifications. The simulation category includes many tools ranging from circuit-level simulation (SPICE or its derivatives), timing level simulation, logic level simulation, and behavioral simulation. Many other simulation tools have also been developed for device level simulation and process simulation for technology development. The design rules checking CAD category includes the tools for layout rules checking, electrical rules checking, and reliability rules checking. The layout rules checking program has been highly effective in weeding out potential yield problems and circuit malfunctions.

References

1. A. D. Lopez and H.-F. S. Law, "A dense gate matrix layout method for MOS VLSI," *IEEE Transactions on Electron Devices*, vol. ED-27, no. 8, pp. 1671-1675, August 1980.

2. S. M. Kang, R. H. Krambeck, H.-F. S. Law, and A. D. Lopez, "Gate matrix layout of random logic in a 32-bit CMOS CPU chip adaptable to evolving logic design," *IEEE Transactions on Computer-Aided Design*, vol. CAD-2, no. 1, pp. 18-29, January 1983.

3. E. Horst, C. Muller-Schloer, and H. Schwartzel, *Design of VLSI Circuits*, Heidelberg: Springer-Verlag, 1987.

4. T. C. Hu and E. S. Kuh, *VLSI Circuit Layout: Theory and Design*, IEEE Press, 1985.

5. C. Sechen and A. Sangiovanni-Vincentelli, "The TimberWolf placement and routing package," *IEEE Journal of Solid-State Circuits*, vol. SC-20, no. 2, pp. 510-522, April 1985.

6. Y. Leblebici and S. M. Kang, *Hot-Carrier Reliability of MOS VLSI Circuits*, Norwell, MA: Kluwer Academic Publishers, 1993.

7. S. S. Sapatnekar and S. M. Kang, *Design Automation for Timing-Driven Layout Sythesis*, Norwell, MA: Kluwer Academic Publishers, 1993.

8. B. T. Murphy, "Cost-size optima of monolithic integrated circuits," *Proceedings of IEEE*, vol. 52, pp. 1937-1945, December 1964.

9. A. V. Ferris-Prabhu, "On the assumptions contained in semiconductor yield models," *IEEE Transactions on Computer-Aided Design*, vol. 11, pp. 955-965, August 1992.

10. T. E. Dillinger, *VLSI Engineering*, Englewood Cliffs, NJ: Prentice-Hall, Inc., 1988.

11. C. F. Fey, "Custom LSI/VLSI chip design complexity," *IEEE Journal of Solid-State Circuits*, vol. SC-20, no. 2, April 1985.

12. E. E. Hollis, *Design of VLSI Gate Array ICs*, Englewood Cliffs, NJ: Prentice Hall, Inc., 1987.

Exercise Problems

13.1   An ADD/SUBTRACT logic circuit is shown below. It performs the ADD operation for $P = 0$ and SUBTRACT for $P = 1$.

(a)   Draw an equivalent CMOS logic diagram by noting that most CMOS gates, except for the transmission gate and XOR, are inverting. For example, the AND gate is implemented with NAND followed by an inverter.

(b)   By using the gate array platform given on page 533, implement the CMOS circuit as compactly as possible with the aspect ratio, which is the ratio of vertical dimension to horizontal dimension, as close to 1 as possible.

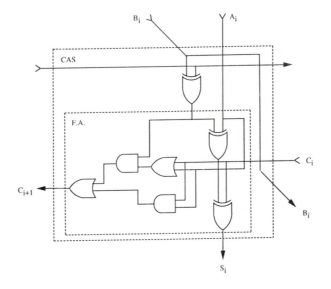

13.2   For the CMOS circuit in Problem 13.1,

(a)   First develop a small library of CMOS cells.

(b)   Place the cells into a single row and interconnect them with proper ordering such that the total interconnection wire length is minimized.

13.3   A measure of design productivity predicts the required engineer-months in terms of design implementation styles, such as repeated transistors (RPT); non-repeatable unique transistors (UNQ); PLA, RAM, and ROM transistors; the experience level of engineers (yr); the productivity improvement per year (D); and the design complexity (H). The formula proposed by Fey is

$$Engineer\text{-}Months\ (EM) = (1+D)^{-yr}\left[A + Bk^{H}\right]$$

where the number k of equivalent transistors in the design is expressed by

$$k = UNQ + C \cdot RPT + E \cdot PLA + F\sqrt{RAM} + G\sqrt{ROM}$$

In this formula, the transistor count for each style is in units of thousands and the coefficients A, B, C, D, E, F, G, and H are model parameters which depend on the designers' experience and CAD tool support. The parameter (yr) represents the number of years since the extraction time of the model parameters. A set of sample values for these parameters are $A = 0$, $B = 12$, $C = 0.13$, $D = 0.02$, $E = 0.37$, $F = 0.65$, $G = 0.08$, and $H = 1.13$.

(a)   Discuss how one would extract the model parameters within a design organization.

(b)   A 24-bit floating-point processor has been designed using 20,500 repeated transistors, 10,500 unique transistors, 105,500 RAM transistors, and 150,200 ROM transistors. Calculate the expected engineer-months (EM) by assuming the experience year value of $yr = 3$. Note that the transistor counts in the formula are in units of thousands, for instance, $UNQ = 10.5$, not 10,500.

13.4   A large-scale fast prototyping system has been produced by using a very large array of field programmable logic arrays (FPGAs).

(a)   Discuss the pros and cons of such prototyping systems for proof of design concepts and verification in view of effort and speed performance of the design.

(b)   How would you compare the hardware prototyping method with the computer simulation method?

13.5   As the design complexity increases with increasing number of on-chip transistors, the on-chip noises become more pronounced. Discuss the impact of packaging in suppressing on-chip noises in view of the numbers and strategic placement of ground and power pads and the numbers of ground and power planes.

13.6   The testing of VLSI chips at speed has become increasingly more difficult due to undesirable parasitic effects in a testing environment. Also the cost of high-speed testing machines has become very high and, hence, in reality it has become difficult for smaller manufacturers to procure such equipment. Discuss what problem chip testing only at lower speed causes for systems houses that take such chips to develop systems at speed. What alternative ways can be used to ease the problem in the absence of at-speed testers?

13.7   Draft plans for developing a chip as a function of design turnaround time and development cost. In particular, what specific design styles would be chosen when the customer requires that the chip be delivered in one month, six months, and one year, respectively?

CMOS Gate Array Platform (for Problem 13.1)

CHAPTER 14

DESIGN FOR MANUFACTURABILITY

14.1. Introduction

Digital circuits should be designed such that fabricated circuits meet the specifications for their performances, such as speed and power dissipation, under all operating conditions. However, random fluctuations in fabrication processes cause an undesirable spread in the circuit performance. Also, random variations in the circuit operating conditions, such as the power supply voltage V_{DD} and the operating temperature, result in variations in the circuit performance. Excessive deviations of performance can lead to a significant yield loss and thus can increase the unit cost of the product. It is, therefore, important that the effect of these inevitable statistical variations in the processing and environmental conditions be considered early on during the design of the circuit. The circuit performance should be made least sensitive to these variations and should have enough margins that a large fraction of the manufactured circuits pass the acceptability criteria. This is the essential motivation for *design for manufacturability* (DFM).

The term design for manufacturability (also called *statistical design* in computer-aided design literature) encompasses many methodologies and techniques. In this chapter, we briefly discuss some important issues in design for manufacturability which have an impact on digital circuit design techniques. In particular, these issues are parametric yield estimation, parametric yield maximization, worst-case analysis, and variability minimization. We will discuss formulation of the problems and some simple solution strategies for each of these issues.

The relationship between processing and device parameters and their effect on circuit and system performances are shown in Fig 14.1.

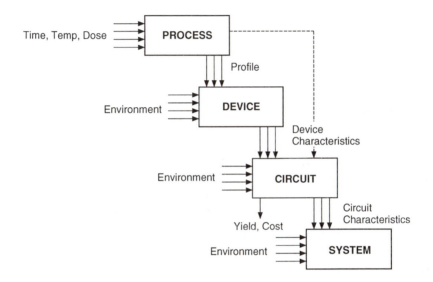

Figure 14.1.   Relationship between process and device parameters and circuit and system performance.

14.2. Process Variations

Figure 14.2 shows a plot of SPICE simulation results for the output waveform of a 4-bit adder circuit for nominal $V_{DD} = 5$ V at room temperature. It can be seen that the output varies significantly due to variations in device parameters caused by process fluctuations. Although a common set of masks is used to fabricate integrated circuits, some chips will have short delay times while others will have long delay times. In other words, circuit performances can vary significantly due to uncontrollable variations in the fabrication lines. Thus, an important design task is to minimize the impact of process variations on the circuit performances.

As discussed in Chapter 2, the fabrication procedure of CMOS integrated circuits is highly complex. Most submicron MOS technologies use more than 10 masks for over 100 steps of chemical processes to deposit oxide layers and photoresist materials, to transfer mask patterns to wafers with optical lithography, followed by chemical etchings. Even with computer-controlled high-precision fabrication steps, some errors in mask alignment, doping or implantation of targeted amounts of impurities, chemical etching of polysilicon gate lengths of MOS transistors, and thickness control of the thin-gate oxide layer are inevitable.

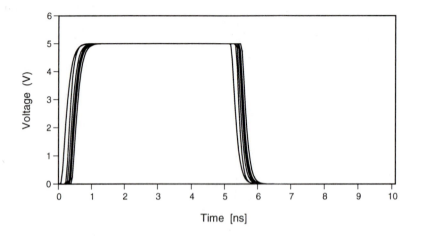

Figure 14.2.   Process-related variation in output waveform of a 4-bit adder.

The performances of digital circuits critically depend on the I-V characteristics of MOS transistors and the parasitics in interconnection lines, both of which can vary due to process variations. A particular transistor on a chip can have a much higher or lower drain current than its counterpart on a different chip on the same wafer. This variation can be even larger for chips on different wafers. A similar statement can be made for the parasitic resistances and capacitances in interconnection structures. In this chapter, we will concentrate on the variation of transistor characteristics because it has a larger impact, although the variation of interconnection parasitics is becoming non-negligible.

We recall that the drain current of an MOS transistor can be described by

$$I_d = \mu C_{ox} \frac{W}{L} f(V_{DS}, V_{GS}, V_T) \tag{14.1}$$

where μ is the mobility of electrons in nMOS, holes in pMOS,
$C_{ox} = \varepsilon_{ox}/t_{ox}$ is the gate oxide capacitance per unit area,
W/L is the ratio of the channel width to channel length, and
V_T is the threshold voltage of the transistor.

The mobility of majority carriers in the surface channel depends on the doping level in the substrate or tub. The threshold voltage depends on the flatband voltage of the MOS system and the substrate doping. Also, C_{ox} is inversely proportional to the gate oxide layer thickness t_{ox}. Thus, random fluctuations in μ, t_{ox}, W/L, and V_T will cause corresponding variations in the drain current even under the same biasing conditions. These random variations in the drain current may be translated into random variations in circuit performances such as delay times, power consumption, and logic threshold voltage. Among the quantities in (14.1), the only designable parameter under the jurisdiction of the designer is the nominal value of the aspect ratio, W/L. Thus, in order to make the circuit performance less sensitive to process variations, the most obvious design choice is to determine the optimal values of W and L for various MOS transistors in the circuit. Also, a more careful layout with proper orientation of transistors can be done to make the processed circuit less sensitive to process variations (this is even more important for analog circuits). In digital circuits, the value of L is usually chosen to be the minimum value allowed, except for sensitive circuits such as the level shifter described in Chapter 12 and memory cells for which the channel leakage is a great concern. In analog circuits, the L value may be made an order of magnitude larger than the minimum value to minimize the sensitivity of circuit performances to process variations.

14.3. Basic Concepts and Definitions

In this section, we introduce some basic concepts and define terms commonly used in design for manufacturability.

Circuit Parameters

Due to random variations in manufacturing processes and operating conditions, it is expected that the actual values of circuit parameters will be different from their nominal or target values. For instance, the actual channel width W of a MOS transistor can be decomposed into a statistically varying component ΔW and a nominal component W^o, i.e., $W = W^o + \Delta W$. In general, any circuit parameter can be considered to have a nominal component and an uncontrollable statistically varying component as shown in Table 14.1.

In this table, the geometrical parameters have a nominal component which can be set to particular values by the circuit designer. Such a nominal component is said to be *designable* or *controllable*, e.g., W^o and L^o. The statistically varying component of the geometrical parameters is called the *noise* component, and it represents the uncontrollable fluctuation of a circuit parameter about its designable component, e.g.,

	Actual	=	Nominal	+	Random
Geometrical Parameters					
MOS channel width	W	=	W^o	+	ΔW
MOS channel length	L	=	L^o	+	ΔL
Device Model Parameters					
Threshold voltage	V_T	=	V_T^o	+	ΔV_T
Gate oxide thickness	t_{ox}	=	t_{ox}^o	+	Δt_{ox}
Mobility	μ	=	μ^o	+	$\Delta \mu$
Operating Conditions					
Power supply voltage	V_{DD}	=	V_{DD}^o	+	ΔV_{DD}
Temperature	T	=	T^o	+	ΔT

Table 14.1.   Circuit parameters as sum of the nominal and random components.

ΔW and ΔL. For the device model parameters and operating conditions, the nominal component is not under the control of the designer and is set by nominal processing and operating conditions. For these parameters, the nominal and random components are together called the noise component, e.g., V_T and V_{DD}. In general, we can express any circuit parameter x_i as

$$x_i = d_i + s_i \qquad (14.2)$$

where d_i is the designable component and s_i is the random noise component. For circuit parameters which do not have a designable component, d_i is set to zero. Similarly, for circuit parameters which are completely controllable, s_i is set to zero.

All of the designable components are commonly grouped to form a set of parameters called *designable parameters*. These parameters are denoted by the vector **d**. Similarly, all of the noise components are grouped to form the set of *noise parameters*. These are denoted by the random vector **s**. Vectorially, (14.2) can be written as

$$\mathbf{x} = \mathbf{d} + \mathbf{s} \qquad (14.3)$$

Other commonly used terminology classifies the noise parameters based on whether their variation is related to fluctuations in manufacturing processes or operating conditions. The former are referred to as *internal noise parameters* and the latter as *external noise parameters*. For instance, V_T is considered to be an internal noise parameter, while V_{DD} is an external noise parameter.

Noise Parameter Distributions

A noise parameter is treated as a random variable. Since each circuit parameter consists of a designable component and a noise component, it can also be treated as a random variable. Any random variable is characterized by a *probability density function* (and by a mean and a standard deviation which depends on the density function). The vector of noise parameters **s** can be considered to be a random vector and is characterized by a *joint probability density function (jpdf)*. We will denote the *jpdf* of the noise parameters by $f(\mathbf{s})$. The vector of circuit parameters x is also a random vector and its *jpdf* is denoted by $f(\mathbf{x})$ or $f(\mathbf{d},\mathbf{s})$. The second notation emphasizes that the variability of circuit parameters comes from the noise components, but the *jpdf* may depend on the designable parameters.

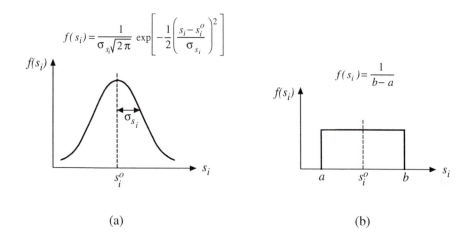

$$f(s_i) = \frac{1}{\sigma_{s_i}\sqrt{2\pi}} \exp\left[-\frac{1}{2}\left(\frac{s_i - s_i^o}{\sigma_{s_i}}\right)^2\right]$$

$$f(s_i) = \frac{1}{b-a}$$

(a)   (b)

Figure 14.3.   Probability density function for (a) a Gaussian and (b) a uniform random variable.

The statistical distributions of the internal noise parameters can be obtained via test structure measurements and parameter extraction. However, to simplify the analysis, a commonly used assumption is that the internal noise parameters have a Gaussian distribution, while the external noise parameters are uniformly distributed random variables (Fig. 14.3). The internal noise parameters are *statistically correlated* due to the sequential nature of the processing steps. The external noise parameters are, however, *statistically independent* random variables. Hence, the internal noise parameters are treated as a correlated Gaussian random vector and the external noise parameters as an independent uniform random vector. When noise vector **s** is multivariate Gaussian, then its distribution can be completely characterized by its mean vector μ and the covariance matrix **Q**, i.e., $\mathbf{s} \sim \text{MVG}(\mu, \mathbf{Q})$.

Circuit Performance Measures

A circuit performance measure is used to monitor the functionality of the circuit. For example, the propagation delay of an inverter or the signal skew between various branches of a clock distribution tree can be considered to be performance measures.

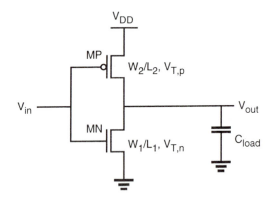

Figure 14.4. A CMOS inverter circuit.

Consider the simple CMOS inverter circuit shown in Figure 14.4. Let us make the following assumptions about the circuit parameters to illustrate some basic concepts:

(i) The widths and lengths of the MOS transistors are fixed and are not subject to statistical variations, i.e., W_1, L_1, W_2, and L_2 are designable parameters.

(ii) The internal noise parameters are the threshold voltages of MN and MP, $V_{T,n}$ and $V_{T,p}$, and the common gate oxide thickness t_{ox}. We assume for simplicity that these random variables are uncorrelated (independent) and Gaussian. The threshold voltage $V_{T,n}$ has a mean of 0.8 V and a standard deviation of 0.17 V, $V_{T,p}$ has a mean of −0.9 V and a standard deviation of 0.17 V, and t_{ox} has a mean of 27.5 nm and a standard deviation of 4.17 nm.

(iii) The external noise parameters are the power supply voltage V_{DD}, which is uniformly distributed in the range [4.8 V, 5.2 V], and the operating temperature T, which is uniformly distributed in the range [30 °C, 90 °C]. Moreover, V_{DD} and T are considered to be independent random variables.

Suppose that we are interested in the propagation delay of the inverter as the performance measure. The propagation delay of a CMOS inverter τ_p is given by (6.4), while the expressions for the high-to-low and low-to-high propagation delays are given in (6.22a) and (6.23a) and are repeated here for convenience.

$$\tau_{PHL} = \frac{C_{load}}{k_n(V_{DD} - V_{T,n})} \left[\frac{2V_{T,n}}{V_{DD} - V_{T,n}} + \ln\left(\frac{4(V_{DD} - V_{T,n})}{V_{DD}} - 1 \right) \right] \qquad (14.4a)$$

$$\tau_{PLH} = \frac{C_{load}}{k_p\left(V_{DD} - |V_{T,p}|\right)} \left[\frac{2|V_{T,p}|}{V_{DD} - |V_{T,p}|} + \ln\left(\frac{4\left(V_{DD} - |V_{T,p}|\right)}{V_{DD}} - 1 \right) \right] \qquad (14.4b)$$

It can be seen from the above equations that the two propagation delays are dependent on the widths and lengths of the MOS transistors, as well as on their threshold voltages and gate oxide thicknesses. The dependence on the power supply voltage is also evident. Although the dependence on the operating temperature is not shown explicitly, the temperature influences the propagation delay by affecting device parameters such as the mobility and threshold voltage of the MOS transistors. One can generalize from the above observation that a performance measure r is a function of the designable, internal noise and external noise parameters of the circuit, i.e.,

$$r = r(\mathbf{d}, \mathbf{s}) = r(\mathbf{x}) \qquad (14.5)$$

In some cases (as in the propagation delays above), it may be possible to express a performance measure by a closed-form analytical equation in terms of the circuit parameters of interest. In many cases, however, especially for large circuits, it is not possible to express the performances as explicit functions of the circuit parameters. In such cases, the value of a circuit performance for given values of the circuit parameters can be obtained by circuit simulation.

When the internal and external noise parameters are assumed to be fixed at their mean values, the corresponding circuit performance value is called the *nominal value*. Since it depends on the designable parameters alone, the nominal value of a performance r is denoted by

$$r^o(\mathbf{d}) = r(\mathbf{d}, \mathbf{s}^o) \qquad (14.6)$$

where $\mathbf{s}^o$ is the mean vector of the noise parameters. For the inverter example, the nominal value of the propagation delay τ_p is computed to be $\tau_p^o = 0.186$ ns for a load capacitance of $C_{load} = 0.5$ pF.

Due to inevitable statistical variations, there is a spread in the performance values about the nominal value. To demonstrate the variability in circuit performances, we fix the designable parameters (MOS channel lengths and widths) and vary the internal and external noise parameters. First, we vary $V_{T,n}, V_{T,p}$, and t_{ox} according to their statistical distribution while keeping the external noise parameters fixed. For each of the 1000 samples obtained in this manner, the τ_p value is computed from SPICE simulation results; its histogram is plotted in Figure 14.5. The histogram shows that the τ_p values are distributed mostly in the range [0.1 ns, 0.25 ns]. Figure 14.6 compares the transient waveforms for the best (smallest delay) and worst (largest delay) cases among these samples with the nominal case. Next, we vary the external noise parameters V_{DD} and T, while keeping the internal noise parameters fixed at their nominal values. The histogram for this case is plotted in Figure 14.7 and shows that the τ_p values are distributed in a much smaller range of [0.17 ns, 0.19ns].

The circuit performance measure, which is a function of the random circuit parameters, is also a random variable. Therefore, a performance measure will have a mean value and a standard deviation. The histograms in Figs. 14.5 and 14.7 show the approximate form of the distribution of τ_p. The relationship between a performance measure and the circuit parameters is often not known explicitly. Thus, the probability distribution of a performance is also not known explicitly, and its mean and standard deviation have to be estimated. For the inverter example and for the case of internal noise parameter variations alone (i.e., the case corresponding to the histogram of Fig. 14.5), the mean and standard deviation of τ_p are estimated to be 0.184 ns and 0.023 ns, respectively. Note that the nominal value of a performance is not the same as its mean. However, like the nominal value, the mean and standard deviation of a performance measure are functions of the designable parameters alone, since the effects of the noise parameters have been "averaged out."

Parametric Yield and Performance Variability

Another important concept related to circuit performance is the *yield*. Minimization of yield loss is the central notion of design for manufacturability. There are a large number of factors which lead to yield losses: material defects, lithography misalignments, processing variations, and insufficient design margins. *Catastrophic faults* are shorts and opens which cause malfunctioning of the circuit, whereas *parametric faults* are shifts in device and circuit performances. A circuit under parametric faults may be logically functional but may fail to meet some performance specifications. Each circuit performance is associated with a specification which determines its acceptability. A circuit is said to be *acceptable* if each of its performances satisfies the corresponding specifications. The yield (also called *parametric yield* to distinguish it from the functional yield) is defined as the fraction of the total number of manufactured circuits that is acceptable, i.e.,

$$\text{parametric yield} = \frac{\text{total number of acceptable circuits}}{\text{total number of manufactured circuits}} \quad (14.7)$$

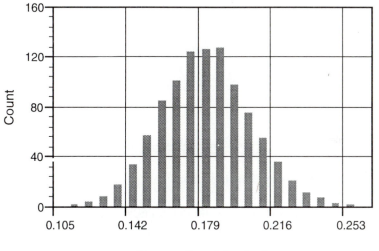

Propagation delay [ns]

Figure 14.5. Histogram showing variation of τ_p with $V_{T,n}$, $V_{T,p}$, and t_{ox}.

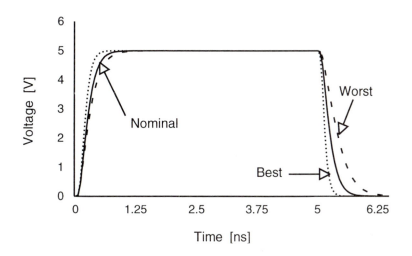

Time [ns]

Figure 14.6. Comparison of waveforms for the nominal, best, and worst cases.

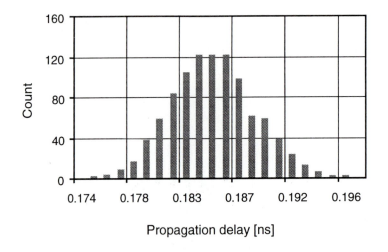

Figure 14.7.   Histogram showing variation of τ_p with V_{DD} and T.

The yield of a circuit directly determines the profitability of the product, hence, the maximization of yield is the primary motivation of design for manufacturability. In the preceding inverter example, suppose that we are considering variations in the internal noise parameters alone and that the acceptability criterion is that τ_p should be less than 0.19 ns. From the histogram of Fig. 14.5, it can be seen that a large fraction of the sampled circuits would fail the acceptability criterion, which would lead to a poor yield of 61.2%. Note that a poor parametric yield may result even when the nominal value of the performance meets the specifications.

The *variability* of a circuit performance is a measure of the spread in the performance values due to uncontrollable statistical variations in the circuit parameters. The minimization of variability is another important task in design for manufacturability, since it results in uniform products which are more desirable. Several measures of variability are used in practice: the standard deviation or variance, the ratio of the standard deviation and the mean, the range of the performances, etc.

To summarize the various points and observations made in this section:

- Circuit parameters are composed of designable and noise components.
- Noise parameters represent the statistical variations due to manufacturing and environmental fluctuations.
- Designable parameters are deterministic, while noise parameters are random.

- Circuit performances are functions of designable and noise parameters.
- Circuit performances are characterized by a probability distribution, which usually has to be estimated.
- A circuit with acceptable nominal performances may have low parametric yield.

14.4. Design of Experiments and Performance Modeling

Suppose that there are n circuit parameters of interest denoted by $\mathbf{x} = (x_1, x_2, ..., x_n)$. These circuit parameters can be designable parameters or noise parameters. As noted earlier, a circuit performance r is a function $r(\mathbf{x})$ of these parameters. Usually, this function is not known explicitly, and for particular values of $\mathbf{x}$, r has to be evaluated using a circuit simulator such as SPICE. Circuit simulations are computationally expensive, especially if the circuit size is large and transient simulations are required. An attractive alternative would be to construct a compact model of the circuit performance in terms of the parameters $\mathbf{x}$ and then use the performance model instead of the circuit simulator to evaluate the performance. The utility of such an approach depends on two criteria. First, the model should be computationally efficient to construct and evaluate so that substantial computational savings can be achieved. Second, the model should be accurate. Clearly, these two features are conflicting requirements, and some optimal trade-offs should be made in order to develop a good model.

The next question is how to build such a model. Since the model is to act as a surrogate to the circuit simulator, it should be built from values of r obtained by circuit simulations. Figure 14.8 shows the procedure for developing such a model. The model building procedure consists of four steps. First, m *training points* are selected in the $\mathbf{x}$-space. The ith training point is denoted by $\mathbf{x}_i = (x_{1i}, x_{2i}, ..., x_{ni})$. In the second step, the circuit is simulated at the m training points, and the values of the performance measure are obtained from the simulation results as $r(\mathbf{x}_1),...,r(\mathbf{x}_m)$. In the third step, a pre-assigned function of r in terms of $\mathbf{x}$ is "fitted" to the data. In the final step, the model is validated for accuracy. If the model is not sufficiently accurate, then the modeling procedure is repeated with a larger number of training points or with different models.

The model is called the *response surface model* (RSM) of the performance. The computational cost of modeling depends on m, the number of training points, and on the procedure of fitting the model to the simulation data. The accuracy of the model is calibrated by computing error measures which quantify the "goodness of fit." The accuracy of the derived model would be influenced greatly by the manner in which the training points are selected from the $\mathbf{x}$-space. *Experimental design* techniques are systematic means of selecting the training points in the "best" possible manner, in the sense that the smallest number of training points will be required and the most accurate model can be obtained from them.

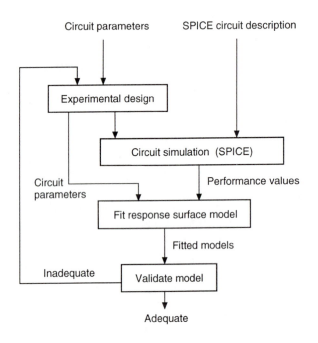

Figure 14.8.   Performance modeling procedure.

Design of experiments (DOE) is an established branch of statistics, and has been used successfully in many manufacturing fields since the 1920s. In this chapter, we provide a brief glimpse into this discipline by discussing some commonly used DOE techniques in the context of design for manufacturability of integrated circuits. To illustrate certain points, it is convenient to assume that the RSM of the performance measure r is a quadratic polynomial in the circuit parameters x_i, $i=1, 2,..., n$. In particular,

$$r'(\mathbf{x}) = \alpha_0 + \sum_{i=1}^{n} \alpha_i x_i + \sum_{i=1}^{n} \sum_{j=i}^{n} \alpha_{ij} x_i x_j \qquad (14.8)$$

is the RSM used, where the coefficients α_0, α_i, and α_{ij} are the fitting parameters in the model. Note, however, that the discussion is valid for any other RSM as well.

Factorial Design

In this experimental design technique, each of the parameters $x_1, x_2, ..., x_n$ is quantized into two levels or settings (the smallest and largest values in its range). Without any loss of generality, we can assume that these values are -1 and $+1$ for each parameter after normalization. A full factorial design consists of all possible combinations of values for the n parameters. Therefore, a full factorial design for n parameters has 2^n training points or experimental runs. The design matrix for the case of $n = 3$ is shown in Table 14.2 and the design is depicted in Fig.14.9. The perfomance

	Parameters			Interactions				
Run	x_1	x_2	x_3	$x_1 \times x_2$	$x_1 \times x_3$	$x_2 \times x_3$	$x_1 \times x_2 \times x_3$	r
1	-1	-1	-1	$+1$	$+1$	$+1$	-1	r_1
2	-1	-1	$+1$	$+1$	-1	-1	$+1$	r_2
3	-1	$+1$	-1	-1	$+1$	-1	$+1$	r_3
4	-1	$+1$	$+1$	-1	-1	$+1$	-1	r_4
5	$+1$	-1	-1	-1	-1	$+1$	$+1$	r_5
6	$+1$	-1	$+1$	-1	$+1$	-1	-1	r_6
7	$+1$	$+1$	-1	$+1$	-1	-1	-1	r_7
8	$+1$	$+1$	$+1$	$+1$	$+1$	$+1$	$+1$	r_8

Table 14.2.   Full factorial design for $n = 3$.

data for the kth run are denoted by r_k. The full factorial design provides much information on the relationship between the parameters x_i and the performance r.

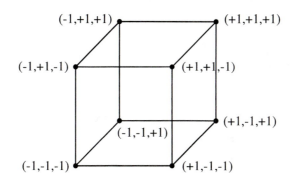

Figure 14.9.   Pictorial representation of full factorial design for $n = 3$.

For example, one can evaluate the *main* or *individual effect* of a parameter x_i, which quantifies how much the parameter affects the performance. The main effect is the difference between the average performance value when the parameter is at the high level (+1) and the average performance value when the parameter is at the low level (−1). The main effect of x_i is the coefficient of the x_i term in the RSM of (14.8). One can also determine the *interaction effects* of two or more parameters which quantify how those factors jointly affect the performance. Two factor interactions can be computed as the difference between the average performance value when both parameters are at the same level and the average performance value when the parameters are at different levels. The two-factor interaction effect of parameters x_i and x_j is the coefficient of the $x_i x_j$ term in (14.8). Higher-order multifactor interactions can be computed in a recursive manner.

| | Parameters | | | Interactions | | | | |
Run	x_1	x_2	x_3	$x_1 \times x_2$	$x_1 \times x_3$	$x_2 \times x_3$	$x_1 \times x_2 \times x_3$	r
1	−1	−1	+1	+1	−1	−1	+1	r_1
2	−1	+1	−1	−1	−1	−1	+1	r_2
3	+1	−1	−1	−1	+1	+1	+1	r_3
4	+1	+1	+1	+1	+1	+1	+1	r_4

Table 14.3.   Half-fraction of full factorial design for $n = 3$.

Thus, the full factorial design allows us to estimate all the first-order and cross-factor second-order coefficients in the RSM of (14.8). However, it does not allow us to estimate the coefficients of the pure quadratic terms x_i^2. Moreover, the number of experimental runs increases exponentially with the number of parameters. In most modeling situations, the information about the high-order multifactor interaction effects is often unnecessary and unimportant. It is possible to reduce the size of a full factorial design without compromising the accuracy of the main effects and the low-order interaction effects that are of primary interest. This is accomplished by considering only a fraction of the original full factorial design by systematically eliminating some of the runs. Such designs are called *fractional factorial designs*. If k is the degree of the fraction ($k = 1$ is a half fraction, $k = 2$ is a quarter fraction, etc.), then such designs have 2^{n-k} experimental runs and are called 2^{n-k} designs. One of the half-fractions of the full factorial design of Table 14.2 is shown in Table 14.3. We observe that some of the columns of Table 14.3 are identical. It is not possible to distinguish between effects that correspond to identical columns. Such effects are said to be *confounded* or *aliased* with each other. In the fractional design of Table 14.3, we see that the main effect of x_3 is confounded with the interaction effect of x_1 and x_2.

Moreover, the column $x_1 \times x_2 \times x_3$ is identical to a column of 1s. This implies that the three-factor interaction effect is confounded with the grand average of the performances. Confounding, however, is not really a problem since in most applications, high-order multifactor interactions are negligible and may be ignored. In the quadratic RSM in (14.8), only main effects and two-factor interaction effects are important, and it can be assumed that all higher-order interactions are absent. One of the most important attributes of fractional factorial designs is that these designs are orthogonal, which allows the model coefficients to be estimated with minimum errors.

Central Composite Designs

As mentioned above, one of the problems of factorial designs in regard to the RSM of (14.8) is that the coefficients of the pure quadratic terms cannot be estimated. This can be done with a *central composite design*, which is a combination of a factorial (full or fractional) and a "star" design. Figure 14.10 shows the central composite design for the case of $n = 3$. The factorial design is the "cube" design shown in dotted lines,

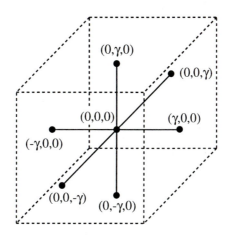

Figure 14.10. Central composite design for $n = 3$.

while the star design is shown in solid lines. Each parameter in this design can take on five levels, 0, ±1, ±γ, where $0 < \gamma < 1$, and the star section of the design consists of $(2n + 1)$ runs, which includes

- one *center* point, where all parameters are set to 0, and
- $2n$ *axial* points, with the axial-pair values set to $-\gamma$ and $+\gamma$ while all other parameters are set to 0.

The parameter γ is usually selected by the user. The main advantage of the central composite design is that all the coefficients of (14.8) can be estimated using a reasonable number of simulations.

Taguchi's Orthogonal Arrays

Another popular experimental design technique is Taguchi's method using *orthogonal arrays* (OAs). An orthogonal array is a fractional factorial design matrix which allows a balanced and fair comparison of levels of any parameter or interaction of parameters. These orthogonal arrays are readily available in a table format and they are of two types. The first category of OAs corresponds to parameters which have been quantized to two levels, and the second corresponds to OAs where the parameters are quantized to three levels each. These arrays are often available as tables in textbooks which discuss Taguchi techniques. As an example, we show the L18 OA in Table 14.4. The number in the designation of the array indicates the number of trials or experimental runs. The L18 OA belongs to the second category of designs, i.e., each parameter has three levels. In the Taguchi technique, the experimental design matrix for the designable or controllable parameters is called the *inner array*, while that for the noise parameters is called the *outer array*. The L18 design of Table 14.4 can be applied to the inner array as well as to the outer array.

Latin Hypercube Sampling

The factorial and central composite experimental designs described above set the parameters to certain levels or quantized values within their ranges. As a result, most of the parameter space remains unsampled. A more "space filling" sampling strategy would be more desirable. The most obvious method of obtaining more complete coverage of the parameter space would be to use the simplest sampling strategy, *random sampling*, to obtain the training points in the parameter space. Technically, a random sample is drawn from a probability density function. We have already noted that the internal and external noise parameters are random variables and have associated density functions. The designable parameters are, however, deterministic variables. For the purpose of sampling, we assume for a designable parameter that any value within its range is equally likely. In other words, we assume that the designable parameters are independent and uniformly distributed random variables for sampling. Once the random samples are generated, the circuit is simulated and the values of the performance measure are computed. Random samples are theoretically convenient and easy to generate, and many inferences can be drawn about the probabilistic distribution of the performance. This sort of random sampling is also known as *Monte Carlo sampling*. The problem with random sampling, however, is that a large sample is required to estimate quantities with sufficiently small errors.

Parameters

Run	1	2	3	4	5	6	7	8
1	1	1	1	1	1	1	1	1
2	1	1	2	2	2	2	2	2
3	1	1	3	3	3	3	3	3
4	1	2	1	1	2	2	3	3
5	1	2	2	2	3	3	1	1
6	1	2	3	3	1	1	2	2
7	1	3	1	2	1	3	2	3
8	1	3	2	3	2	1	3	1
9	1	3	3	1	3	2	1	2
10	2	1	1	3	3	2	2	1
11	2	1	2	1	1	3	3	2
12	2	1	3	2	2	1	1	3
13	2	2	1	2	3	1	3	2
14	2	2	2	3	1	2	1	3
15	2	2	3	1	2	3	2	1
16	2	3	1	3	2	3	1	2
17	2	3	2	1	3	1	2	3
18	2	3	3	2	1	2	3	1

Table 14.4.   Taguchi's L18 orthogonal array.

Moreover, a random sample may not be very space-filling either. This can be understood by considering the bell-shaped Gaussian density function of Fig. 14.3(a). In random sampling, values near the peak of the bell-shaped curve would be more likely to be selected in the sample since such values have a greater probability of occurrence. Stated differently, values away from the central region of the distribution would not be well represented in the sample.

   Latin hypercube sampling (LHS) is a sampling strategy that alleviates this problem. It ensures that each parameter x_i has all portions of its distribution represented by sample values. If S is the sample size, then the range of each x_i, $i=1, 2, ..., n$, is divided into S non-overlapping intervals of equal marginal probability $1/S$. Each such interval is sampled once to obtain S values for each of the parameters. The S values for one parameter are then randomly paired with the S values of another, and so on. The process is illustrated in Fig. 14.11, where x_1 is a uniform random variable, x_2 is a Gaussian random variable, and $S = 5$. Note that the (marginal) probability is the area under the individual probability density curves. Therefore, intervals with equal probability are those with equal areas under the probability density curve. For the uniform random variable x_1, such intervals are of equal length. For x_2, the central intervals have smaller lengths (since the density there is higher) than the intervals away from the center (where the density is lower). Figure 14.11 also shows that $S = 5$ values are chosen from each region for both x_1 and x_2 (shown as open circles). These values are then randomly paired so as to obtain the sample points (shown as filled dots). This example also illustrates some of the advantages of Latin hypercube sampling. First, LHS provides more uniform coverage of the input parameter space than other experimental design techniques. Second, a sample of any size can be easily generated and all types of probability densities are handled.

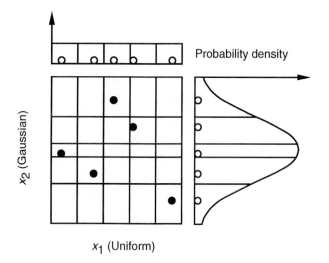

Figure 14.11.   Latin hypercube sampling for a uniform and a Gaussian parameter.

Model Fitting

Once the experimental design is done to select the training points in the **x**-space, the circuit is simulated at the training points and the values of the performance measure are extracted from the simulation results. Let S denote the number of training points. Now, we have S points in the x-space and corresponding values of r. We will describe the technique of fitting the quadratic RSM of (14.8) with the understanding that the same techniques can be used to fit any RSM. The aim in model fitting is to determine the coefficients in the model such that the model fits the data as accurately as possible, i.e., the fitting error is minimized. The number of coefficients in the quadratic RSM of (14.8) is $C = (n+1)(n+2)/2$, where n is the number of model parameters. There are interpolation methods which can be used to fit the quadratic RSM with a smaller number of data points than coefficients, i.e., with $S < C$. But we will restrict ourselves to the case of $S \geq C$. If $S = C$, there are as many equations as unknowns, and simple simultaneous equation solving will provide the values of the coefficients. In such a case, there is no attempt to minimize any fitting error. It is, therefore, a better practice to collect more data points than the number of coefficients so that the optimal set of model coefficients can be determined. This method is called *least-squares fitting* (in a numerical analysis context) or *linear regression* (in a statistical context). The error measure used is called the sum of squared errors and is given as

$$\varepsilon = \sum_{k=1}^{S} (r_k - r_k')^2 \tag{14.9}$$

where r_k is the simulated performance value at the kth data point, and r_k' is the model-predicted value. Note that r_k', and therefore ε, depends on the model coefficients. The aim of least-squares fitting is to obtain the values of the coefficients such that the error is minimized. Formally stated, least-squares fitting is the following optimization problem:

$$\underset{\alpha_i}{\text{minimize}} \quad \left[\varepsilon = \sum_{k=1}^{S} (r_k - r_k')^2 \right] \tag{14.10}$$

The error ε is used to determine the adequacy of the model. If the model is considered to be inaccurate, the modeling procedure has to be repeated with a larger number of training points or a different design strategy, or a different model of the performance may have to be used. There are several measures of accuracy of a model, but these are available in many statistics textbooks and are not discussed here.

14.5. Parametric Yield Estimation

The parametric yield is used to characterize the manufacturability of a circuit. As defined in (14.7), parametric yield is the fraction of manufactured circuits that meet all acceptability criteria. Let $\mathbf{r} = (r_1, r_2, ..., r_p)$ denote the p circuit performance measures of interest. Each performance has acceptability specifications which can be expressed as

$$a_k \leq r_k \leq b_k, \quad k = 1, 2, ..., p \tag{14.11}$$

where a_k denotes the lower limit and b_k the upper limit of acceptability for the kth performance. The specifications on the circuit performances define an *acceptability region* in the p-dimensional *performance space* denoted by A_r and expressed as

$$A_r = \left\{ \mathbf{r} \big| a_k \leq r_k \leq b_k, \ k = 1, 2, ..., p \right\} \tag{14.12}$$

For example, consider an adder circuit where the performances are the power dissipation P_d and the propagation delay τ_p. Suppose the specifications are as follows:

$$P_d \leq 0.5 \text{ mW}$$
$$\tau_p \leq 0.16 \text{ ns} \tag{14.13}$$

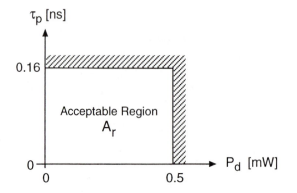

Figure 14.12.   Region of acceptability in performance space for adder example.

Figure 14.12 illustrates the acceptability region in the performance space for the above case. The parametric yield is defined as the fraction of manufactured circuits that are acceptable, and since the circuit performances are random variables, the parametric yield can be expressed as

$$Y = \Pr(\mathbf{r} \in A_r) \tag{14.14}$$

Since the probability density functions of the circuit performances are not known explicitly, the above probability is difficult to evaluate. Therefore, an alternative approach is used to estimate the yield.

In addition to the performance specifications, the circuit parameters may also be restricted to a subset of the parameter space. These restrictions may arise from physical considerations such as nonnegative resistance values. Let the allowed circuit parameter space be represented by $\mathbf{X}$. Then, we can define a region in the circuit parameter space called the *acceptable region* or *feasible region* A_x as follows:

$$A_x = \left\{ \mathbf{x} \mid a_k \le r_k(\mathbf{x}) \le b_k, \ k = 1, 2, ..., p \text{ and } \mathbf{x} \in \mathbf{X} \right\} \tag{14.15}$$

Note the distinction between A_r and A_x: A_r denotes the acceptable region in the performance space, while A_x denotes the acceptable region in the circuit parameter space. Clearly, A_x is a subset of $\mathbf{X}$, i.e., $\mathbf{X} \supset A_x$. The mapping $\mathbf{x} \to \mathbf{r}$ from the circuit parameter space to the performance space is determined by the functions $r_k(\mathbf{x})$, $k = 1$, $2,..., p$. These functions are known implicitly (i.e., they have to be evaluated by circuit simulation) or can be approximated by explicit response surface models as explained in the previous section. The inverse mapping, $\mathbf{r} \to \mathbf{x}$, from the performance space to the circuit parameter space is generally not known. Therefore, the boundaries that define the acceptable region in the parameter space cannot be determined easily.

Figure 14.13 shows an example of a hypothetical acceptable region in the parameter space for $p = 2$ and $n = 2$. Consider the designable parameter vector (also referred to as the *design point*) marked $P = (d_1, d_2)$ on the figure. The circle around this point represents the actual values (x_1, x_2) of the circuit parameters that may be realized as a result of statistical variations. It is easy to see that the circuit parameters of acceptable circuits for this design point lie in the intersection of the circle and the acceptable regions A_x. In other words, the parametric yield Y of the circuit for this design point is the area (in a probabilistic sense) of the intersection. Note that for a given process technology, Y can be represented as a function of the designable parameters alone, since the noise components are averaged out when the area is computed. This can be understood from Fig. 14.13 by realizing that as the design point (which is the center of the circle that represents parameter variations) is moved toward the center of the acceptable region, the parametric yield increases.

Given that the acceptable region A_x consists of circuit parameter values for which the circuit performances meet specifications, the parametric yield Y at a design

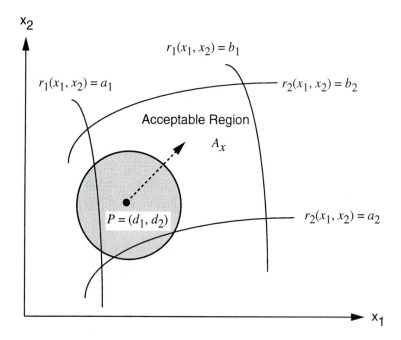

Figure 14.13.   Acceptable region in the circuit parameter space for $p = 2$ and $n = 2$.

point $\mathbf{d}$ can be defined as the probability that the actual circuit parameter $\mathbf{x}$ belongs to A_x. Mathematically,

$$Y(\mathbf{d}) = \Pr(\mathbf{x} \in A_x) = \Pr(\mathbf{d} + \mathbf{s} \in A_x) = \int_{A_x} f(\mathbf{d}, \mathbf{s}) ds \tag{14.16}$$

We define an indicator function $I(r_1, r_2, ..., r_p)$ as follows:

$$I(r_1, r_2, ..., r_p) = \begin{cases} 1, & \text{if } a_k \le r_k \le b_k \text{ for all } k \\ 0, & \text{otherwise} \end{cases} \tag{14.17}$$

Then the yield expression can be rewritten as

$$Y(\mathbf{d}) = \int I\big(r_1(\mathbf{d}, \mathbf{s}), r_2(\mathbf{d}, \mathbf{s}), ..., r_p(\mathbf{d}, \mathbf{s})\big) f(\mathbf{d}, \mathbf{s}) ds \tag{14.18}$$

Next, we discuss two simple methods for parametric yield estimation.

Direct Monte Carlo Method

Monte Carlo or simple random sampling is the most widely used method for parametric yield estimation. The various steps are outlined below.

Step 1: Generate a (large) number of samples of the noise parameters s_i, $i = 1, 2,...,$ N_{MC}, from the joint probability distribution of the noise parameters $f(s)$.

Step 2: Simulate the circuit performances $r_1(\mathbf{d}, s_i), r_2(\mathbf{d}, s_i),...,r_p(\mathbf{d}, s_i), i = 1, 2, ..., N_{MC}$.

Step 3: Calculate the estimated yield as the fraction of samples that are acceptable.

$$Y'(\mathbf{d}) = \frac{1}{N_{MC}} \sum_{i=1}^{N_{MC}} I\left(r_1(\mathbf{d}, s_i), r_2(\mathbf{d}, s_i),..., r_p(\mathbf{d}, s_i)\right) \qquad (14.19)$$

The error of the yield estimate is given by

$$\sigma(Y') = \left[\frac{Y'(1-Y')}{N_{MC}-1} \right]^{1/2} \qquad (14.20)$$

Thus, the accuracy of the yield estimate is inversely proportional to the square root of the sample size. The Monte Carlo method has the following advantages:

- It provides the error of the yield estimate.
- The sample size N_{MC} is independent of the number of circuit parameters.
- It makes no restrictive assumptions regarding the nature of the joint probability density function $f(s)$.
- It makes no restrictive assumptions on the nature of the relationship between the circuit parameters and the circuit performances.

Although various strategies to reduce the sample size, such as variance reduction techniques, importance sampling, control variates, and stratified sampling, have been considered, Monte Carlo sampling is still expensive in terms of computational cost due to the large number of samples and the high cost of circuit simulations.

Performance Model Method

In this method, the response surface models of the circuit performance measures are used to evaluate the performance values for a given set of circuit parameter values. This avoids expensive circuit simulations and makes yield analysis feasible. Since the designable parameter vector $\mathbf{d}$ is fixed during yield estimation, the response surface models of the circuit performances are expressed in terms of the noise parameters alone. These response surface models are denoted by $r_k'(\mathbf{s})$, $k = 1, 2,..., p$. For example, the quadratic RSMs of (14.8) can be used with x_i replaced by s_i. The various steps of the yield estimation procedure are enumerated below.

Step 1: Design an experiment and simulate the circuit at the training points.

Step 2: Fit the RSMs of the performances $r_1'(\mathbf{s})$, $r_2'(\mathbf{s})$,..., $r_p'(\mathbf{s})$, and improve model accuracy if required.

Step 3: The parametric yield Y is obtained by Monte Carlo sampling using the RSMs. Generate a (large) sample of the noise parameters, i.e., $\mathbf{s}_i$, $i = 1, 2,..., N_{MC}$, from the joint density function of $\mathbf{s}$.

Step 4: Compute $r_k'(\mathbf{s}_i)$, $k = 1, 2,..., p$ and $i = 1, 2,..., N_{MC}$.

Step 5: The estimated parametric yield is given by (14.19) with r_k replaced by r_k'.

This method preserves the advantages of the Monte Carlo method of yield estimation and at the same time considerably reduces the computational cost.

A Simple Example of Parametric Yield Estimation

Let us consider the simple inverter example of Fig. 14.4. To illustrate the yield estimation procedure, we make the following assumptions:

(i) The operating conditions are not subject to statistical variations, i.e., there are no external noise parameters.

(ii) The channel lengths of MN and MP are also not subject to statistical variations. Moreover, we assume that both channel lengths are fixed at $0.8 \, \mu m$. The channel width of an MOS transistor is the sum of a designable (nominal) component and a noise component. The designable components are W_1^o and W_2^o for transistors MN and MP, respectively. The corresponding noise components are ΔW_1 and ΔW_2.

(iii) Two additional noise parameters arise from random variations in the threshold voltages of MN and MP ($V_{T,n}$ and $V_{T,p}$). We assume that all four noise parameters are independent Gaussian random variables. Their means and standard deviations (s.d.) are as follows:

$$
\begin{array}{llll}
\Delta W_1: & \text{mean} = 0 \, \mu m\,, & \text{s.d.} = 0.03 \, \mu m \\
\Delta W_2: & \text{mean} = 0 \, \mu m, & \text{s.d.} = 0.06 \, \mu m \\
V_{T,n}: & \text{mean} = 0.8 \, V, & \text{s.d.} = 0.067 \, V \\
V_{T,p}: & \text{mean} = -0.9 \, V, & \text{s.d.} = 0.067 \, V
\end{array}
$$

(iv) There are two performance measures of interest. The first is the propagation delay τ_p defined earlier. The second is the area of the circuit. Since it is not possible to evaluate the area accurately without layout data, we assume that the area is given by the sum of the products of the widths and lengths of all MOS transistors in the circuit. Furthermore, since the transistor length is assumed to be fixed, the area measure can be simplified to be the sum of all transistor widths. Therefore, the area measure A_m is given by

$$A_m = W_1 + W_2 = W_1^o + \Delta W_1 + W_2^o + \Delta W_2 \qquad (14.21)$$

(v) The specifications on the two performances are: $\tau_p \leq 0.172$ ns and $A_m \leq 35$ μm.

With these assumptions, the parametric yield is computed using the two methods outlined above at the design point $\mathbf{d}_{init} = (W_1^o = 10, W_2^o = 20)$, where all of the widths are shown in units of μm. The value of N_{MC} used in both methods is 1000. In the direct Monte Carlo method, N_{MC} circuit simulations are used to estimate the parametric yield at $\mathbf{d}$ to be 79.5% (the error of the estimate, according to (14.20), is 1.28%). In the performance modeling method, 10 circuit simulations are used to construct the following linear RSM for the propagation delay τ_p in terms of the noise parameters:

$$\tau_p' = 0.169 + 0.0069 V_{T,n} - 0.0071 V_{T,p} - 0.0007 \Delta W_1 - 0.0008 \Delta W_2 \qquad (14.22)$$

Note that an RSM for the area measure is not required since (14.21) already expresses it explicitly in terms of the circuit parameters. Furthermore, as discussed previously, the RSM for the propagation delay is given in terms of the noise parameters alone since the design point is fixed. Based on this RSM, the parametric yield is estimated to be 79.6% (with an error of 1.27%). 10 circuit simulations were required in the performance modeling approach as opposed to 1000 simulations used in the direct Monte Carlo method. This example shows that the performance modeling approach can be quite accurate and provides a substantial saving in the number of circuit simulations.

14.6. Parametric Yield Maximization

As noted in the previous section, the parametric yield is a function of the designable parameter values. Therefore, the designable parameters may be adjusted to maximize the yield. This is the basic idea of parametric yield maximization. Various yield maximization approaches have been proposed over the years. These approaches can be classified into two categories: *Monte Carlo-based methods* and *geometrical methods*.

Monte Carlo-Based Methods

In these approaches, the yield integral of (14.16) or (14.18) is computed by the Monte Carlo method as before (or improved variants of the Monte Carlo methods), and the yield is then numerically maximized. Several choices of optimization techniques are available in the maximization of yield. Some techniques do not require the derivatives of the yield with respect to the designable parameters. These are, however, slower than those that require derivative information. Since the yield is a multi-dimensional integral involving statistical distributions, analytical formulae for the derivatives are usually not available. There are many sophisticated methods for approximating these derivatives, but the simplest one uses finite difference methods. In the Monte Carlo-based approaches, the acceptable region is not explicitly charac-terized. Rather, the decision as to whether a set of circuit parameter values belongs to the acceptable region is made by evaluating the performances and verifying whether they meet the specifications. The evaluation of circuit performances can be done both by actual circuit simulation as well as by constructing analytical response surface models for the performances.

Geometrical Methods

These methods build an approximation to the acceptable region A_x and this approximation is used in the yield maximization. There are two techniques that can be used to approximate A_x. The first technique progressively builds a geometric approxi-mation, such as a simplex, to A_x. This technique is called *simplicial approximation*. The simplicial approximation method has one important drawback: the cost of constructing the approximation to A_x and maximizing the yield grows exponentially as the number of circuit parameters increases. This problem is often called the "curse of dimension-ality." The second technique uses analytical models of the circuit performances, such as the response surface models introduced earlier. Then the boundaries of A_x are given by the constraint equations, $r_k' = a_k$ and $r_k' = b_k$, where r_k' denotes the performance model for the kth performance and a_k and b_k are the lower and upper acceptability limits, respectively. Many methods use analytically approximated acceptability regions as they are relatively inexpensive for a large number of circuit parameters. Once the acceptability region approximation has been constructed, yield maximization involves a technique called *design centering*, which attempts to move the design point towards the center of the acceptability region. Design centering is an attractive method and many approaches to design centering have been proposed.

A Simple Yield Maximization Method

Since a detailed description of the yield maximization approaches is beyond the scope of this book, a simple method for parametric yield maximization is outlined below. This method belongs to the class of Monte Carlo-based methods described above.

Step 1: Assume models for all circuit performances in terms of both the designable and the noise parameters. These are denoted by $r_k'(\mathbf{x})$, or alternatively by $r_k'(\mathbf{d}, \mathbf{s})$, for $k = 1, 2,..., p$.

Step 2: Design an experiment and simulate the circuit at the training points.

Step 3: Fit the models for the performances and validate the models.

Step 4: The parametric yield of a circuit at a design point $\mathbf{d}$ is obtained by drawing a Monte Carlo sample of the noise parameters and using (14.19) to estimate the yield.

Step 5: Maximize the estimated yield $Y'(\mathbf{d})$ with respect to $\mathbf{d}$ using some optimization algorithm. The same Monte Carlo sample of noise parameters is used at each new design point encountered during the optimization. Let the final design point be $\mathbf{d}^*$. Obtain a confirmatory yield estimate at $\mathbf{d}^*$.

In the above procedure, the experimental design of Step 2 and the model of Step 3 are given in terms of both the designable and the noise parameters. However, the Monte Carlo sample in the yield estimation of Step 4 is given in terms of the noise parameters alone.

A Simple Example of Parametric Yield Maximization

Let us consider the familiar inverter example. We make the same assumptions that were used to illustrate the yield estimation procedures in Section 14.5. The designable parameters are the nominal channel widths of MN and MP, W_1^o and W_2^o, respectively. Earlier, we had estimated the parametric yield at the point $\mathbf{d}_{init} = (W_1^o = 10, W_2^o = 20)$ to be $Y'(\mathbf{d}_{init}) = 79.6\%$. The yield maximization procedure described above is used. We construct an RSM for τ_p in terms of both of the designable parameters W_1^o and W_2^o, and the noise parameters $V_{T,n}$, $V_{T,p}$, ΔW_1, and ΔW_2. At each design point, the yield is estimated from a Monte Carlo sample of the four noise parameters. The final design point obtained from the optimization is $\mathbf{d}_{final} = (W_1^o = 11, W_2^o = 22)$. At $\mathbf{d}_{final}$, the parametric yield is estimated to be 100%. Figure 14.14 shows a comparison of the scatter of τ_p and A_m values before and after yield maximization. This plot shows that at $\mathbf{d}_{init}$, the τ_p specification was violated at many of the points while the area measure satisfied its specifications at all the points.

14.7. Worst-Case Analysis

Worst-case analysis is the most commonly used technique in industry for considering manufacturing process tolerances in the design of digital integrated circuits. These

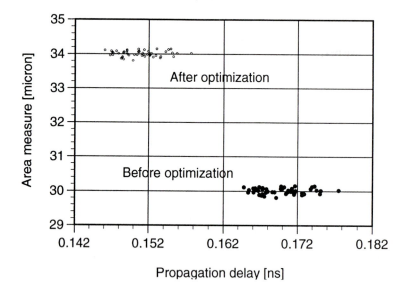

Figure 14.14.   Comparison of circuit performances before and after optimization. During the course of yield maximization, delay and area are traded off so that at $\mathbf{d}_{final}$, all points satisfy both the delay and the area specifications.

approaches are relatively inexpensive compared to the yield maximization approaches in terms of computational cost and designer effort, and they also provide high parametric yields. At any design point, uncontrollable fluctuations in the circuit parameters cause circuit performances to deviate from their nominal designed values. The goal of worst-case analysis is to determine the worst values that the performances may have under these statistical fluctuations. In addition to finding the worst-case values of the circuit performances, this analysis also finds the corresponding worst-case values of the noise parameters. The worst-case noise parameter vector is used in circuit simulation to verify whether circuit performances are acceptable under these conditions. Similar to worst-case analysis, one can also perform *best-case analysis*. In fact, industrial designs are often simulated under best, worst, and nominal noise parameter conditions, which provides designers with quick estimates of the range of variation of circuit performances.

For some circuit performances, e.g., delay, the larger their values are, the worse they become. For other performances, e.g., power dissipation, the smaller the values are the better they become. For each performance r, we can therefore define a *worst-case direction* as follows:

$$w = \begin{cases} +1, & \text{if larger values are worse than smaller values} \\ -1, & \text{if smaller values are worse than larger values} \end{cases} \qquad (14.23)$$

The worst-case performance is defined as a value that bounds some desired percentage of the population of possible performance values. This percentage or probability is called the *worst-case probability* p. Therefore, the worst-case circuit performance value r^{wc} can be defined as the value for which

$$\rho = \begin{cases} \Pr\!\left(r \ge r^{wc}\right), & \text{if } w = +1 \\ \Pr\!\left(r \le r^{wc}\right), & \text{if } w = -1 \end{cases} \qquad (14.24)$$

f(r): Probability density function of performance r

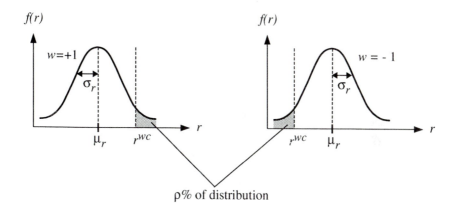

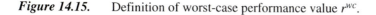

p% of distribution

Figure 14.15.   Definition of worst-case performance value r^{wc}.

The definition is illustrated in Fig. 14.15 where the probability density function of the performance r is denoted by $f(r)$. Here, the worst-case performance value is the value for which a fraction p of the distribution is worse than that value. Note that for any random variable, there exists a *probability distribution function* (to be distinguished from the probability density function) which expresses the probability that the random variable is less than some specified value. For the performance r, we can define a probability distribution function $F()$ as

$$F(a) = \Pr(r \le a) \qquad (14.25)$$

The worst-case performance value r^{wc} can now be defined in terms of the probability distribution function

$$\rho = \begin{cases} 1 - F(r^{wc}), & \text{if } w = +1 \\ F(r^{wc}), & \text{if } w = -1 \end{cases} \tag{14.26}$$

One may raise a question about the relationship between worst-case analysis and the parametric yield of a circuit. Traditional worst-case analysis does not solve the yield maximization problem. Instead, it provides the designer with a single-ended test for the variability in a performance measure. If a circuit is designed such that all the worst-case performance values satisfy their specifications, i.e., if

$$a_k \leq r_k^{wc} \leq b_k, \text{ for } k = 1, 2, ..., p \tag{14.27}$$

the parametric yield of the circuit is at least $(1 - \rho)\%$. However, if all of the specifications are not met under worst-case conditions, no definitive statement can be made regarding the parametric yield of the circuit.

We will describe two techniques for worst-case analysis. The first, often called the "corners" technique, is a simplistic and commonly used procedure. In many cases, however, this technique is too conservative (overly pessimistic) and produces design bottlenecks. The second technique is introduced for a more accurate and realistic analysis. The reader should keep in mind that the designable parameters are fixed during worst-case analysis and therefore do not appear in the discussion below. Further, since the procedures discussed apply uniformly to all performance measures, only one performance measure r is used in the discussion.

The Corners Technique

In this technique, each noise parameter is *independently* set to its extreme value. For a noise parameter s_i with mean s_i^o and standard deviation σ_i, a typical extreme value is

$$s_i = s_i^o \pm 3\sigma_i, \ i = 1, 2, ..., n_s \tag{14.28}$$

where n_s is the number of noise parameters. The direction of deviation of a noise parameter from its mean (i.e., whether $+$ or $-$ is used in the above formula) depends on whether increasing or decreasing the noise parameter s_i makes the performance worse. This is determined by *sensitivity analysis*, i.e., by computing the derivative of the performance with respect to a noise parameter and by considering the sign of that derivative. For each of the n_s noise parameters, the sensitivity is computed at its nominal (or mean) value. Since this technique sets each noise parameter to its extreme or corner value, it is called the "corners" or the one-at-a-time technique.

Once the worst-case values of the noise parameters have been obtained, the worst-case performance value r^{wc} is computed by simulating the circuit with those values (at a fixed design point). This worst-case analysis technique is very conservative, since all of the noise parameters are independently set at their extreme values. The probability that such a combination of noise parameter values will actually occur in a fabricated circuit is extremely small. Therefore, the predicted worst-case value of the circuit performance is overly pessimistic. If the circuit can be designed such that all of the specifications are met even under these pessimistic conditions, then the parametric yield is guaranteed to be very high. In many cases, however, such a design cannot be obtained, which causes severe bottlenecks.

A More Realistic Worst-Case Analysis Technique

In this technique, the worst-case performance value is computed first. How does one compute it? We have two possible cases depending on whether the performance is Gaussian or not.

Gaussian performance measure

If the performance r is Gaussian, its density function is completely characterized by its mean μ_r and its standard deviation σ_r. Moreover, given a worst-case probability ρ, one can refer to readily available tables for the distribution function of a standard Gaussian random variable to determine the value of r^{wc}. A standard Gaussian random variable has mean 0 and standard deviation 1. The random variable r can be converted into a standard Gaussian random variable q via the transformation $q = (r - \mu_r)/\sigma_r$. $\Phi()$ is the standard notation for the distribution function of a standard Gaussian random variable, and $\Phi^{-1}()$ is its inverse. In this case, it can be shown that the solution of (14.26) is

$$r^{wc} = \mu_r + \Phi^{-1}(1 - \rho) \, w \, \sigma_r \qquad (14.29)$$

Non-Gaussian performance measure

If the performance measure r is not Gaussian, (14.26) cannot be used to compute r^{wc}. In this case, the distribution function $F()$ can be estimated using Monte Carlo techniques. Since a large number of samples are required in the Monte Carlo technique, an RSM for the performance measure is usually used. Given the value of ρ and the estimated $F(\ )$, (14.26) is numerically solved for r^{wc} using the well-known Newton-Raphson procedure.

After r^{wc} has been obtained, one has to compute the corresponding set of noise parameter values that would result in the worst-case performance. Recall that the designable parameters are fixed and that the circuit performance is a function of the noise parameters alone, i.e., $r(\mathbf{s})$. Therefore, the worst-case noise parameter vector $\mathbf{s}^{wc}$ is the solution to the following equation

$$r(\mathbf{s}^{wc}) = r^{wc} \qquad (14.30)$$

Note that equation (14.30) above has n_s unknowns and there are infinitely many solutions (or a surface of solutions) for the worst-case noise vector. Any combination of noise parameter values on that surface produces the worst-case performance; thus, an additional condition is required to uniquely identify $\mathbf{s}^{wc}$. The most intuitively appealing choice for the worst-case vector would be the solution of (14.30) which is most probable. This corresponds to the most likely combination of noise parameter values that would be seen in a population of manufactured circuits which would produce the worst-case performance value. Recall that the noise parameter vector is characterized by the joint probability density function (*jpdf*) $f(\mathbf{s})$. Therefore, one can formulate the following constrained maximization problem for determining the worst-case noise vector:

$$\begin{aligned}\text{maximize } \ & f(\mathbf{s}) \\ \text{subject to } \ & r(\mathbf{s}) = r^{wc}\end{aligned} \qquad (14.31)$$

In the above problem, the maximization of the density is equivalent to the maximization of probability, and the constraint ensures that the worst-case performance value is produced.

If the *jpdf* of the noise parameters is Gaussian, the optimization has a simple implication. In this case, the solution $\mathbf{s}^{wc}$ of (14.31) is the point on the surface (14.30) which is (in a probabilistic sense) closest to the mean vector $\mathbf{s}^o$. This is illustrated in Fig. 14.16 for the case of $n_s = 2$. If we assume that the noise parameters s_1 and s_2 are independent, then the contours of equal probability density $f(s_1, s_2)$ are circles centered at the mean point (s_1^o, s_2^o). The solution of (14.31) is akin to moving along the surface $r(s_1, s_2) = r^{wc}$, and looking for the point that is closest to the mean point.

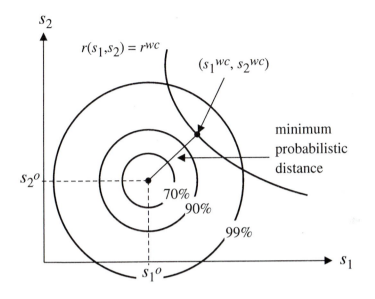

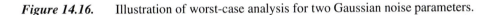

Figure 14.16.   Illustration of worst-case analysis for two Gaussian noise parameters.

If, in addition to the fact that the noise parameters are Gaussian, the performance r is linear with respect to the noise parameters, (14.31) can be solved analytically to provide a closed-form solution for the worst-case noise vector. In other cases, the constrained maximization problem has to be solved numerically using an optimization method.

A Simple Worst-Case Analysis Example

To illustrate the worst-case analysis procedure, we will again consider the example of the inverter shown in Fig. 14.4. We have four noise parameters $V_{T,n}$, $V_{T,p}$, ΔW_1, and ΔW_2, whose statistical distributions have been stated in Section 14.5. The propagation delay τ_p is a smaller-the-better performance measure and so its worst-case direction is $w = +1$. For this circuit, the propagation delay τ_p is relatively insensitive to variations in ΔW_1 and ΔW_2. This is also evident from the RSM of (14.22), where the coefficients of the linear ΔW_1 and ΔW_2 terms are small. Hence, for this example, we drop these two noise parameters. This is an example of a more general technique called *parameter screening*, which is utilized in many design-for-manufacturability methods. Screening helps to reduce the number of parameters being considered and thereby the computational cost.

The fixed design point used in the worst-case analysis is the final design point $\mathbf{d}_{final}$ obtained in the parametric yield maximization example of the previous section: $\mathbf{d}_{final} = (W_1^o = 12, W_2^o = 22)$. The following RSM is constructed for the propagation delay τ_p:

$$\tau_p' = 0.151 + 0.0056 V_{T,n} - 0.0073 V_{T,p} \tag{14.32}$$

When τ_p is expressed as a linear function of the independent Gaussian random variables $V_{T,n}$ and $V_{T,p}$, it becomes a Gaussian random variable. Its mean and variance are estimated to be 0.151 ns and 0.0028 ns, respectively. For a worst-case probability ρ of 0.13%, we can compute the worst-case value using (14.29) to be $\tau_p^{wc} = 0.159$ ns. The corresponding worst-case values of $V_{T,n}$ and $V_{T,p}$ are found by solving (14.31) to be $V_{T,n}^{wc} = 0.91$ V and $V_{T,p}^{wc} = -1.06$ V. Note that these values are different from those that would be obtained from the corners method.

The results from the worst-case analysis using the linear response surface of (14.32) must be verified. For verification, the worst-case value of the propagation delay is obtained by solving (14.26) numerically using the procedure described earlier for a non-Gaussian performance measure. This procedure yields $\tau_p^{wc} = 0.159$ ns. To find the corresponding set of worst-case noise parameter values, we first draw a large Monte Carlo sample of the noise parameters and simulate the inverter circuit to obtain the propagation delay value for each sample. Next, we isolate the sample point for which the performance is within a small tolerance of the worst-case value (we use 0.0005 ns) and for which the probability density is the largest. Then, another Monte Carlo sample is drawn in a small region around this point and the process is repeated until sufficient accuracy in the estimate is achieved. A plot of the values of the propagation delay and the probability density is shown for the last set of sample points in Figure 14.17. The point with the largest value of the probability density is chosen (rightmost point on Fig. 14.17) as the worst-case noise parameter vector: $V_{T,n}^{wc} = 0.89$ V and $V_{T,p}^{wc} = -1.07$ V. The worst-case analysis using the response surface method is therefore verified.

14.8. Performance Variability Minimization

Variabilities in circuit performances are caused by processing and environmental fluctuations. Minimization of performance variability is another important topic in design for manufacturability. Circuits with low variabilities will result in products which are more uniform and have higher quality. As mentioned earlier, there are several quantities which may be used to measure the variability in a performance. We will use the variance of a performance r, denoted by σ^2, to quantify its variability. Two points which were made in Section 14.3 are reiterated here. First, the mean and variance

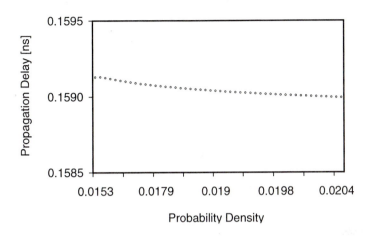

Figure 14.17.   Monte Carlo confirmation of worst-case analysis.

of a performance measure are functions of the designable parameters alone since the effects of the noise parameters have been averaged out. This implies that the variability of a performance measure can be minimized by properly choosing the values of the designable parameters. This is the essence of variability minimization. Second, the mean and variance of a performance have to be estimated since the probability density of the performance is not known. To estimate these two quantities at a design point, a Monte Carlo sample of the noise parameters is drawn: s_i, $i = 1, 2, ..., N_{MC}$. Then the mean μ is estimated by the sample mean given by

$$\mu(\mathbf{d}) = \frac{1}{N_{MC}} \sum_{i=1}^{N_{MC}} r(\mathbf{d}, \mathbf{s}_i)$$ (14.33)

and the variance σ^2 is estimated by the sample variance given by

$$\sigma^2(\mathbf{d}) = \frac{1}{N_{MC}-1} \sum_{i=1}^{N_{MC}} r(\mathbf{d}, \mathbf{s}_i)^2 - \frac{N_{MC}}{N_{MC}-1} \mu^2$$ (14.34)

The specifications on the variability of the circuit performances can be expressed as

$$\sigma_k^2 \le c_k, \text{ for } k = 1, 2, ..., p$$ (14.35)

where σ_k^2 denotes the variance of performance r_k and p is the number of circuit performances. In addition to the variability specifications, there may also be specifications on the nominal values of the circuit performances:

$$a_k \le r_k^o \le b_k, \text{ for } k = 1, 2, ..., p \qquad (14.36)$$

Thus, the variability minimization problem is to determine the optimal set of designable parameter values such that the variability and nominal performance specifications are met. In the parametric yield maximization methods, we had only one objective function to optimize, namely the yield. In the variability minimization problem, however, there may be multiple objective functions since for each performance there are two objectives, and for a particular circuit, multiple performances may be of interest. Such optimization problems are called *multi-criteria optimization problems*. There is a wide body of literature on methods to solve such problems; in this section, we provide a simple (although not the most efficient) solution strategy.

For each performance measure r_k, we define a *variability penalty* A_k using the variance σ_k^2 as follows:

$$A_k(\mathbf{d}) = 100 \left(\frac{\sigma_k^2(\mathbf{d})}{c_k} \right) \qquad (14.37)$$

This linear relationship is shown in Fig. 14.18(a). The best value of σ_k^2 is zero, and the corresponding penalty is zero; the penalty is higher for larger values of σ_k^2. Next, we define a *performance penalty* B_k which depends on the nominal value of the performance, r_k^o, according to one of the relationships shown in Fig. 14.18(b)-(d). Figure 14.18(b) shows a definition for B_k that can be used if smaller values of performance are better than larger values, and Fig. 14.18(c) shows a definition for the opposite case. If all values within the specification are equally desirable, then the definition of Fig. 14.18(c) may be used. Note that each of the above definitions for the penalty is linear; various other nonlinear relationships may also be used, based on the relative desirability of the values. The variability and performance penalties for each performance measure are defined as above.

Next, we define a *circuit penalty* at a design point, $Z(\mathbf{d})$, as

$$Z(\mathbf{d}) = \max_{1 \le k \le p} \left\{ A_k(\mathbf{d}), B_k(\mathbf{d}) \right\} \qquad (14.38)$$

The circuit penalty expresses the overall quality of the design with respect to the variability and nominal performance specifications. Now, we can formulate the

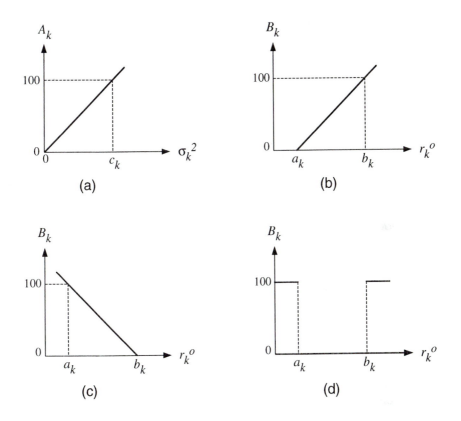

Figure 14.18. Definitions of (a) variability penalty, (b) performance penalty for smaller-the-better case, (c) performance penalty for larger-the-better case, and (d) performance penalty when all acceptable values are equally good.

variability minimization problem as the minimization of the circuit penalty. Formally, the variability minimization problem is to

$$\text{minimize} \quad Z(\mathbf{d}) \tag{14.39}$$

Thus, we have converted the multi-criteria optimization into an optimization involving a single objective function. In so doing, however, our objective function is no longer differentiable. Therefore, the choice of optimization methods to perform the minimization of (14.39) must be restricted to those that do not require the objective functions to be differentiable.

An Example of Variability Minimization

Let us consider the CMOS clock driver circuit shown in Fig. 14.19. The top branch has three inverters and the lower branch has two inverters to illustrate the problem of signal skew that is present in many clock trees.

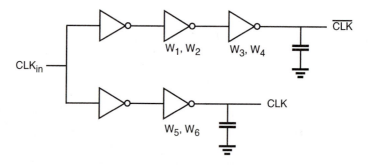

W_1, W_2 W_3, W_4

$\overline{CLK}$

CLK_{in}

W_5, W_6

CLK

Figure 14.19.   Clock distribution circuit.

For our purpose, we shall define the skew to be the difference in the times at which CLK and its complement cross $0.5\ V_{DD}$. The definition is illustrated in Fig. 14.20, where we show two skews, one corresponding to the rising edge of CLK (ΔS_r) and the other corresponding to the falling edge of CLK (ΔS_f). The signal skew ΔS is defined as the larger of the two skews and is the performance measure of interest in this circuit.

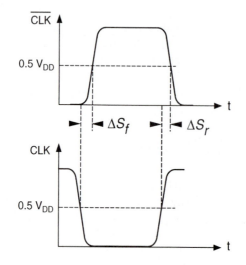

Figure 14.20.   Definition of rising and falling clock skews.

The variability minimization problem is to ensure that

(i)   the *range* of ΔS is less than 0.05 ns, and
(ii)   the nominal value of ΔS is less than 0.5 ns.

The range is defined as the difference between the maximum and minimum values of ΔS.

The designable parameters in the circuit are the nominal channel widths of the two transistors in the second and third inverters of the top branch and the last inverter of the lower branch. The designable parameters are marked on Fig. 14.19. The internal noise parameters are the noise components of the channel widths and lengths of all of the nMOS and pMOS transistors in the circuit, and the common gate-oxide thickness of all the transistors in the circuit. The external noise parameters are the power supply voltage V_{DD} and the operating temperature. To perform this variability minimization, we construct a response surface model for ΔS in terms of the designable, internal, and external noise parameters. We estimate the range and the nominal value of ΔS using the RSM. The initial design point is $d_{init} = (W_1 = 3, W_2 = 6, W_3 = 3, W_4 = 6, W_5 = 3, W_6 = 6)$, where all the widths are shown in μm. The range and nominal value of ΔS at d_{init} are estimated to be 0.311 ns and 1.398 ns, respectively. At the end of variability minimization, we obtain the final design point as $d_{final} = (1.7, 11.1, 16.3, 26.9, 2.5, 38.6)$. The range and nominal value of ΔS at d_{final} are estimated to be 0.04 ns and 0.384 ns, respectively. If the acceptability criteria for the circuit are taken to be the specifications on the range and nominal value of ΔS, then the parametric yield at d_{init} is 0%, which implies that the nominal performance at the initial design point is not acceptable. At d_{final}, the parametric yield is computed to be 100%. We have achieved a 72.5% reduction in the nominal value of ΔS, with a simultaneous reduction of 87% in the range of ΔS.

References

1.   A. Strojwas, Ed., *Selected Papers on Statistical Design of Integrated Circuits*, IEEE Press, 1987.

2.   G. E. P. Box, W. G. Hunter, and J. S. Hunter, *Statistics for Experimenters: An Introduction to Design, Data Analysis and Model Building*, New York: John Wiley and Sons Inc., 1978.

3.   G. E. P. Box and N. R. Draper, *Empirical Model Building and Response Surfaces*, New York: John Wiley and Sons Inc., 1987.

4.   I. Miller and E. Freund, *Probability and Statistics for Engineers*, second edition, Englewood Cliffs, NJ: Prentice Hall Inc., 1977.

5.   W. Maly and S. W. Director, Ed., *Statistical Approach to VLSI*, Amsterdam: North Holland, 1994.

6.   P. J. Ross, *Taguchi Techniques for Quality Engineering*, New York: Mc-Graw Hill,1988.

Exercise Problems

14.1   Consider a simple RC circuit model for a point-to-point interconnect in CMOS chips. The 50% delay time for a step input pulse is

$$\tau_{50\%} = 0.38 \, R \, C$$

Both R and C values are subject to random fluctuations due to process variations.

(a)   Express the sensitivities of $\tau_{50\%}$ with respect to R and C.

(b)   Express the percentage change in the delay in terms of percentage changes in both R and C values under the assumption that their values can vary independently.

(c)   Determine the largest delay increase possible due to $\pm10\%$ fluctuations in both R and C values.

14.2   An empirically determined relationship between the bulk mobility of majority carriers and the doping level is

$$\mu = \mu_{min} + \frac{\mu_{max} - \mu_{min}}{1 + \left(\dfrac{N}{N_0}\right)^{\alpha}} \; [\text{cm}^2/\text{V} \cdot \text{s}]$$

where $\mu_{max} = 1360$, $\mu_{min} = 92$, $N_0 = 1.3 \times 10^{17} /\text{cm}^3$, $\alpha = 0.91$ for electrons, and the corresponding numbers for holes are $495, 48, 6.3 \times 10^{16}$ and 0.76.

(a)   Express the sensitivity of the mobility μ with respect to the doping level N.

(b)   Determine the percentage change in μ for 10% change in doping level from the nominal value for electrons and holes, respectively.

14.3   The dependence of the mobility on temperature is determined empirically as

$$\mu(T) = \frac{\mu(300\ K)}{\left(\dfrac{T}{300}\right)^{1.5}}$$

(a)   Derive an expression for $\Delta\mu/\mu$ in terms of $\Delta T/T$.

(b)   Calculate the percentage change in μ due to a 5% increase in temperature from 300 K for $\mu(300\ K) = 980\ cm^2/V \cdot s$.

14.4   Let us consider the device transconductance parameter

$$k = \mu C_{ox} \frac{W}{L}$$

It is known that all four variables μ, C_{ox}, W, and L are random variables due to process variations.

(a)   Derive an expression for the percentage change in k due to the percentage changes in μ, C_{ox}, W, and L.

(b)   If you assume that all four parameters change independently (corners technique), what would be the maximum percentage change expected due to 5% changes in all four parameters for nominal values of $(\mu^o, C_{ox}{}^o, W^o, L^o) = (980\ cm^2/V \cdot s, 35 \times 10^{-8}$ F/cm^2, 5 μm, 1.0 μm)?

14.5   The logic threshold voltage V_{th} of a CMOS inverter from (5.87) is

$$V_{th} = \frac{V_{T0,n} + \sqrt{\dfrac{1}{k_R}}\left(V_{DD} + V_{T0,p}\right)}{1 + \sqrt{\dfrac{1}{k_R}}}$$

where

$$k_R = \frac{k_n}{k_p} = \frac{\mu_n C_{ox} \dfrac{W_n}{L_n}}{\mu_p C_{ox} \dfrac{W_p}{L_p}}$$

(a)   Assuming that $\mu_n = 2.5\ \mu_p$, $L_n = L_p$ for simplicity, derive an expression for the percentage change in V_{th} in terms of percentage changes in $V_{T0,n}$, $V_{T0,p}$, and W_p/W_n.

(b)   For nominal values of $V_{T0,n} = 0.8$ V, $V_{T0,p} = -0.8$ V, and $W_p = 2.5\ W_n$, determine the maximum (worst-case) deviation of V_{th} for a ±0.1 V change in both $V_{T0,n}$ and $V_{T0,p}$, and a ±15% change in W_p/W_n.

14.6   A CMOS chip with three signal input terminals, one output terminal, one power supply terminal, and one ground terminal is presented to you. The input signal voltages can vary from 0 to 5 V, the power supply voltage can vary from 4.5 V to 5.5 V, and the device operating temperature can range from 0 to 85 °C.

We want to design an experiment so that the DC response of the chip can be described by a quadratic response surface model for three inputs, power supply voltage, and temperature.

(a)   Design a full factorial experiment.
(b)   Design a Latin hypercube experiment.
(c)   Design Taguchi's orthogonal array.

14.7   Design a CMOS full adder circuit and its layout using $W/L = 5/2$ for nMOS transistors and $W/L = 10/2$ for pMOS transistors. For a capacitive load of 1 pF, determine the worst-case delay of your particular design when the range of power supply variations is 4.5 to 5.5 V and the operating temperature ranges from 25 °C to 85 °C. Also assume that the magnitude ranges of the threshold voltages are from 0.8 to 1.2 V for both transistor types. All other parameters are assumed to be at their nominal values in this problem.

14.8   Determine the worst-case power dissipation for the design in Problem 14.7.

14.9 Suppose that the noise parameters are jointly Gaussian. The joint probability density function of a Gaussian random vector s with mean s^o and variance/covariance matrix Q is given by

$$f(\mathbf{s}) = \frac{1}{\left(\sqrt{2\pi}\right)^{n_s}\left(\sigma_1\sigma_2\cdots\sigma_{n_s}\right)}\exp\left[-\frac{1}{2}\left(\mathbf{s}-\mathbf{s}^o\right)^T Q^{-1}\left(\mathbf{s}-\mathbf{s}^o\right)\right]$$

Consider the RC circuit model of Problem 14.1 with the performance measure being the 50% delay time $\tau_{50\%}$. Suppose R and C are independent Gaussian random variables. R has a mean of $R^o=10$ kΩ and a standard deviation of $\sigma_R=1$ kΩ. C has a mean of $C^o=10$ pF and a standard deviation of $\sigma_C=1$ pF. The worst-case direction for $\tau_{50\%}$ is +1.

(a) Determine the mean and standard deviation of $\tau_{50\%}$.

(b) $\tau_{50\%}$ is not a Gaussian random variable. To verify this, draw a Monte Carlo sample of R and C, use the formula given in Problem 14.1 to compute $\tau_{50\%}$, and plot the density of $\tau_{50\%}$. Compare the density to that of a Gaussian random variable having the mean and standard deviation of $\tau_{50\%}$ that you computed in (a).

(c) The worst-case value of $\tau_{50\%}$ can be computed using the procedure for a non-Gaussian performance measure given in the discussion of worst-case analysis. Suppose the value is denoted by $\tau_{50\%}{}^{wc}$. To find the worst-case values of R and C, one must solve (14.31). Using the formula for the (*jpdf*) of a Gaussian random vector, reformulate the maximization of (14.31) into a minimization. The objective function for this minimization is called the "probabilistic distance." For this example, the minimization problem can be solved analytically using pencil and paper. Obtain the condition for minimization.

(d) Use the fact that $R^o = C^o$ and $\sigma_R = \sigma_C$ to obtain R^{wc} and C^{wc} in terms of $\tau_{50\%}{}^{wc}$.

14.10 (a) List the pros and cons of using the pessimistic worst-case analysis results in the design of integrated circuits in view of parametric yield, design effort, and project schedule.

(b)   Would the use of the simple-minded, pessimistic worst-case analysis technique such as the corners technique always shorten the development time?

(c)   What can be the most serious problem that can be faced by the pessimistic analysis in a big project which involves many different organizations?

14.11 Worst-case models even for the same circuit are different for different performances. For instance, the worst-case MOS model for delay time will be different from the worst-case MOS model for power dissipation. Yet, most design practices have been carried out using *slow, medium, fast* transistor models for both nMOS and pMOS transistors. Discuss how one would simulate the clock skew in a CMOS clock distribution circuit under such circumstances. What would be the correct way of performing circuit simulation of clock skews, if the transistor models can be custom-ordered by designers?

14.12 Discuss the strategy for developing manufacturable specifications for chip processing in terms of required control on threshold voltages, variations in channel lengths and widths, gate oxide thickness, and substrate and tub doping profiles.

14.13 The so-called "throw-over-the-wall" approach overstrains the effort needed by the counterpart organization. Discuss how one can avoid such overstraining requirements with shared responsibility in order to make the whole technology development process more cost-effective.

14.14 Taguchi's orthogonal array approach appeals to designers due to its simplicity; the design of the experiment is readily available in table form. Apply the L18 in MOS circuit design by using an example of your choice to demonstrate the principle of statistical design.

14.15 From (14.31) and (14.32) derive the expression for the worst-case noise vector for two independent Gaussian random variables, each with a mean of 0 and standard deviation of 1.

CHAPTER 15

DESIGN FOR TESTABILITY

15.1. Introduction

The task of determining whether fabricated chips are fully functional is highly complex and can be very time-consuming. However, when faulty chips pass an improperly designed test, they can cause system failures and enormous difficulty in system debugging. It is known that the debugging cost increases by about tenfold from chip level to board level, and also from board level to system level. Thus, it is of great importance to detect faults as early as possible. As the number of transistors integrated into a single chip increases, the task of chip testing to ensure correct functionality becomes increasingly more difficult. However, in a production environment, many chips must be tested within a short time for timely delivery to customers. To overcome such difficult issues, *design for testability* has become ever more critical. In this chapter, we discuss types of faults, the corresponding fault models, design of testable circuits, and self-testing circuits. The testability will be defined in terms of observability and controllability, which are also commonly used in control and system theory. Introductory material in this chapter is largely based on a tutorial by Patel. For in-depth treatment of this topic, the reader is encouraged to consult Abramovici et al.

15.2. Fault Types and Models

Chip testing, in the conventional sense, is usually multi-purpose and attempts to detect faults in fabrication, design, and failures due to stressful operating conditions, namely the reliability problems. Input test vectors are devised and applied to the device under test (DUT) or circuit under test (CUT) as its stimuli. Then the measured outputs are compared with the expected correct responses to determine whether DUT is good (go) or bad (no-go). The main difficulty in testing is caused by the fact that only input and output pins of DUT are accessible, although at the test bench in the development laboratory, the internal nodes of unpackaged chips can be probed at the topmost metal level before passivation is done. As the operating clock frequency of the chips increases beyond several tens of megahertz, the at-speed test has also become a difficult problem. The difficulty stems from the signal integrity (transient ringing) problem in sending test signals from the tester to DUT and in detecting response signals from DUT due to impedance mismatch and transmission line problems in the tester interconnects. The impedance mismatch problem has been partially addressed in chip I/O design or by using a table lookup technique to correct delay measurement errors. In addition to the tester problem, the generation of correct test vectors to detect all modeled faults and design errors in complex chips, either manually or through an automatic test patter generator (ATPG), has become a difficult task. In this chapter, we will limit our discussion to faults caused by physical defects.

Examples of *physical defects* include:

• defects in silicon substrate
• photolithographic defects
• mask contamination and scratches
• process variations and abnormalities
• oxide defects

The physical defects can cause electrical faults and logical faults. The *electrical faults* include:

• shorts (bridging faults)
• opens
• transistor stuck-on, stuck-open
• resistive shorts and opens
• excessive change in threshold voltage
• excessive steady-state currents

The electrical faults in turn can be translated into logical faults. The *logical faults* include:

• logical stuck-at-0 or stuck-at-1
• slower transition (delay fault)
• AND-bridging, OR-bridging

The relationships between physical defects, electrical faults, and logical faults can be explained using a simple NOR2 gate as shown in Fig. 15.1. A metallic blob (physical defect) between the common drain terminal in the n-diffusion region and the ground bus line shown in Fig. 15.1(a) can be modeled as a *resistive short* between the output node Z and the ground as shown in Fig. 15.1(b), and also by a *stuck-at-0* (s-a-0) fault of output Z when the resistance is low or a *pull-up delay fault* when the resistance is high, as shown in Fig. 15.1(c).

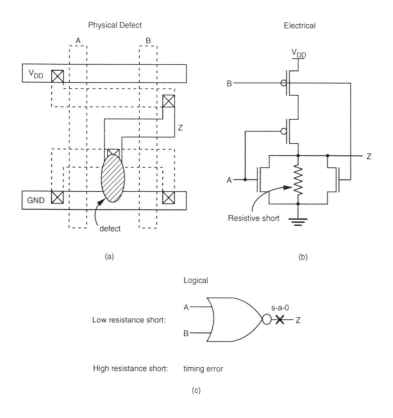

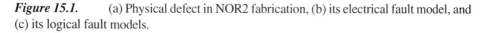

Figure 15.1.   (a) Physical defect in NOR2 fabrication, (b) its electrical fault model, and (c) its logical fault models.

Figure 15.2 shows other types of faults in a CMOS circuit consisting of NOR2, NAND2, and inverter gates. In this circuit, the input line B can be stuck-at-1 (s-a-1), since some part of the input line is shorted to the power rail. The pMOS transistor of the first stage NOR2 gate is stuck-on due to a process problem that causes a short between its source and drain terminals. The top nMOS transistor in the NAND2 gate, on the other hand, is stuck-open due to either an incomplete contact (open) of the source or drain node or due to a large separation of drain or source diffusion from the gate, which causes permanent turn-off of the transistor regardless of the input C value. The stuck-on and stuck-open faults are elaborated on in Fig. 15.3. The bridging fault between the output line of the inverter and the input line C can be due to a fabrication defect which causes a short between any two parts of the two lines. Although in the circuit diagram, these two lines are seemingly far apart, in the actual layout, some parts of these two lines can be close to each other. In such a layout, these two lines can be shorted due to underetching in the line patterning process.

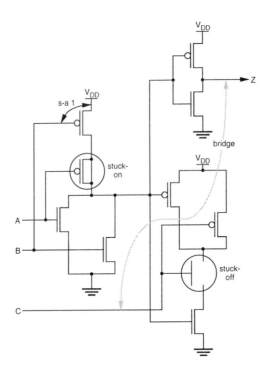

Figure 15.2.   Some process-related defects in a CMOS circuit consisting of NOR2, NAND2, and inverter gates.

The single stuck-at fault models are used frequently, although the DUT can have defects that do not map to a single stuck-at fault. Some of the reasons are:

- Complexity of test generation is greatly reduced.
- Single stuck-at fault is independent of technology, design style.
- Single stuck-at tests cover a large percentage of multiple stuck-at faults.
- Single stuck-at tests cover a large percentage of *unmodeled* physical defects.

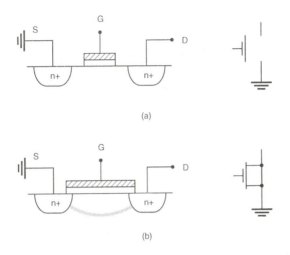

Figure 15.3.   MOS transistor with (a) stuck-open (off) fault and (b) stuck-on (short) fault.

In fact, it has been shown that in a two-level circuit with no redundancy, any complete test set for all single stuck-at faults can cover all stuck-at faults. Multiple stuck-at fault models find applications for fuse or anti-fuse based programmable designs such as programmable gate arrays, field programmable gate arrays (FPGAs), and RAMs.

The *delay fault* which causes timing failures at target speed can be due to several factors. To name a few,

- improper consideration of on-chip interconnect delays and other timing considerations,
- excessive variations in the fabrication process which cause significant variations in circuit delays and clock skews,
- opens in metal lines connecting parallel transistors which make the effective transistor size much smaller,
- aging effects such as hot-carrier induced delay increase.

The task of detecting delay faults is even more subtle than detecting functional faults in steady state. The functional test is usually done at speeds lower than the target speed due to the limitations of the testers. Special clocking is used to apply delay tests on a slow tester. The fault models mentioned above are used in fault simulation aimed at

- test generation,
- construction of fault dictionaries, or
- circuit analysis in the presence of faults.

Each fault dictionary stores the expected output response of every faulty circuit to a particular test vector corresponding to a particular simulated fault.

15.3. Controllability and Observability

The *controllability* of a circuit is a measure of the ease (or difficulty) with which the controller (test engineer) can establish a specific signal value at each node by setting values at the circuit input terminals. The *observability* is a measure of the ease (or difficulty) with which one can determine the signal value at any logic node in the circuit by controlling its primary input and observing the primary output. Here the term *primary* refers to the I/O boundary of the circuit under test. The degree of controllability and observability and, thus, the degree of testability of a circuit, can be measured with respect to whether test vectors are generated deterministically or randomly. For example, if a logic node can be set to either logic 1 or 0 only through a very long sequence of random test vectors, the node is said to have a very *low random controllability* since the probability of generating such a vector in random test generation is very low. There exist time constraints in practice, and in such cases the circuit may not be considered testable. There are deterministic procedures for test generation for combinational circuits, such as the D-algorithm which uses a recursive search procedure advancing one gate at a time and backtracking, if necessary, until all the faults are detected. The D-algorithm requires a large amount of computer time. To overcome such shortcomings, many improved algorithms such as Path-Oriented DEcision Making (PODEM) and FAN-out-oriented test generation (FAN) have been introduced. Sequential circuit test generation is several orders of magnitude more difficult than these algorithms. To ease the task of ATG, design-for-test (DFT) techniques are routinely employed.

Let us now consider the simple circuit in Fig. 15.4 consisting of four simple logic gates. To detect any defect on line 8, the primary inputs A and B must be set to logic 1. However, such a setting forces line 7 to logic 1. Thus, any stuck-at-1 (s-a-1)

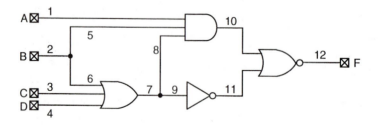

Figure 15.4.   A simple circuit consisting of four gates with four primary inputs and one primary output.

fault on line 7 *cannot be tested* at the primary output, although in the absence of such a fault, the logic value on line 7 can be fully controllable through primary inputs B, C, and D. Therefore, this circuit is *not fully testable*. The main cause of this difficulty in this circuit is the fact that input B fans out to lines 5 and 6, and then after the OR3 gate, both line signals are combined in the AND3 gate. Such a fanout is called *reconvergent fanout*. Reconvergent fanouts make the testing of the circuit much more difficult.

If a large number of input vectors are required to set a particular node value to 1 or 0 (fault excitation) and to propagate an error at the node to an output (fault effect propagation), then the testability is low. The circuits with poor controllability include those with feedbacks, decoders, and clock generators. The circuits with poor observability include sequential circuits with long feedback loops and circuits with reconvergent fanouts, redundant nodes, and embedded memories such as RAM, ROM, and PLA.

15.4. Ad Hoc Testable Design Techniques

One way to increase the testability is to make nodes more accessible at some cost by physically inserting more access circuits to the original design. Listed below are some of the ad hoc testable design techniques.

Partition-and-Mux Technique

Since the sequence of many serial gates, functional blocks, or large circuits are difficult to test, such circuits can be partitioned and multiplexors (muxes) can be inserted such that some of the primary inputs can be fed to partitioned parts through multiplexers with accessible control signals. With this design technique, the number of accessible nodes can be increased and the number of test patterns can be reduced. A case in point would be the 32-bit counter. Dividing this counter into two 16-bit parts would reduce the testing time in principle by a factor of 2^{15}. However, circuit partitioning and addition of multiplexers may increase the chip area and circuit delay. This practice is not unique and is similar to the divide-and-conquer approach to large, complex problems. Figure 15.5 illustrates this method.

Initialize Sequential Circuit

When the sequential circuit is powered up, its initial state can be a random, unknown state. In this case, it is not possible to start the test sequence correctly. The state of a sequential circuit can be brought to a known state through *initialization.* In many designs, the initialization can be easily done by connecting asynchronous preset or clear-input signals from primary or controllable inputs to flip-flops or latches.

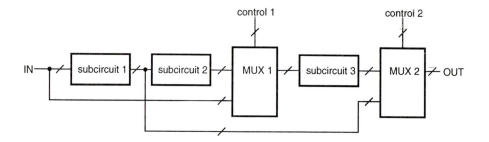

Figure 15.5.   Partition-and-mux method for large circuits.

Disable Internal Oscillators and Clocks

To avoid synchronization problems during testing, internal oscillators and clocks should be disabled. For example, rather than connecting the circuit directly to the on-chip oscillator, the clock signal can be ORed with a disabling signal followed by an insertion of a testing signal as shown in Fig. 15.6.

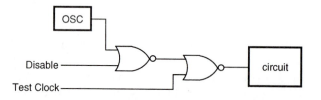

Figure 15.6.   Avoid synchronization problems via disabling of the oscillator.

Avoid Asynchronous Logic and Redundant Logic

The enhancement of testability requires serious tradeoffs. The speed of an asynchronous logic circuit can be faster than that of the synchronous logic circuit counterpart. However, the design and test of an asynchronous logic circuit are more difficult than for a synchronous logic circuit, and its state transition times are difficult to predict. Also, the operation of an asynchronous logic circuit is sensitive to input test patterns, often causing race problems and hazards of having momentary signal values opposite to the expected values. Sometimes, designed-in logic redundancy is used to mask a static hazard condition for reliability. However, the redundant node cannot be observed since the primary output value cannot be made dependent on the value of the redundant node. Hence, certain faults on the redundant node cannot be tested or detected. Figure 15.7 shows that the bottom NAND2 gate is redundant and the stuck-at-1 fault on its output line cannot be detected. If a fault is undetectable, the associated line or gate can be removed without changing the logic function.

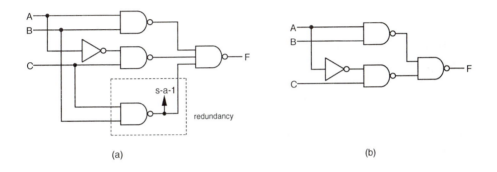

$$F = AB + BC + \overline{A}C$$
$$= AB + \overline{A}C$$

Figure 15.7.   (a) A redundant logic gate example. (b) Equivalent gate with redundancy removed.

Although it is unessential to test redundant nodes when they are designed in as backup parts, either to enhance the circuit reliability or to increase the fabrication yield, the use of redundant circuits can make test generation much more complex and difficult. In fact, test generators, especially the random or deterministic test generators, would not be able to recognize such design intent. Some redundancy in circuits may be unintentional due to lack of design efficiency.

Avoid Delay-Dependent Logic

Chains of inverters can be used to design in delay times and use AND operation of their outputs along with inputs to generate pulses, as shown in Fig. 15.8.

Most automatic test pattern generation (ATPG) programs do not include logic delays to minimize the complexity of the program. As a result, such delay-dependent logic is viewed as redundant combinational logic, and the output of the reconvergent gate is always set to logic 0, which is not correct. Thus, the use of delay-dependent logic should be avoided in design for testability.

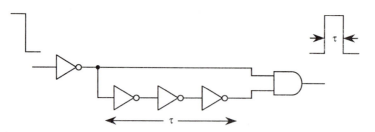

Figure 15.8.   A pulse-generation circuit using a delay chain of three inverters.

15.5. Scan-Based Techniques

As discussed earlier, the controllability and observability can be enhanced by providing more accessible logic nodes with use of additional primary input lines and multiplexors. However, the use of additional I/O pins can be costly not only for chip fabrication but also for packaging. A popular alternative is to use scan registers with both shift and parallel load capabilities. The scan design technique is a structured approach to design sequential circuits for testability. The storage cells in registers are used as observation points, control points, or both. By using the scan design techniques, the testing of a sequential circuit is reduced to the problem of testing a combinational circuit.

In general, a sequential circuit consists of a combinational circuit and some storage elements. In the scan-based design, the storage elements are connected to form a long serial shift register, the so-called *scan path,* by using multiplexors and a mode (test/normal) control signal, as shown in Fig. 15.9.

In the test mode, the scan-in signal is clocked into the scan path, and the output of the last stage latch is scanned out. In the normal mode, the scan-in path is disabled and the circuit functions as a sequential circuit. The testing sequence is as follows:

Step 1:   Set the mode to *test* and let latches accept data from scan-in input.
Step 2:   Verify the scan path by shifting in and out the test data.
Step 3:   Scan in (shift in) the desired state vector into the shift register.
Step 4:   Apply the test pattern to the primary input pins.
Step 5:   Set the mode to *normal* and observe the primary outputs of the circuit after sufficient time for propagation.
Step 6:   Assert the circuit clock for one machine cycle to capture the outputs of the combinational logic into the registers.
Step 7:   Return to *test* mode; scan out the contents of the registers, and at the same time scan in the next pattern.
Step 8:   Repeat steps 3-7 until all test patterns are applied.

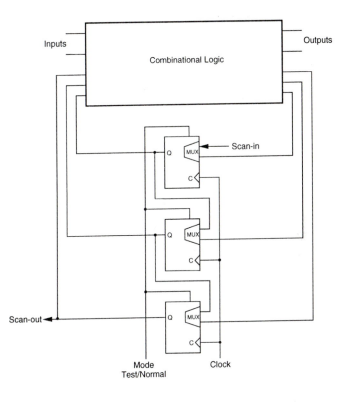

Figure 15.9.   The general structure of scan-based design.

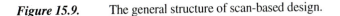

The storage cells in scan design can be implemented using edge-triggered D flip-flops, master-slave flip-flops, or level-sensitive latches controlled by complementary clock signals to ensure race-free operation. A detailed discussion of such latches and flip-flops is given in Chapter 8. Figure 15.10 shows a scan-based design of an edge-triggered D flip-flop. In large high-speed circuits, optimizing a single clock signal for skews, etc., both for normal operation and for shift operation, is difficult. To overcome this difficulty, two separate clocks, one for normal operation and one for shift operation, are used. Since the shift operation does not have to be performed at the target speed, its clock is much less constrained.

An important approach among scan-based designs is the level sensitive scan design (LSSD), which incorporates both the level sensitivity and the scan path approach using shift registers. The level sensitivity is to ensure that the sequential circuit response is independent of the transient characteristics of the circuit, such as the component and wire delays. Thus, LSSD removes hazards and races. Its ATPG is also simplified since tests have to be generated only for the combinational part of the circuit.

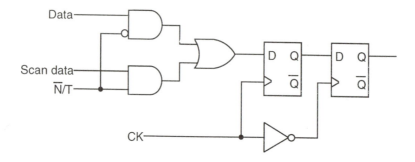

Figure 15.10.   Scan-based design of an edge-triggered D flip-flop.

The boundary scan test method is also used for testing printed circuit boards (PCBs) and multichip modules (MCMs) carrying multiple chips. Shift registers are placed in each chip close to I/O pins in order to form a chain around the board for testing. With successful implementation of the boundary scan method, a simpler tester can be used for PCB testing.

On the negative side, scan design uses more complex latches, flip-flops, I/O pins, and interconnect wires and, thus, requires more chip area. The testing time per test pattern is also increased due to shift time in long registers.

15.6. Built-In Self Test (BIST) Techniques

In built-in self test (BIST) design, parts of the circuit are used to test the circuit itself. On-line BIST is used to perform the test under normal operation, whereas off-line BIST is used to perform the test off-line. The essential circuit modules required for BIST include:

• Pseudo random pattern generator (PRPG)
• Output response analyzer (ORA)

The roles of these two modules are illustrated in Fig. 15.11. The implementation of both PRPG and ORA can be done with Linear Feedback Shift Registers (LFSRs).

Pseudo Random Pattern Generator

To test the circuit, test patterns first have to be generated either by using a pseudo random pattern generator, a weighted test generator, an adaptive test generator, or other means. A pseudo random test generator circuit can use an LFSR, as shown in Fig. 15.12.

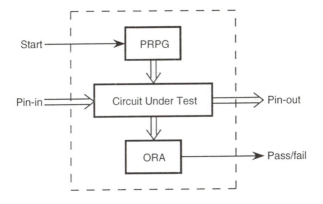

Figure 15.11.   A procedure for BIST.

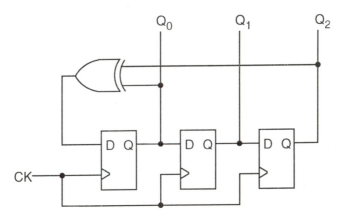

Figure 15.12.   A pseudo-random sequence generator using LFSR.

Linear Feedback Shift Register as an ORA

To reduce the chip area penalty, data compression schemes are used to compare the compacted test responses instead of the entire raw test data. One of the popular data compression schemes is the *signature analysis*, which is based on the concept of cyclic redundancy checking. It uses polynomial division, which divides the polynomial representation of the test output data by a characteristic polynomial and then finds the remainder as the signature. The signature is then compared with the expected signature to determine whether the device under test is faulty. It is known that compression can cause some loss of fault coverage. It is possible that the output of a faulty circuit can match the output of the fault-free circuit; thus, the fault can go undetected in the signature analysis. Such a phenomenon is called *aliasing*.

In its simplest form, the signature generator consists of a single-input linear feedback shift register (LFSR), as shown in Fig. 15.13, in which all the latches are edge-triggered. In this case, the signature is the content of this register after the last input bit has been sampled. The input sequence $\{a_n\}$ is represented by polynomial $G(x)$ and the output sequence by $Q(x)$. It can be shown that $G(x) = Q(x) P(x) + R(x)$, where $P(x)$ is the characteristic polynomial of LFSR and $R(x)$ is the remainder, the degree of which is lower than that of $P(x)$. For the simple case in Fig. 15.13, the characteristic polynomial is

$$P(x) = 1 + x^2 + x^4 + x^5$$

For the 8-bit input sequence $\{1\ 1\ 1\ 1\ 0\ 1\ 0\ 1\}$, the corresponding input polynomial is

$$G(x) = x^7 + x^6 + x^5 + x^4 + x^2 + 1$$

and the remainder term becomes $R(x) = x^4 + x^2$, which corresponds to the register contents of $\{0\ 0\ 1\ 0\ 1\}$.

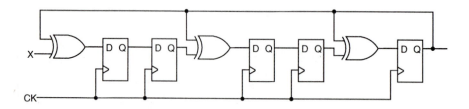

Figure 15.13.   Polynomial division using LFSR for signature analysis.

Output Response Analyzer

The on-chip storage of a fault dictionary containing all test inputs with the corresponding outputs is prohibitively expensive in terms of the chip area. A simple alternative method is to compare the outputs of two identical circuits for the same input, with one of them regarded as reference. However, if both circuits have the same faults, their outputs can still match. Such faults cannot be detected with this technique, although the probability of two identical circuits having exactly the same faults would be very low.

In addition to the above circuits for built-in self test, self-checking design techniques can be used to detect faults autonomously during on-line operation. Usually a checker circuit is inserted such that the checker generates and sends out a signal when on-line faults occur. The distribution of checkers throughout a very large digital circuit or system can provide prompt detection of the fault location by tracing the checker which sent the fault signal. The use of self-checking circuits simplifies the development of software diagnostic programs. However, some additional hardware is required, and the checker itself needs to have self-checking capability. When self-checking capability of the checker itself is required, a single-output checker is not sufficient since that output may have a stuck-at fault, thus preventing the detection of actual faults in the circuit under test. A checker with a *pair* of outputs can be used instead to overcome this problem.

Built-In Logic Block Observer

The built-in logic block observer (BILBO) register is a form of ORA which can be used in each cluster of partitioned registers. A basic BILBO circuit is shown in Fig. 15.14, which allows four different modes controlled by C_0 and C_1 signals.

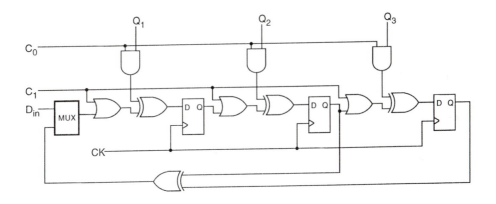

Figure 15.14.   3-bit BILBO example.

C_0	C_1	Mode
0	0	linear shift
1	0	signature analysis
1	1	data (complemented) latch
0	1	reset

The BILBO operation allows monitoring of circuit operation through exclu-sive-ORing into LFSR at multiple points, which corresponds to the signature analyzer with multiple inputs.

15.7 Current Monitoring I_{DDQ} Test

An often-used technique for testing fabrication defects is the I_{DDQ} test. Under a bridging fault, the static currents drawn from the power supply in CMOS circuits can be noticeably high, well beyond the expected range of leakage currents. For example, if the drain node of the pMOS transistor in a CMOS inverter is shorted to the power supply rail due to a bridging fault, its I_{DDQ} current can be very high even when the input is high. It can also detect other fabrication defects not easily detected by other test methods, including:

- gate oxide short
- channel punch-through
- p-n diode leakage
- transmission-gate defect

The I_{DDQ} test consists of applying the test vector and then monitoring the current drawn from the power supply rail in DC steady state. Although this test requires more testing time, the fault detection capability is greatly improved with small circuit overhead required to monitor the I_{DDQ} in various parts of DUT.

While stuck-at tests require both fault sensitization and fault effect propaga-tion, the I_{DDQ} test requires only fault sensitization. However, its performance in open drain and open gate test is less effective. The I_{DDQ} fault coverage is relatively easy to obtain and may potentially offer a full-chip coverage capability for large designs.

The design guidelines for I_{DDQ} testability are as follows:

- low static current states, e.g., full CMOS is preferred
- no active pull-ups or pull-downs
- no internal drive conflicts, e.g., drivers share a bus
- no floating nodes in the circuit
- no degraded voltages, e.g., must have $V_{OH} = V_{DD}$ and $V_{OL} = 0$

References

1.   M. Abramovici, M. A. Breuer, and A. D. Friedman, *Digital Systems Testing and Testable Design*, New York, NY: Computer Science Press, 1990.

2.   N. H. E. Weste and K. Eshraghian, *Principles of CMOS VLSI Design*, second edition, Reading, MA: Addison-Wesley Publishing Co., 1993.

3.   A. Osseiran, *Design for Testability*, Swiss Federal Institute of Technology (EPFL) Intensive Summer Course Note, 1993.

4.   N. Jha and S. Kundu, *Testing and Reliable Design of CMOS Circuits*, Norwell, MA: Kluwer Academic Publishers, 1990.

5.   E. J. McClusky, *Logic Design Principles with Emphasis on Testable VLSI Circuits*, Englewood Cliffs, NJ: Prentice-Hall, 1986.

6.   J. H. Patel, *ECE443 Class Notes*, University of Illinois at Urbana-Champaign, Spring 1994.

7.   M.R. Barber, "Fundamental timing problems in testing MOS VLSI on modern ATE," *IEEE Design and Test*, pp. 90-97, August 1984.

8.   M.E. Mokhari-Bolhassan and S.M. Kang, "Analysis and correction of VLSI delay measurement errors due to transmission-line effects," *IEEE Trans. Circuits and Systems*, vol. 35, pp. 19-25, January 1988.

9.   M.A. Breuer and A.D. Friedman, *Reliable Design of Digital Systems*, Rockville, MD: Computer Science Press, 1976.

10.   R.L. Wadsack, "Fault modeling and logic simulation of CMOS and MOS integrated circuits," *Bell System Technical Journal*, vol. 57, no. 5, pp. 1449-1474, May-June 1978.

Exercise Problems

15.1   Give a logic circuit example in which stuck-at-1 fault and stuck-at-0 fault are indistinguishable.

15.2   Show that the remainder of LFSR in Fig. 15.13 is indeed $R(x) = x^4 + x^2$.

15.3   Explain the merits or demerits of the bus structure in relation to testability. How would the bus structure impact the chip area overhead?

15.4   Determine whether the leakage current test for chips should be done prior to or after the functional test. What can you say about the test frequency of chips containing dynamic circuits designed to operate at very high frequency? Can it fail the functional test at much lower frequency? If so, explain why.

15.5   Show a few logic circuit examples whose logical fault coverage is dependent on the test vector sequence.

15.6   Find the set of all test vectors which detects the stuck-at-0 fault in line B in Fig. 15.2. Repeat for the stuck-at-1 fault in line C.

15.7   Show that if there are *undetectable* stuck-at faults in a combinational circuit then the circuit can be reduced according to the following rules (the rule set for OR gates is given below; prove it and find the rules for AND, NOR, NAND, and XOR gates).

Undetectable Fault	Reduction Rule for OR Gates
Input x_i s-a-0	Remove input x_i
Input x_i s-a-1	Remove OR gate, connect output to 1
Output s-a-0	Remove OR gate, connect output to 0
Output s-a-1	Remove OR gate, connect output to 1

15.8   Apply the rules in Problem 15.7 to the circuit shown in Fig. 15.7.

APPENDIX

DESIGN AND IMPLEMENTATION OF A 16-BIT ADDER

A.1. Introduction

Here we present a full-custom design and implementation of a CMOS 16-bit binary adder circuit using generic 2 μm CMOS design rules. This circuit design problem was assigned as a team project to the students registered in ECE 382 - *Large Scale Integrated Circuit Design*, a senior-level course at the University of Illinois at Urbana-Champaign, during the 1994 spring semester. The circuit design presented in the following sections reflects but one approach to the problem, as completed by one of the several student teams who undertook this project. The authors wish to thank their students for their willing participation and sincere efforts in active learning, in particular the design team consisting of Eugene Hodges, Teresa Johnson, Wayland Middendorf, Eugene Zeldin, and Charles Hall, whose design is presented here.

A.2. Project Description

Students who take the senior-level course ECE382 - *Large Scale Integrated Circuit Design* at the Department of Electrical and Computer Engineering of the University of Illinois at Urbana-Champaign are expected to have a working knowledge of circuit simulation tools such as SPICE. Students are requested to complete a full-custom layout of their designs using a graphic layout editor which is installed on advanced engineering workstations in design laboratories.

The abbreviated project description is as follows:

> *A full-custom CMOS 16-bit binary adder circuit will be designed using 2 μm generic CMOS design rules. The competition is on speed, chip area, and power; the best design should be the fastest in speed (50% weight), the smallest in chip area (30% weight), and it should have the lowest power consumption (20% weight).*
>
> *Each design team, consisting of at most five students, should investigate and choose the best CMOS 16-bit adder circuit. Design issues to be considered include circuit type (static or dynamic), carry propagation arrangements, buffering, and chip floorplanning. Both input latches and output latches will be required to hold data. The adder circuit should be capable of driving a capacitive load of 10 pF.*
>
> *Each team should partition the task to its members evenly so that the deadline can be met with successful completion of the design. All layout plots, circuit diagrams, and SPICE simulation results on speed and power should be submitted with a final report by the end of the project deadline.*

The student teams were given six weeks to complete this open-ended design project. In the following sections, excerpts of the selected team's final project report are presented.

A.3. Background and Comparisons

Many different designs have been proposed for full adders, with different ranges of speed and area costs. The simplest adder is the ripple carry adder, composed of a cascade of full-adder blocks linked together by connecting the carry bits. Each full adder cell performs an exclusive-OR (XOR) operation on A, B, and C_{in} to compute the sum (S) and then computes C_{out}. However, the speed is limited by the speed of the carry bit rippling through the carry chain. If the delay of one stage is τ_i, then the overall delay of an n-bit adder is:

$$\tau_{total} = n \cdot \tau_i$$

This delay becomes non-trivial with a 16-bit implementation. To improve this delay, the carry lookahead adder computes the carry bits separately, parallelizing this computation as much as possible. Two intermediate signals, propagate (P) and generate (G), are computed for each bit of the addition and used to compute the sum and carry bits (Fig. A.1).

$$P = A \oplus B$$
$$G = A \cdot B$$
$$S = P \oplus C_{in}$$
$$C_{out} = G + P \cdot C_{in}$$

Since the propagate and generate signals corresponding to all preceding inputs must be used to compute the carry signals of subsequent stages, the fan-in of the gates used by successive stages easily gets out of hand. For this reason, each carry lookahead

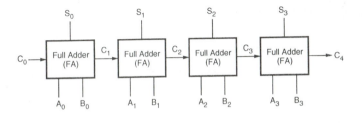

(a)

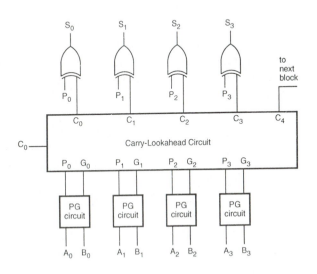

(b)

Figure A.1.   (a) Conventional ripple carry adder circuit. (b) 4-bit carry lookahead adder.

addition usually spans no more than four stages. These carry lookahead (CLA) stages are then cascaded by connecting their carry-in and carry-out bits. Under this scheme, the total delay is

$$\tau_{total} \approx (n/4) \cdot \tau_i$$

Carry lookahead adders are usually implemented with domino CMOS circuits to make them faster. The Manchester carry adder is an improvement of the carry lookahead adder that greatly reduces the area and complexity of the carry chain. Let C_i denote the carry signal which is necessary to produce the i_{th} sum bit. Then, it can be shown that the circuits generating the carry signals C_1, C_2, and C_3 are all *contained* in the circuit generating the C_4 signal, because each C_i expression is a part of the C_{i+1} expression. The Manchester carry chain is actually the domino CMOS circuit generating C_4, with the other three carry bits obtained from its internal nodes. (*Author's note: Further information on the Manchester carry circuit can be found in Section 9.5.*) Note, however, that the domino CMOS circuit can now have six transistors in series in the worst case, which can significantly slow down the circuit. A carry-bypass signal can be used to improve the performance of this worst case, corresponding to all P_i high.

The carry-select adder improves the speed of the adder at the expense of significantly increasing the area. Since C_4 can be logic 0 or 1, two sets of the C_5 - C_8 calculation hardware are constructed, one with a C_{in} of 1, the other with a C_{in} of 0. The outputs of the two sets are multiplexed with the actual value of C_4 controlling the multiplexer, which selects the correct output. Thus C_1 - C_4 and C_5 - C_8 are computed in parallel. Successive stages of this circuitry can be cascaded to create larger adders such that all the carry bits are computed in parallel with the correct one being selected when the correct previous carry becomes available. However, there is a large increase in area due to the duplicated hardware. For a 16-bit adder, this extra hardware seems too expensive except in cases where speed is the entire design goal. This project, however, weights speed with 50% importance and area with 30% importance. Thus the choice was made not to use the carry select and instead use the Manchester, which seems to achieve a nicer balance. Furthermore, our improvements on the above Manchester design, which will be discussed later, reduced the Manchester area and improved upon the speed.

Table 1 below gives our worst-case delay time, total area, and power estimation. As mentioned earlier, the worst case for the Manchester with carry chain occurs when all 16 propagate signals (P_i) are high. However, since we used the carry bypass enhancement, the worst case occurs when P_1 is low and G_1 is high, and all other P_i are high and G_i low. Simulating the entire 16-bit adder was deemed prohibitively time consuming and is not necessary to simulate the worst-case delay. To simulate the

A.2. The propagate signal (P) determines whether a previous carry in the Manchester carry chain should propagate up the chain, and the generate signal (G) determines whether a new carry is generated in the carry chain.

The propagate circuit uses a six-transistor implementation of the exclusive-OR (XOR) operation. This static CMOS design consists of one transmission gate and two inverters. However, the generate circuit is a domino CMOS circuit implementing the AND operation. The circuit is made dynamic for two reasons: (1) the dynamic design is much smaller than a functionally equivalent static design, and (2) it helps optimize the design of the Manchester carry chain. The output of this circuit is precharged low because the input to the inverter is precharged high. This effect allows the elimination of one nMOS transistor per G_i input on the Manchester carry chain. This will be discussed further in the Manchester carry chain section below. Also, the charge-sharing problem in this circuit is reduced by the fact that the pMOS transistor on the inverter is larger than normal to decrease V_{th}. Simulation was used to verify that charge sharing did not cause any incorrect circuit operation. The series nMOS transistors for A, B, and CLK were made larger to increase the speed of the circuit. The circuit diagrams of the propagate-generate circuits and the layout of the PG module are shown in Fig. A.2.

Manchester Carry Chain

The Manchester carry chain is the most important module since its delay time is the dominant delay in the adder. The original Manchester design shown below can be significantly improved. First, the clocked nMOS transistor in series with the C_0 nMOS transistor can be removed due to the fact that the C_0 signal is guaranteed to be low in our implementation. This guarantee is satisfied as follows:

a)   The C_0 input to the Manchester chain at the least significant end is obtained from output of the dynamic carry-input latch, which is precharged low initially.

b)   The C_0 input for higher Manchester chains is guaranteed low due to the fact that the output C_4 of the previous Manchester chain is effectively precharged low, which is a *side effect* of the Manchester design.

Additionally, since the G_i inputs are effectively precharged low as discussed above in the generate circuit design, the clocked nMOS in series with each G_i can be removed. Thus our implementation has both fewer transistors and is made faster due to the fact that there are now fewer transistors in series for any given path to ground.

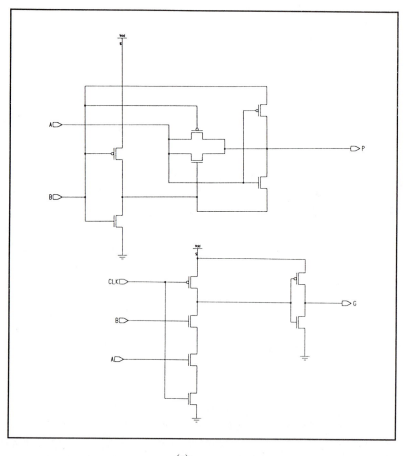

(a)

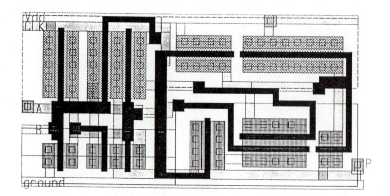

(b)

Figure A.2.   (a) Circuit diagrams of the propagate (P) and generate (G) circuits. (b) Mask layout of the PG module.

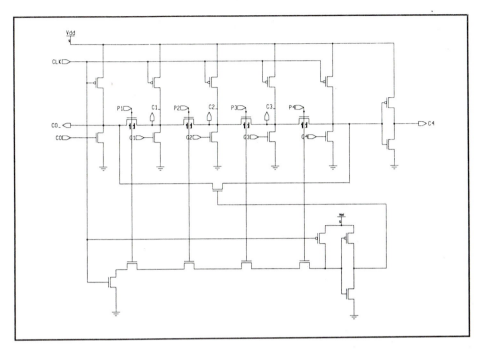

(a)

(b)

Figure A.3.   (a) Circuit diagram of the enhanced Manchester carry chain circuit.
(b) Mask layout of the Manchester carry chain module.

        Although these improvements help a great deal, the circuit still suffers from
the problem of having to discharge the internal capacitances through five series nMOS
transistors. To improve this worst case, two optimizations were performed. First, the

(*W/L*) ratios of the pMOS and nMOS transistors were optimized through simulation. The pMOS transistors for precharging could be made smaller since they work in parallel, allowing the pMOS transistor on the inverter to be significantly larger and thus pull up faster. The nMOS transistor (*W/L*) ratios were also optimized through simulation to provide the fastest possible discharge of the internal capacitances within a given layout height. Second, a carry bypass signal was added to improve the discharging of the gate input to the inverter. This made the carry output C_4 react faster if the P_i were all high and C_0 was high. Simulation showed that all of these optimizations decreased the delay of the carry chain by more than a factor of 5 over the initial Manchester implementation. The longest delay in the carry chain was less than 3 ns. The circuit diagram and the layout of the Manchester carry chain is shown in Fig. A.3.

Sum Circuit

The sum circuit implements an XNOR operation and is the same as the propagate circuit with the addition of an inverter on the output. All considerations made for the propagate are replicated for this circuit.

Output Latch

The output latch design is critically important to achieving a low overall delay time for the adder since it must drive a 10 pF load. The 10 pF load is extremely large relative to the internal circuit capacitances that are measured in femtofarads. Thus, a minimum transistor (*W/L*) ratio layout will produce a very long delay time, which was verified through simulation. In order to decrease this delay time, a buffer consisting of four successively larger inverters was used (Fig. A.4). The sizes of the four inverters were determined from SPICE simulations. The size ratios of the successive inverters in the buffer circuit are (1/1), (2/1), (6/1), and (15/1). Simulation showed that this achieved a delay time of 5 ns. The layout with minimum-sized transistors had a delay of more than an order of magnitude larger (about 80 ns). The delay of 5 ns could be decreased further by using even larger inverters, but there is a delay versus area tradeoff. The current latch implementation occupies about 30% of the total area, due to the large inverter buffers.

Four-Bit Adder

A four-bit adder was constructed using the modules described above; it was used to obtain detailed SPICE simulation results. This is due to the fact that full simulation of the 16-bit adder was deemed too time consuming and tedious. The actual 16-bit adder circuit simply consists of four 4-bit adder modules. Test cases were used in SPICE simulation to verify that the circuit performed additions correctly. Exhaustive testing was not feasible, so several diverse test-input cases were chosen and all

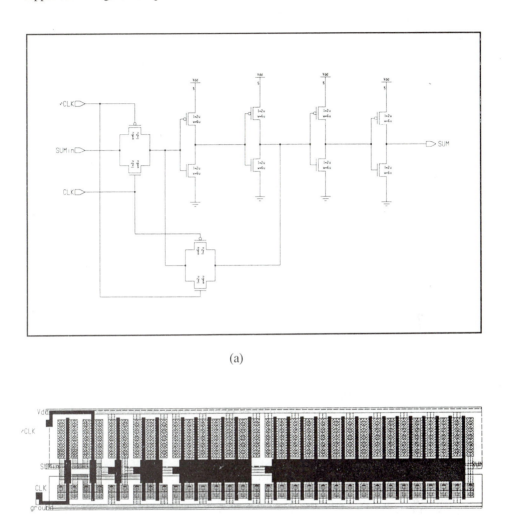

(a)

(b)

Figure A.4. (a) Circuit diagram of the tapered output latch circuit. (b) Mask layout of the output latch.

worked properly. Furthermore, as with all modules, the layout was compared to the schematic through the use of the Layout Versus Schematic (LVS) check. Lastly, this module was used to compute the power dissipation for the entire 16-bit adder, which was found to be 58.9 mW (maximum). The power was actually measured for the four-bit adder cell and then multiplied by 4.25, to account for a total of four of these cells plus the carry-in and carry-out latches. The mask layout of the 4-bit adder is shown in Fig. A.5.

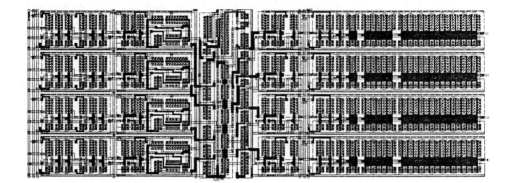

Figure A.5.   Mask layout of the 4-bit adder module.

Complete 16-Bit Adder

The 16-bit adder circuit is constructed of four identical 4-bit adder cells, a dynamic input latch for carry in, an output latch for carry out, and two inverters, one for each of the clocks, to supply an inverted clock signal (Fig. A.6). This complete circuit was assembled from individually tested modules. Also, the critical path was simulated to get the worst-case delay time, which is approximately 20 ns. The complete layout for the 16-bit adder is shown in Fig. A.7.

A.6. SPICE Simulations

Each sub-module of the 16-bit adder, a 4-bit adder, and the critical path of the 16-bit adder were simulated using SPICE. In these SPICE simulations, it is important to include as many parasitic effects as practical in the transistor models so that simulation results are as accurate as possible. The transistor model statements used in the simulations are as follows:

```
.model mn nmos (vto=0.8 lambda=0 gamma=0 cj=1.4e-4 cjsw=4.5e-10
uo=580 ld=0.1e-6 phi=0.6 tox=100e-9 rsh=3.0)

.model mp pmos (vto=-0.8 lambda=0 gamma=0 cj=5.6e-4 cjsw=7.1e-
11 uo=175 ld=0.1e-6 phi=0.6 tox=100e-9 rsh=3.0)
```

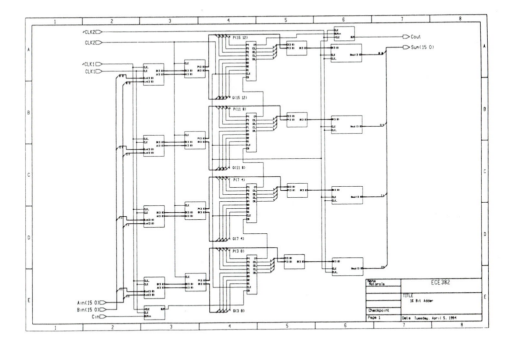

Figure A.6.   Top-level circuit diagram of the 16-bit adder, which consists of individual modules (PG modules, input and output latches, Manchester carry chains) as described earlier.

These model statements include the parameters as specified for the fabrication process. Also, the transistor description statements must accurately reflect the layout used to implement the circuit. A sample transistor description for pMOS and nMOS from the **Double_Latch** file listing is as follows:

```
m0 5 ~CLK Bin Vdd mp l=2u w=22u ad=110p as=110p pd=54u ps=54u
nrd=0.2273 nrs=0.2273
```

```
m6 5 CLK Bin ground mn l=2u w=8u ad=40p as=40p pd=26u ps=26u
nrd=0.625 nrs=0.625
```

This SPICE file was automatically generated from the layout through *circuit extraction*. Thus, these transistor statements accurately reflect the parameters for each transistor. If the parameters for the source and drain areas and perimeters, for instance, were not specified, the simulation would incorrectly show a much faster circuit. Not including the source and drain parameters **ad**, **as**, **pd**, and **ps** is equivalent to not specifying a **cj** and **cjsw** in the model statement since they default to 0.

Figure A.7.   Mask layout of the complete 16-bit adder circuit.

Although SPICE simulation is only an approximation, every effort was made to ensure that the results were as accurate as possible. As many parasitic capacitances and resistances as possible were included in the simulations. The capacitances due to the interconnect metal layers were included as a lumped capacitance at each node in every SPICE file. In the simulation of the four-bit adder cell, the resistance of the polysilicon interconnect was included and had no noticeable effect on the results.

INDEX